T0199268

N. K. Fageria
V. C. Baligar
R. B. Clark

Physiology of Crop Production

More pre-publication
REVIEWS, COMMENTARIES, EVALUATIONS . . .

"With the explosion in world population there is great demand for increasing food production. Urban growth and anthropogenic-induced factors of land degradation have reduced available land for crop production. Abiotic and biotic stresses have become major constraints for food production. Therefore, there is great need for production of more crop yield per unit area. Knowledge of crop physiology is crucial to understand the biological process and functions that determine the growth and development of crops, which are major components of crop yield.

This book includes eight chapters that cover extensively both theoretical and practical aspects of crop physiological processes that have profound effects on production potentials of major agricultural crops. The book includes chapters on plant canopy and root architecture, growth and yield components that determine yield, photosynthesis, source-sink relationships, carbon dioxide influence on crop yield, and the physiology of drought and mineral nutrition. Authors have given extensive and comprehensive discussion on various physiological processes that contribute to overall yield expression in crops. This book provides us with important information on harnessing crop physiology principles in achieving desired crop yield under range of management and ecosystems. The book has considerable value to students, scientists, and professionals working in the fields of agronomy, crop production, and plant breeding for abiotic stresses and high nutritional quality. It is easy to read, with plenty of illustrations, data tables, and excellent references. Rich sources of information given on crop physiology will serve as excellent materials for teaching real physiology for improved efficient crop production."

Moustafa A. Elrashidi
Soil Scientist,
United States Department of Agriculture,
National Resources Conservation Service,
National Soil Survey Center

Food Products Press®
An Imprint of The Haworth Press, Inc.
New York • London • Oxford

Physiology of Crop Production

FOOD PRODUCTS PRESS®
Crop Science
Amarjit S. Basra, PhD
Editor in Chief

Plant-Derived Antimycotics: Current Trends and Future Prospects edited by Mahendra Rai and Donatella Mares

Concise Encyclopedia of Temperate Tree Fruit edited by Tara Auxt Baugher and Suman Singha

Landscape Agroecology by Paul A. Wojtkowski

Concise Encyclopedia of Plant Pathology by P. Vidhyasekaran

Molecular Genetics and Breeding of Forest Trees edited by Sandeep Kumar and Matthias Fladung

Testing of Genetically Modified Organisms in Foods edited by Farid E. Ahmed

Fungal Disease Resistance in Plants: Biochemistry, Molecular Biology, and Genetic Engineering edited by Zamir K. Punja

Plant Functional Genomics edited by Dario Leister

Immunology in Plant Health and Its Impact on Food Safety by P. Narayanasamy

Abiotic Stresses: Plant Resistance Through Breeding and Molecular Approaches edited by M. Ashraf and P. J. C. Harris

Teaching in the Sciences: Learner-Centered Approaches edited by Catherine McLoughlin and Acram Taji

Handbook of Industrial Crops edited by V. L. Chopra and K. V. Peter

Durum Wheat Breeding: Current Approaches and Future Strategies edited by Conxita Royo, Miloudi M. Nachit, Natale Di Fonzo, José Luis Araus, Wolfgang H. Pfeiffer, and Gustavo A. Slafer

Handbook of Statistics for Teaching and Research in Plant and Crop Science by Usha Rani Palaniswamy and Kodiveri Muniyappa Palaniswamy

Handbook of Microbial Fertilizers edited by M. K. Rai

Eating and Healing: Traditional Food As Medicine edited by Andrea Pieroni and Lisa Leimar Price

Handbook of Plant Virology edited by Jawaid A. Khan and Jeanne Dijkstra

Physiology of Crop Production by N. K. Fageria, V. C. Baligar, and R. B. Clark

Plant Conservation Genetics edited by Robert J. Henry

Introduction to Fruit Crops by Mark Rieger

Sourcebook for Intergenerational Therapeutic Horticulture: Bringing Elders and Children Together by Jean M. Larson and Mary Hockenberry Meyer

Agriculture Sustainability: Principles, Processes, and Prospects by Saroja Raman

Introduction to Agroecology: Principles and Practice by Paul A. Wojtkowski

Handbook of Molecular Technologies in Crop Disease Management by P. Vidhyasekaran

Handbook of Precision Agriculture: Principles and Applications edited by Ancha Srinivasan

Dictionary of Plant Tissue Culture by Alan C. Cassells and Peter B. Gahan

Handbook of Potato Production, Improvement, and Postharvest Management edited by Jai Gopal and S. M. Paul Khurana

Physiology of Crop Production

N. K. Fageria
V. C. Baligar
R. B. Clark

Food Products Press®
An Imprint of The Haworth Press, Inc.
New York • London • Oxford

For more information on this book or to order, visit
http://www.haworthpress.com/store/product.asp?sku=5148

or call 1-800-HAWORTH (800-429-6784) in the United States and Canada
or (607) 722-5857 outside the United States and Canada

or contact orders@HaworthPress.com

Published by

Food Products Press®, an imprint of The Haworth Press, Inc., 10 Alice Street, Binghamton, NY 13904-1580.

PUBLISHER'S NOTE
The development, preparation, and publication of this work has been undertaken with great care. However, the Publisher, employees, editors, and agents of The Haworth Press are not responsible for any errors contained herein or for consequences that may ensue from use of materials or information contained in this work. The Haworth Press is committed to the dissemination of ideas and information according to the highest standards of intellectual freedom and the free exchange of ideas. Statements made and opinions expressed in this publication do not necessarily reflect the views of the Publisher, Directors, management, or staff of The Haworth Press, Inc., or an endorsement by them.

Cover design by Jennifer M. Gaska.

Library of Congress Cataloging-in-Publication Data

Fageria, N. K., 1942-
  Physiology of crop production / N. K. Fageria, V. C. Baligar, R. B. Clark.
      p. cm.
  Includes bibliographical references.
  ISBN-13: 978-1-56022-288-0 (hard : alk. paper)
  ISBN-10: 1-56022-288-3 (hard : alk. paper)
  ISBN-13: 978-1-56022-289-7 (soft : alk. paper)
  ISBN-10: 1-56022-289-1 (soft : alk. paper)
  1. Crops—Physiology. 2. Crop yields. I. Baligar, V. C. II. Clark, R. B. (Ralph B.) III. Title.

SB112.5.F34 2005
571.2—dc22

                                                        2005020381

CONTENTS

ABOUT THE CO-AUTHORS

N. K. Fageria, DSc, Ag, is Senior Research Soil Scientist at the National Rice and Bean Research Center, Empresa Brasileira de Pesquisa Agropecuaria (EMBRAPA), Santo Antônio de Goiás, Brazil. Dr. Fageria is the author or co-author of several books and more than 200 scientific journal articles, book chapters, and review articles, and has been an invited speaker to several national and international symposiums and meetings.

V. C. Baligar, PhD, serves as a Research Scientist with USDA-ARS in Beltsville, Maryland. Dr. Baligar is co-author or co-editor of several books and over 200 scientific journal articles, book chapters, and review papers, and is an elected Fellow of the American Society of Agronomy and Soil Science Society of America.

Ralph B. Clark, PhD, worked as a research chemist/plant physiologist for over thirty-seven years. He is the author or co-author of more than 280 scientific journal articles, book chapters, and review articles. He is an elected Fellow of the American Society of Agronomy, the Soil Science Society of America, and the Crop Science Society of America. Dr. Clark is now retired.

Preface and Acknowledgments

A major global success of the last half of the twentieth century was the increase in crop production exceeding population growth. At present, 1.2 billion people in the world live in a state of absolute poverty, of which 800 million people live under uncertain food security and 160 million preschool children suffer from malnutrition. In addition, the land available for crop production is decreasing steadily due to urban growth and land degradation. The trend is expected to be much more dramatic in developing than in developed countries. Natural and human-induced abiotic and biotic stresses have become major constraints for global food production. Under these situations, increasing crop yields per unit area is crucial to meet world food demand. Increasing food production is possible only with efficient use of natural resources, especially land and water, and use of better crop production technology.

Crop physiology is one of the most important subjects to understand biological processes and functions. Through the science of crop physiology, it is possible to understand crop growth and development, components of plant yield, and their interactions. Here an attempt has been made to compile a book that gives a comprehensive coverage of theoretical and practical aspects of crop physiological processes that impact production potentials of major agricultural crops. The eight chapters included in this book are related to plant canopy and root architecture, growth and yield components which determine yield, photosynthesis, source-sink relationships, influence of carbon dioxide on crop yields, and the physiology of drought and mineral nutrition. This unified reference provides an extensive coverage of the physiology of crop production directed toward maintaining or increasing crop productivity. Discussion in each of the chapters is supported by experimental results to make the book as practical as possible. References from different regions of the world have been cited to provide a broader perspective of the various subjects covered.

* * *

Preparing a book of this nature involves the assistance and cooperation of many people, to whom we are especially grateful. Two outstanding scientists selected to review the first draft of this book include Dr. C. D. Foy, USDA-ARS, Beltsville, Maryland, and Dr. David C. Martens, Professor of Agronomy, VPI, Blacksburg, Virginia. These reviewers made valuable suggestions for improving the quality of the book and the authors gratefully acknowledge their contributions. Many scientific societies and publishers have granted permission to reproduce figures and tables from copyright materials, and we sincerely thank all of them. Dr. N. K. Fageria expresses his sincere thanks to the National Rice and Bean Research Center of EMBRAPA, Brazil, for granting sabbatical leave during the course of preparing this book. Dr. V. C. Baligar thanks the USDA-ARS, Beltsville, for providing necessary facilities in preparing the book. We express our appreciation to the publisher and share in their pride of a job well done.

Finally, we express sincere appreciation to our families for their understanding and patience, without which this book could not have been accomplished.

Chapter 1

Plant Canopy Architecture

INTRODUCTION

Canopy architecture can be viewed as a visible but momentary expression of plant form (Tomlinson, 1987). When used to describe plants, *architecture* refers to the spatial configuration of the above-ground plant organs and implies that this configuration has functional significance (Campbell and Norman, 1990; Lynch, 1995). The concept of plant canopy architecture is, in many ways, an abstraction that relates plant form to an orderly, genetically determined "growth plan." The relationship between genetic encoded information and the resulting canopy architecture is analogous to that between a blueprint and the building it represents. Unlike buildings, plant canopies are dynamic and constantly changing with time and space in response to their environment.

A primary function of plant canopies is to intercept radiation to drive photosynthesis and other metabolic processes. This interception occurs with varying degrees of efficiency, which is determined chiefly by leaf area, configuration relative to the sun, and to a lesser extent the spatial arrangement of leaves (Duncan, 1975). Other important morphological canopy characteristics include plant height, tillering, branching patterns, and sizes and shapes of individual leaves. At least 90 percent of the biomass of higher plants is derived from CO_2 assimilated through photosynthesis (Zelitch, 1982). Because crop growth and yield are determined largely by photosynthesis, they are also highly dependent upon the amount of light intercepted by the canopy. Therefore, in environments in which light is the most limiting factor to crop growth and production, canopy properties that lead to greater interception of sunlight generally increase photosynthesis, and consequently grain or dry matter yield. Conversely, those that lead to poor light interception result in lower growth and yield.

1

When speaking of plant canopies or other plant features, it is often useful to have a mental picture of an ideal form. Early humans preferentially selected individuals with certain desirable attributes (Stoskopf, 1981), which was the earliest form of plant domestication. This process continues today with modern plant breeders, who often have a mental concept of an ideal plant, or *ideotype,* that guides their breeding programs. The goal of "ideotype breeding" is to define theoretically the most efficient plant type for a particular crop and environment, and then to breed toward this goal. Donald (1968) defined ideotype as a biological model that is expected to perform or behave in a predictable manner within a defined environment. More specifically, a "crop ideotype" is a plant model that is expected to yield a greater quantity and/or quality of grain, oil, or other useful product in a particular environment when developed as a cultivar. *Environment* from a plant breeding point of view can be defined as the integrated influence of all nongenetic variables affecting phenotypic expression of crop species or genotypes (Saeed and Francis, 1984).

It should be apparent that what is ideal for one environment would not be ideal for another. For example, a plant for hot and dry environments may not be ideal for cold and wet environments. Therefore, development of a crop ideotype has to take into account what is termed "genotype by environment interaction," or G×E interaction (Kang, 1998; Rao et al., 2002). The term G×E interaction stems largely from elementary statistical analysis of variance for yield among different crop genotypes for a particular species grown under a variety of environments. These environments often range from low stress to high stress. An elementary statistical yield model that can be used to describe yield response to genotype and environment is

$$Y = \mu + G + E + G \times E + \varepsilon \qquad (1.1)$$

where Y equals yield for a particular treatment combination (yield of a particular genotype in a particular environment); μ equals overall mean yield; G equals the genotypic effect upon yield; E equals the environmental effect upon yield; and ε equals random error. Depending upon the crop, physiological trait, and range of environments, G×E can be very large or practically nonexistent. An understanding of the importance of G×E therefore necessitates physiological studies on the response of different plant processes to different environmental factors. Only then can an ideotype, or model, be developed. Several workers considered G×E interaction as linear functions of environment and proposed regression of yield of a genotype on the mean yield of all genotypes in each environment to evaluate genotype performance stability (Eberhart and Russell, 1966; Finlay and Wilkinson, 1963; Perkins and Jinks, 1968).

On the basis of several physiological studies, Donald (1968) described a wheat ideotype for a favorable environment that had a short stem, relatively few, small, erect leaves, a large ear with many florets per unit of dry matter of the crop, an erect ear, awns, and a single culm. Asana (1968) also described a wheat ideotype suitable for dryland conditions.

One of the major benefits of ideotype breeding is that breeders are forced to define their goals and strategies in advance (Ortiz and Langie, 1997). For example, based upon the morphological design postulated by Donald (1968), it was concluded that a successful crop plant needed high competitive ability relative to its mass and high efficiency relative to its use of environmental resources. It has been shown that high-yielding cultivars of barley and rice are suppressed or eliminated in mixtures (Sinha and Swaminathan, 1984). Therefore, a breeder would have to envisage different ideotypes when breeding crops for monoculture as opposed to mixed cropping systems, since the two systems represent two different kinds of competition. Using the ideotype approach, breeders select for and not against specific phenotypes.

Many examples exist for achieving yield increases through the use of ideotype breeding for improved canopy architecture. In 1989, the International Rice Research Institute (IRRI), located in the Philippines, conceptualized and developed new plant ideotypes for annual upland, rainfed lowland, and direct-seeded irrigated rice (Table 1.1). Figure 1.1 shows the architecture of IRRI's ideotype for irrigated rice, which was designed to produce fewer tillers than the improved cultivars commonly grown today, but almost every tiller will bear panicles. The new rice ideotype, which has a vigorous root system to draw nutrients from the soil, will be appropriate for both direct seeding and transplanting.

Ideotypes have also been suggested for the development of higher-yielding dry or common bean cultivars. Denis and Adams (1978) suggested an ideotype that consists of a relatively large plant with

TABLE 1.1. Characteristics of ideotypic rice plants.

Upland rice	Rainfed lowland rice	Direct-seeded irrigated rice
130 cm tall	130 cm tall	90 cm tall
Very sturdy stems	Very sturdy stems	Very sturdy stems
5-8 panicles per plant	6-10 panicles per plant	3-4 panicles per plant
150-200 grains per panicle	150-200 grains per panicle	200-250 grains per panicle
Erect upper leaves, droopy lower leaves	Dark green, erect or moderately droopy leaves	Dark green, erect, thick leaves
Deep, thick roots	Extensive root system	Vigrous root system
100 days growth cycle	120-150 days growth cycle	100-130 days growth cycle
Multiple disease and insect resistance	Multiple disease and insect resistance	Multiple disease and insect resistance
3-4 t·ha^{-1} yield potential	5-7 t·ha^{-1} yield potential	13-15 t·ha^{-1} yield potential
	Strong submergence tolerance	0.6 harvest index
	Strong grain dormancy	

Source: Adapted from IRRI, 1989.

numerous nodes, leaves, and reproductive structures. They also suggested ideotypes with more open canopy architecture to facilitate more uniform light interception throughout the canopy.

For maize, Mock and Pearce (1975) defined an ideotype with a leaf area index (LAI, the ratio of leaf surface to ground surface) > 4, with stiff, vertically oriented leaves located above the ear, and horizontally oriented leaves below the ear, in order to maximize light interception by the entire canopy. Ideotypes have also been described for several other crops such as barley (Rasmusson, 1987).

An analysis of canopy architecture of a crop or variety helps to explain how plants utilize their aerial environments. The interaction of the aboveground vegetation of a crop with the environment is largely controlled by canopy architecture (Welles and Norman, 1991) in terms of the distribution, area, and shape of leaves, stems, and inflorescences. This information can be useful to crop scientists interested

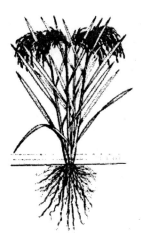

FIGURE 1.1. Schematic ideotype of direct-seeded irrigated rice plant. *Source:* Reproduced by permission of International Rice Research Institute, Los Baños, Philippines, 1989.

in improving morphological traits of annual crop cultivars for higher productivity. The objective of this chapter is to characterize ideal plant architecture, in terms of height, tillering, and leaf characteristics. We also discuss management strategies for ideal plant architecture and relate these to yield enhancement.

PLANT HEIGHT

Plant height is the distance from ground level to the tip of the tallest leaf for seedlings or juvenile plants. For mature plants, it is the distance from ground level to the tip of the tallest panicle, ear, or head in cereals and branch legumes. Short and sturdy culms, more than any other character, favor lodging resistance. Lodging is the permanent displacement of the stems from their upright position. There are three types of lodging: breaking of the stem, bending of the stem, and rolling, in which the whole plant is uprooted from the ground and falls over. A tall crop variety has greater bending momentum than a short one because of culm height. Early lodging of long, thin culms disturbs leaf arrangement, increases mutual shading, interrupts transport of nutrients and photosynthates, causes grains sterility, and reduces yield (Jennings, Coffman, and Kauffman, 1979). Further, strong winds and rains during reproductive and grain filling stages of growth can cause lodging in annual crops. Lodging during the grain filling growth stage can reduce grain quality.

A short and sturdy culm also promotes favorable grain-to-straw ratios, adequate N responses, and high yield capacities. Increased N application is essential for higher yields but causes elongation of the lower internodes, making the crop more susceptible to lodging. In addition, planting date, row spacing, and seeding rate affect lodging in soybean (Willmot, Pepper, and Nafziger, 1989), as well as plant height, branch production, and basal pod height.

A standard evaluation system has been designed by IRRI for plant height, culm resistance, and lodging incidence of rice (Table 1.2). Similarly, Willmot, Pepper, and Nafziger (1989) devised a rating system for lodging for soybean and other legumes to evaluate lodging resistance (Table 1.3). The importance of lodging resistance to high yield capacity has long been recognized, but only in recent years has

TABLE 1.2. Standard evaluation system for rice for plant height, culm strength, and lodging incidence.

Scale rating	Plant height	Culm strength: Plants lodged	Lodging incidence: Area affected
0	Semi-dwarf	Strong	No lodging
1	Semi-dwarf:	Strong:	Less than 20 percent
	Lowland <110 cm and upland <90 cm	No bending	
3		Moderately strong:	20-40 percent
		Most plants bending	
5	Intermediate:	Intermediate:	41-60 percent
	Lowland 110-130 cm and upland 90-125 cm	Most plants moderately bending	
7		Weak:	61-80 percent
		Most plants nearly flat	
9	Tall:	Very weak:	More than 80 percent
	Lowland >130 cm and upland >125 cm	All plants flat	

Source: Adapted from IRRI, 1988.

TABLE 1.3. Rating scale for main stem and branch lodging in soybean.

Lodging score	Visual criteria
1	Almost all plants erect
2	Broken branches on 10 to 30 percent of plants
3	Plants leaning 30° plus 30 to 60 percent with broken branches
4	Plants leaning 45° plus 60 to 90 percent with broken branches
5	All plants leaning 60° plus 100 percent with broken branches

Source: Willmot, Pepper, and Nafziger, 1989. Reproduced by permission from American Society of Agronomy, Madison, WI.

this trait been effectively introduced in rice, wheat, barley, and sorghum. Lodging resistance is principally related to short stature, but it also depends on other characters such as culm diameter, culm wall thickness, and the degree to which leaf sheaths wrap internodes.

The creation of semi-dwarf cultivars spectacularly increased the yielding ability of many crops, such as rice and wheat. For example, Table 1.4 compares yields of older and taller traditional cultivars of upland rice to modern, semi-dwarf cultivars of Brazil. Yields of the modern cultivars are higher, in part, because they are less susceptible to lodging than the old ones. Similarly, lowland rice cultivars used in the Philippines have, over the past 70 years, become shorter, in addition to having smaller, more upright leaves and reduced sensitivity to photoperiod. Panicle weights initially became heavier, but later were made lighter with greatly increased panicle numbers.

A marked increase in harvest index and in grain production per day has been associated with reduced plant height and earlier maturity (Evans, Visperas, and Vergara, 1984). The heritability of dwarfism in cereals is high and is easy to identify, select for, and recombine with other traits. Although yield gains due to the introduction of dwarfing genes into cereals have been remarkable, little evidence exists for concomitant improvements in photosynthetic rate, crop growth rate, or kernel weight.

The introduction of semi-dwarf genes has been advantageous for yield, but extremely short plant stature is not at all advantageous, because leaves are very closely spaced on short culms, which causes high

TABLE 1.4. Plant height, grain yield, and lodging rating of traditional and modern upland rice cultivars under Brazilian conditions.

Cultivar	Plant height (cm)	Grain yield (kg·ha^{-1})	Lodging rating[a]
Traditional			
Rio Paranaíba	108	2780	3
Caiapó	105	2590	2
Guarani	98	2640	4
CNA 8054	108	2470	3
Average	105	2620	3
Modern			
Progresso	86	2620	1
CNA 8172	86	2860	1
CNA 8305	91	2990	1
Canastra	88	2870	1
Average	88	2840	1

Source: Morais, 1998.

[a]Higher values mean relatively more susceptibility to lodging and lower values mean more resistant to lodging.

shading within the plant canopy. Therefore, an optimum plant height is important, which may vary with plant species and environment.

TILLERING

Tillers are the branches that develop from the leaf axils at each unelongated node of the main shoot or from other tillers during vegetative growth. The development of tiller buds after differentiation is greatly affected by environmental conditions, as well as by genotypic characteristics. The environment must be favorable for tiller development. It is essential that sufficient supplies of water, photosynthate, nutrients, and plant hormones are present, and that stress is minimal. The addition of plant nutrients is particularly important when soils have low fertility. Figure 1.2 shows the increase in tiller number with increasing levels of N in lowland rice grown on an Inceptisol of

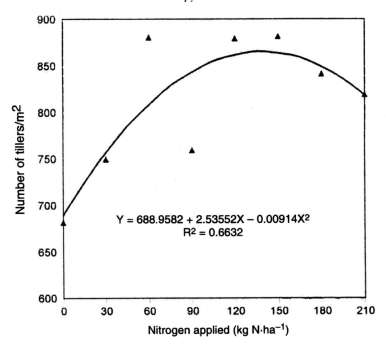

FIGURE 1.2. Relationship between nitrogen application and number of tillers in flooded rice grown on an Inceptisol of central Brazil. *Source:* Reproduced from Fageria, 1998.

central Brazil. Similarly, tiller number of upland rice increased with the application of P at 75 mg·kg^{-1} of soil for an Oxisol of central Brazil (Table 1.5). Nutrients required for the growth of tiller buds of rice must come from the main stem, since tiller buds have neither roots to absorb inorganic nutrients nor leaves to carry out photosynthesis. Once tillers have emerged from the subtending leaf sheaths, they can perform photosynthesis and produce carbohydrates. Tillers can also absorb soil nutrients through their own roots after the third leaf has completely emerged, since roots appear at the prophyll nodes of tillers at this stage of growth (Handa, 1995).

Tiller number is quantitatively inherited. Heritability for tiller number has been low to intermediate, depending on the cultural practices and the uniformity of soil. Although often associated with early vigor in short-stature plants, tiller number is inherited independently

TABLE 1.5. Tiller number per pot under two levels of phosphorus in upland rice genotypes grown on an Oxisol of central Brazil.

Genotypes	0 mg P·kg⁻¹	75 mg P·kg⁻¹
CNA 6187	5	15
CNA 7645	5	16
CNA 7127	5	17
CNA 7680	5	14
CNA 7864	5	18
Rio Paranaiba	5	17
Average	5	16

Source: Adapted from Fageria and Baligar, 1997.

from other major crop traits. In many crosses, tiller erectness or compactness is recessive to spreading culm arrangement (Jennings, Coffman, and Kauffman, 1979).

Whether tillering is governed through genetic, hormonal, or other mechanisms, the full expression of tillering or branching is determined predominantly by the supply of minerals and photoassimilates to the plants. Tiller survival and the production of an inflorescence are paramount to yield. The degree of interplant competition for minerals, photoassimilates, and water will determine how many tillers will reach maturity (Stoskopf, 1981). The proportion of tillers that survive to produce grain depends on the genotype, N fertility, water status, plant density, changes in light quality during crop growth duration, and plant spatial arrangement (Simmons, Rasmusson, and Wiersma, 1982; Simmons and Lauer, 1986; Lauer and Simmons, 1989). In cereals such as rice and wheat, tillering increases as plants develop until a maximum value is reached, after which it decreases. Figure 1.3 illustrates this relationship between plant age and tillering for lowland rice grown on an Inceptisol in central Brazil. Tillering started 18 days after sowing and followed a quadratic relation with crop growth duration. Maximum tillering, based on the quadratic function, was attained at 75 days after sowing, and decreased thereafter.

Decrease in tiller number after reaching a maximum has been attributed to the death of later tillers because of the inability to compete for light and nutrients (Fageria, Santos, and Baligar, 1997). In

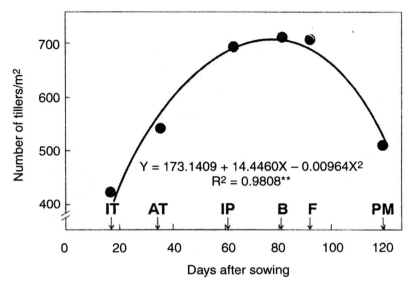

FIGURE 1.3. Relationship between plant age and tiller number in lowland rice. *Source:* Fageria, Santos, and Baligar, 1997. Reproduced by permission of American Society of Agronomy, Madison, WI.

addition, competition for photoassimilates begins between panicles and tillers after panicle development. As growth of many young tillers is suppressed, they may eventually senesce without producing seed (Dofing and Karlsson, 1993). Both shading of tillers and shifts in shoot photoassimilate partitioning patterns have been proposed as factors causing tiller death in grasses (Lauer and Simmons, 1985, 1989). Spiertz and Ellen (1972) reported that light enrichment reduced tiller mortality in perennial ryegrass.

Plant density is another important factor affecting tillering (Counce, Wells, and Gravois, 1992; Wu, Wilson, and McClung, 1998), because it affects competition for light and other resources (Liang, Muo, and Ran, 1986). As plant density and resource competition increase, the proportion of secondary and tertiary tillers in rice decreases (Hoshikawa, 1989). For many plant species, tillering is a mechanism for assuring that the amount of biomass produced is in balance with the amount of plant resources available. For example, Miller, Hill, and Roberts (1991) noted that tiller density in rice increased significantly as plant density increased from 122 to 458 plants/m², while total aboveground mass was not significantly different among the plant populations.

LEAF CHARACTERISTICS

Leaf characters such as erectness, length, width, thickness, toughness, color, and senescence are often associated with yielding ability of plant species or cultivars. Among these characters, erect orientation of leaves is associated with high yielding capacity. Erect leaves permit greater penetration and more even distribution of light into the crop, and thus higher photosynthetic activity. Plants with upright leaves may require less physical space than plants with horizontal leaves and may be adapted to narrow row widths. Leaf angle has been closely correlated with N response in rice, barley, and wheat (Yoshida, 1972). Erect leaves seem to be the result of a pleiotrophic effect of the dwarf gene. Therefore, this trait follows simple recessive modes of inheritance. The erect leaf trait is highly heritable, easily observed at early flowering, and easy to visually rate in pedigree rows of fixed lines (Jennings, Coffman, and Kauffman, 1979).

Leaf thickness is also a desirable leaf trait. Thicker leaves usually have higher densities of chlorophyll per unit leaf area, and hence have greater photosynthetic capacities than thinner leaves (Craufurd et al., 1999). Vigorous leaf growth in crop plants has generally been associated with long-term gains in photosynthetic potential (Blum, Sullivan, and Nguyen, 1997). Plant size may provide substantial yield benefits. For example, rice leaf growth rate is lower compared with maize and barley. A better knowledge of underlying physiological mechanisms involved in limiting leaf growth in rice could facilitate progress in this direction. Lu and Neumann (1999) concluded that the lower growth rate of rice leaves was not associated with comparatively lower rates of epidermal cell production, but instead was associated with the production of smaller mature cells. In addition, this smaller size of rice leaf cells was not associated with less negative osmotic potentials in the cell expansion zone. However, comparative extensibilities of growing leaf tissues and cell walls decreased in the order of barley > maize > rice, and gave good linear correlations with leaf relative growth rates ($r = 0.87^{**}$ and 0.97^{**}, respectively). The data indicated that low cell wall extensibility limits maximum leaf growth rate in rice. Future introduction of increased wall extensibility in rice might lead to increases in leaf growth potential, and conceivably to improved yields.

Leaf size is directly associated with leaf angle, with short leaves tending to be more erect than longer ones. Further, short leaves are usually more evenly distributed throughout the canopy, which permits less mutual shading of leaves and more efficient use of light for photosynthesis. As leaves become longer, they become more difficult for the midrib to support, and therefore tend to droop. Usually, dwarf cereal cultivars have short leaves and tall cultivars have long leaves, although some may have short leaves. These strong associations indicate that leaf length in both dwarf and tall cultivars is a pleiotrophic effect of genes for plant height (Jennings, Coffman, and Kauffman, 1979).

Many examples of higher yield being associated with smaller leaf size and more erect leaf angle distribution are available. Yield of perennial ryegrass was positively correlated with leaf length and LAI, and cultivars with erect foliage were more productive than those with planophile canopies (Rhodes, 1975). Similarly, maize yield has been greater for cultivars with erect canopies than for cultivars with more horizontal orientation (Pendleton et al., 1968). Light entering a canopy of erect leaves was spread over larger photosynthetic areas than in prostrate cultivars, resulting in greater photosynthetic efficiency (Redfearn et al., 1997).

Leaf width is less variable than length. Narrow leaves are more desirable than the wider ones due to more even distribution in the plant canopy. Similarly, thick leaves are associated with high yielding capacities of crop cultivars. However, this character is difficult to measure visually under field conditions and is generally not evaluated in breeding programs. Leaf thickness can be measured microscopically, but it is conveniently expressed as specific leaf area (area per mass) or specific leaf weight (mass per area).

Other leaf characters that have been associated with yield include leaf toughness, color, and senescence. Leaf toughness, which is important in preventing breaking during heavy wind and rain, is related to thickness and lignification of leaf tissues.

Senescence refers to degenerative changes normally associated with the declining phase of leaf, whole plant physiological activities, or both (Wolfe et al., 1988b). Senescence is of economic interest because it affects crop productivity by reducing the active photosynthetic area.

The process of senescence is often considered to be under genetic control, but the role of growth regulators and environmental factors such as temperature, nutrient, and water supply can modify leaf senescence (Nooden, 1980). Water deficit can promote leaf abscission or senescence (Kozlowski, 1973). Jordan (1983) suggested that accelerated postanthesis senescence under conditions of water stress is the result of an insufficient supply of available photosynthate and N in the presence of strong reproductive sinks. Wolfe et al. (1988a) reported that leaf senescence in maize was influenced by the balances between supply and demand for water, which could be affected by canopy size and N, which could also be affected by sizes of the reproductive sinks. Factors that influence the initiation of senescence have been discussed by Thomas and Stoddart (1980).

Extensive studies have demonstrated that water deficits result in early senescence in annual plants (Palta et al., 1994; Zhang et al., 1998; Gebbing and Schnyder, 1999). Early senescence caused by water deficits, however, reduces photosynthesis, shortens the grain-filling period, and finally results in the reduction of grain weight (Brown, Paliyath, and Thompson, 1991; Palta et al., 1994; Zhang et al., 1998). However, Yang et al. (2000, 2001) have reported that water deficit imposed during grain filling enhanced plant senescence, remobilization of prestored C reserves, accelerated grain filling, and improved yield in case where senescence in wheat is unfavorably delayed by heavy use of N. Similarly, plant senescence in monocarpic plants such as rice is the final stage in growth and development (Okatan, Kahanak, and Nooden, 1981; Nooden, 1988; Gan and Amasino, 1997). It is an active, ordered process that involves remobilization of stored food from vegetative tissues to grain (Nooden, Guiamet, and John, 1997; Ori et al., 1999). Delayed senescence results in much nonstructural carbohydrate left in the straw and leads to a low harvest index and reduced grain yield (Yang et al., 2002). Factors that influence the initiation of senescence were discussed by Thomas and Stoddart (1980).

Whole plant relative senescence rates [RSR, $cm^2/(cm^2 \cdot GDD)$ (GDD = growing degree days, base = $10°C$)] can be calculated between two successive measurements performed at t_1 and t_2 using the following formula (Colombo, Kiniry, and Debaeke, 2000):

$$RSR = (GLA_{t1} - GLA_{t2})/(t_2 - t_1) \times (2) / GLA_{t1} + GLA_{t2}) \quad (1.2)$$

where GLA = green leaf area per plant, t_1 = initial time, and t_2 = final time, and GDD = [(maximum temperature + minimum temperature)/2 −10]. Growing degree-days is generally used in defining plant growth stage because it represents a better indication of plant development stage than calendar days (Neild and Seeley, 1977).

BREEDING AND MANAGEMENT STRATEGIES FOR IDEAL PLANT ARCHITECTURE

Ideal plant architecture for higher yield can be achieved through plant breeding and the adoption of adequate cultural practices. It has been recognized that the spectacular yield increase of crops during the second part of the twentieth century have been attributed in almost equal measure to breeding and to the use of inputs.

Breeding

The most important morphological characters that have been bred into high-yielding cereal cultivars, such as rice and wheat, are short, stiff culms for lodging resistance, and erect leaves for increased interception of solar radiation. Several examples of crop improvement through breeding for ideal plant architecture are provided.

Rice

A typical example of breeding semi-dwarf high-yielding cultivars comes from IRRI. The tall tropical cultivar Peta from Indonesia and the subtropical semi-dwarf cultivar De-geo-woo-gen from Taiwan were crossed to produce the semi-dwarf 'IR 8', which produced a record yield of 11 t·ha^{-1} and responded well to N rates up to 150 kg·ha^{-1} at several locations in tropical Asia (Chang, 1976). Dissemination of this improved plant type throughout Latin America was initiated in 1968 by the Colombian-based program of the International Center of Tropical Agriculture together with National Research Institutes in the region (Cuevas-Perez et al., 1995). Scientists at IRRI and several national breeding programs combined most of the desired features in the improved plant type, including reduced height (about 100 cm), leaf erectness, short, dark green leaves, stiff culms, early maturity, photoperiod insensitivity, N responsiveness, and high harvest index (Yoshida, 1981). The wide adoption of 'IR 8' and other high-yielding cultivars, such as 'IR 20' and 'IR 22', made it possible for the semi-dwarf cultivars to become important cultivars in Brazil, Colombia, Peru, Ecuador, Cuba, Mexico, Indonesia, Malaysia, Philippines, India, Pakistan, Bangladesh, and South Vietnam. By 1972-1973, semi-dwarf cultivars occupied a large part of the area planted to high yielding rice cultivars, including about 10 percent of the world total area and 15 percent of the area in tropical Asia (Chang, 1976). Today, high-yielding semi-dwarf cultivars predominate in most lowland rice-producing areas.

Work is in progress at IRRI and many other international and national research centers to further improve plant type, grain quality, and pest resistance (Khush, 1995). Figure 1.4 shows the development of modern high-yielding rice cultivars from formerly prevalent traditional cultivars, and ideotype rice plants of the future. As shown in Figure 1.4, the new plant type for irrigated rice was designed to attain yields of 12 to 13 t·ha^{-1}. Yield improvement beyond 12 t·ha^{-1} will require new plant architectures because of two major problems: The leaves responsible for grain filling will be shaded beneath a dense

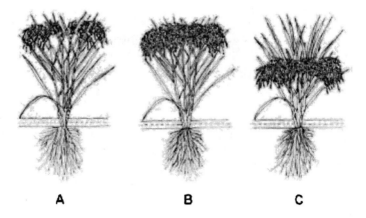

A   **B**   **C**

FIGURE 1.4. New plant types for irrigated rice. A: current generation of new plant types aimed at yields of 12 to 13 t·ha^{-1}; B: same plant type but aimed at yields over 15 t·ha^{-1}; and C: new generation plant type aimed at yields over 15 t·ha^{-1}, but with greater light interception by leaves and resistance to lodging. *Source:* Reproduced by permission of International Rice Research Institute, Los Baños, Philippines, 1993.

cover of panicles, and the immense weight of panicles will result in serious lodging (Figure 1.4). A solution to both problems is to lower the panicle height in the canopy (Figure 1.4). Such adaptations may be relevant to other rice ecosystems as well as to other cereals (IRRI, 1993).

Wheat

Modification of wheat plant architecture through breeding is another example of ideal plant type for improving yield of an important cereal (Byerlee and Curtis, 1988). Wheat yields increased significantly during the twentieth century, with an average global increase of 250 percent during the past 50 years (from 1 to 2.5 t·ha^{-1}). This is remarkable when one considers that wheat yields remained practically unchanged during the first half of the twentieth century (Slafer, Satorre, and Andrade, 1994; Slafer, Calderini, and Miralles, 1996). Better plant architecture through plant breeding and better management practices are responsible for this accomplishment (Calderini and Slafer, 1998).

Detailed analysis of the decade from the mid-1980s to the mid-1990s (Slafer, Satorre, and Andrade, 1994) indicated that worldwide wheat yields might be approaching a ceiling, since average yields did not increase from 1990 to 1995. However, yield potential data of CIMMYT (Centro Internacional de Mejoramento de Maize y Trigo, Mexico) cultivars developed since the 1960s (Figure 1.5) do not indicate a plateau. Indeed, the average increase per year was 0.9 percent (Braun, Rajaram, and Ginkel, 1997). This genetic progress for increasing yield potential was closely associated with increases in photosynthetic activity through the ideal plant canopy (Rees et al., 1993). Both photosynthetic activity and yield potential increased over the 30-year period by 25 percent.

Similarly, an estimated 50 percent of the increase in U.S. wheat yields from 1954 to 1979 can be credited to genetic improvement (Schmidt, 1984). Introduction of semi-dwarf cultivars of wheat had a large impact on productivity of wheat in the Corn Belt and the Great Plains of the United States, and is probably the major source of genetic gain in both regions (Feyerherm, Kemp, and Paulsen, 1988). Semi-dwarf cultivars were planted on more than 90 percent of the

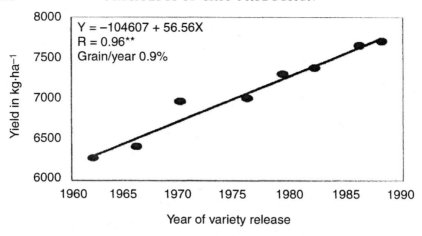

FIGURE 1.5. Mean grain yields for the historical series of bread wheat varieties for the years 1960 to 1990 at Cd. Obregon, Mexico. *Source:* Adapted from Rees et al., 1993.

area planted to wheat in the Corn Belt by 1979, when the genetic gain in productivity was as high as 74 percent (Siegenthaler, Stepanich, and Briggle, 1986). In the Great Plains states (Kansas, Nebraska, and Oklahoma), semi-dwarf cultivars occupied only 9, 1, and 38 percent, respectively, of the area planted to wheat, and genetic improvement for yield was only 45 percent. However, the area planted to semi-dwarf cultivars by 1984 increased to 70, 38, and 76 percent, respectively, in the three states, and genetic improvement increased to 61 percent (Feyerherm, Kemp, and Paulsen, 1988).

Since the beginning of the 1960s, grain yield of winter wheat has increased by about 120 kg·ha^{-1} per year in France. This increase was mainly related to genetic improvement of cultivars and use of adequate levels of N (Gouis and Pluchard, 1997).

Maize

Genetic improvement in grain yield of maize hybrids in North America and Europe during the last three to five decades of the twentieth century has been extensively documented (Tollenaar and Aguilera, 1992). Grain yield improvement of maize hybrids appeared to be the result of increased dry matter accumulation. Increased dry matter may be attributable to either increased absorption of incident photosynthetically active radiation (PAR) and/or improved efficiency of converting absorbed PAR into dry matter. Some evidence indicates that modern hybrids absorb more of the seasonal incidence PAR than the older hybrids. Maximum LAI for modern hybrids was larger than for older ones, and leaves of modern hybrids stay green longer during the final phase of the life growth cycle (Tollenaar and Aguilera, 1992). Full season maize landraces adapted to the lowland tropics are typically tall, leafy, and prone to lodging, and have low harvest indexes (Goldsworthy, Palmer, and Sperling, 1974). During initial stages of maize improvement at CIMMYT, reduction in plant height was a priority (Fischer and Palmer, 1984), and reduced plant height has continued only as a secondary trait in breeding activity. Johnson et al. (1986) reported that 15 cycles of recurrent selection for reduced height in the lowland tropical maize population reduced plant stature by 37 percent and crop duration by 7 percent, and increased the proportional allocation of total biomass to husks, ears, and silks at the 50 percent silking stage. At the same time, researchers observed that grain yield, harvest index, and optimum plant density for grain yield each increased by 50 to 70 percent. Lodging was also substantially reduced. Similarly, Edmeades and Lafitte (1993) reported that lodging in maize declined from 39 to 10 percent with 18 cycles of recurrent selection.

Other Crops

Bridge, Meredith, and Chism (1971) and Bridge and Meredith (1983) reported that yield gains due to genetic improvement of cotton averaged 10.2 and 9.5 kg·ha[-1] per year since about 1910 in the United States. These yield advances have been accompanied by higher lint percentages, smaller seed bolls, and higher micromere values (Meredith and Wells, 1989). Wells and Meredith (1984) indicated that the major component contributing to increased yields was increase in number of fruits. This agrees with Evans's (1980) description of how yield was increased with smaller but more numerous fruits in other major crops.

Vandenberg and Nleya (1999) summarized traits that might optimize canopy structure in common bean at harvest and could be modified through breeding:

1. Long internodes in the lower stem
2. Consistent internode elongation under a wide range of environmental conditions
3. Reduction of stem stunting during early season growth
4. Increased stem length
5. Increased stem strength, particularly in the more basal internode
6. Reduction of pod length without decreasing seed size
7. Increase in pod curvature so that pod tips do not extend below the combine cutterbar
8. Long upright peduncles
9. Commencement of flowering at the upper nodes
10. High fertility at the upper nodes
11. Sufficient number of main stem nodes to maximize productivity in the available growing season

Cultural Practices

In addition to breeding, plant architecture can be modified to a certain extent through cultural practices such as optimal plant spacing, plant density, and fertilization and water management. Miller, Hill, and Roberts (1991) noted that increasing stand density in lowland rice from 120 to 450 plants/m^2 linearly reduced the maximum tillers per plant from seven to three. The level of water management influenced plant survival and tiller appearance. Williams et al. (1990) found that established rice plant stands decreased with increasing water depth from shallow (2.5 to 7.5 cm) to moderate (10.2 to 15.2 cm) and deep (17.8 to 22.8 cm). Similarly, Fageria, Baligar, and Jones (1997) reported that application of N and P to upland, as well as lowland, rice increased tillering in central Brazil. Figure 1.6 shows yield increases of bread wheat as a function of year of cultivar release and N level.

Cultivars of cotton with reduced plant height, short branches, modified leaves, and combinations of these characteristics grown at high plant densities and in narrow-row systems could be a good alternative to increase yields of cotton (Reta-Sánchez and Fowler, 2002). Yield advantages of narrow rows was due to increases in light interception early in the season and increased boll production (Heitholt, Pettigrew, and Meredith, 1992).

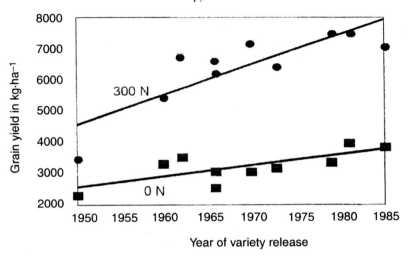

FIGURE 1.6. Grain yield of the historical series of bread wheats at Cd. Obregon, Mexico at 0 and 300 kg N·ha⁻¹. *Source:* Braun, H. J., S. Rajaram, and M. V. Ginkel (1997). CIMMYT's approach to breeding for wide adaptation. In: P. M. A. Tigerstedt (ed.), *Adaptation in plant breeding* (pp. 197-205). Used with kind permission of Kluwer Academic Publishers.

Plant growth regulators such as cycocel (2-chloroethyl-trimethyl-ammonium chloride) are commercially used in Europe as an integral part of intensive cereal management systems to control lodging and maintain high yield potential (Batch, 1981). Khan and Spilde (1992) also reported reduced lodging and improved grain yields of wheat cultivars to ethephon (2-chloroethylphosphoric acid) (an ethylene-producing compound) application.

CONCLUSION

Aboveground plant architecture and its components, including plant height, tillering or branching, leaf size, shape and arrangement, are important plant morphological characters related to yield. The main physiological function of the canopy is interception of light and subsequently plant photosynthetic activity. Distribution of leaf area within a plant canopy is a major factor determining total light interception, which affects photosynthesis, transpiration, and dry matter accumulation. Vertical distribution of leaf area is determined by leaf size, leaf angle, and internodes length. Further, plant height is also related to plant lodging resistance, a character that determines plant yield capacity and grain quality. Some desirable plant architecture characters that are related to high yield are semi-dwarf plant height, thick, short, small and erect leaves, short and stiff culms, upright (compacted) tillering, high N response, and high grain-to-straw ratio. These characteristics can be modified through breeding and cultural practices. Plant breeders have developed high-yielding semi-dwarf cereal cultivars with stiff straw to reduce lodging potential. Complete elimination of lodging, however, has not been achieved without significant reduction in yield potential because of the close association between grain yield, plant height, and total biomass (Austin et al., 1980). High input responsive cultivars (water, fertilizers) of cereals and legumes, permit faster early growth and canopy closure, and reduce the time needed for an adequate root system to develop, thereby permitting the storage phase to begin earlier. Crop plant morphology has been modified to suit monoculture by reduction of branching in legumes or by more upright leaf inclination in cereals. The size of economic organs of plants has been increased substantially, and capacity of transporting assimilates to the various organs has also been increased through breeding during the past few decades (Evans, 1976).

Chapter 2

Root Architecture

INTRODUCTION

Root architecture is another important plant morphological character that deserves special consideration relative to physiological aspects of crop yield, because roots in many instances control development of the entire plant. Root architecture refers to the spatial configuration of root systems, including length, diameter, and hairs. Root architecture is very complex and dynamic in nature. It is related to plant species and cultivars within species, and can be modified by environmental factors (Eghball et al., 1993; Costa et al., 2002). Root systems play a major role in controlling plant growth and development due to their importance in absorption of water and nutrients (Barley, 1970; Lynch, 1995). Availability of nutrients in the soil is determined not only by chemical, physical, and biological factors, but can also be actively influenced by plant roots. A nutrient element has to move to the root and/or be contacted directly by the root surface to be absorbed. For the relatively immobile nutrient such as P, most of the micronutrients, and to some extent K, the amount of roots produced (root surface area) by the plant become a controlling factor in determination of total uptake of these nutrients. Root systems generally contact only very small fractions (1 to 2 percent) of the soil volume. Therefore, the effectiveness of plant roots in absorbing immobile nutrients will be influenced markedly by the extensiveness of root systems, the capacity of a unit root length to absorb contacted nutrients from soil solution, and changes that occur in growth rates and nutrient uptake efficiency of roots as plants grow (Adepetu and Akapa, 1977). Further, roots provide mechanical support to plants from seedling through maturity and development. In addition to their function in water and nutrient absorption and mechanical support, roots are important in N chemistry whereby it can be transported to shoots through the xylem, either as inorganic N or as amino acids and amides. Roots can also synthesize growth substances and hormones such as cytokinins that may be important in leaf function and possibly grain development (Evans and Wardlaw, 1976).

Increased knowledge of root architecture and root development dynamics is crucial for improving productivity of annual crops in agroecosystems. Better understanding of root architecture and growth dynamics of annual crops may lead to more efficient use of applied nutrients and water by crops. The study of plant roots is one of the most promising but least explored areas of research related to plant growth. The aerial portions of plant species have received greater attention and study, probably because of their conspicuousness and easy access, while the subterranean portions have been neglected because of the difficulty for access, the time-consuming effort involved in preparing roots for study, and the disruption of root systems when removed from soil. In addition, information about annual field crop root architecture and growth dynamics as a function of environmental factors is scattered and not often readily accessible. The primary objectives of this chapter are to review the latest scientific advances relative to root architecture and environmental effects on annual crop root growth and function. Information on the rates of growth and distribution of crop roots in soil is also discussed. Our approach should enhance understanding of the contribution of roots to total dry matter of crops, the effects of root size and form on overall crop growth, and the effects of root growth on the environment. Many aspects of root dynamics such as root morphology, top-root ratios, parameters of root growth, methods of root study, root distribution according to soil depth, and growth stages of plants are directly related to environmental factors as well as absorption of water and nutrients, and consequently crop yield.

ROOT MORPHOLOGY

Most root studies have focused on morphology and topology (Hackett and Rose, 1972; Fitter, 1982), and root characteristics such as length, mean diameter, surface area, and mass have been used to quantitatively and qualitatively describe them (Costa et al., 2002). Root morphology is influenced by soil, climatic, and plant factors. Among these factors, plant nutrient and water uptake greatly influence root development and morphology. Root morphology varies with soil moisture content, sowing depth, soil physical, chemical, and biological properties, and genotype. Growth or movement of plant parts respond to gravitational forces known as *geotropism*. While vegetative plant parts are usually negatively geotropic and grow upward, roots are geotropic and grow downward.

Cereals and grasses are commonly known as *monocots* (monocotyledons) and legumes as *dicots* (dicotyledons). Monocots have single cotyledons or leaves arising at the first node of the lead shoot or stem, and dicots have two cotyledons emerging at their first node. Root systems of monocots are fibrous, whereas dicots, such as legumes, have taproots. The fibrous root systems of monocots consist of several root components, including seminal, nodal, and lateral roots. Seminal roots develop from primordia within seeds, and nodal roots develop adventitiously from lower stem nodes. All adventitious roots of stem origin are called nodal roots to distinguish them from other adventitious roots that emerge from the mesocotyl or elsewhere on the plant. Nodal roots are identified by the node number from which they originate. Nodal roots may be functional or nonfunctional (Thomas and Kaspar, 1997). Functional nodal roots are defined as roots that have emerged from stem nodes, entered the soil, and developed lateral roots and/or root hairs. Nonfunctional nodal roots are defined as roots that have emerged from aboveground stem nodes and have not entered the soil or produced lateral roots (Thomas and Kaspar, 1997)

Initial seminal or nodal roots develop laterals, which are classed as roots of the first order. Roots that develop from first-order roots are classed as second-order roots, and additional roots that develop from these laterals are classed as third, fourth, etc., order roots (Yamauchi, Kono, Tatsumi, 1987a,b). Nodal roots are also known as adventitious, coronal, and/or crown roots. Roots of cereals such as rice are formed with mesocotyl, radical (seminal), and nodal or adventitious roots (Yoshida, 1981). Mesocotyl roots emerge from the axis between the coleoptile node and the base of the radical. Mesocotyl roots develop only with certain conditions such as deep seeding or when seeds are treated with chemicals (Yoshida, 1981). Seedlings must rely on roots that initiate on the subcoleoptile internode above the seed or seminal roots below the seed for water uptake until adventitious roots develop.

Adventitious roots are important to seedling establishment because they can conduct more water than smaller-diameter seminal roots. Adventitious roots may develop as early as two weeks after sowing, but require two to four days for available water to initiate them. Seedlings may increase in survival rate when sown at greater soil depths, where larger amounts of soil water availability may increase adventitious root development.

Tiller roots do not form on cereals until tillers have two to three leaves (Klepper, Belford, and Rickman, 1984). Until these roots have developed, parent culms must provide nutrients and water. Parent culms may also have to provide hormonal control so essential for tiller survival. Lag times in root production by tillers and subtillers compared with number of leaves that can be produced likely explains why late tillers and subtillers often have such low potentials for survival (Klepper, Belford, and Rickman, 1984).

Root systems of legumes differ from those of cereals in that dicots are generally of two types: taproot and modified taproot. Taprooted plants have one control root that grows vertically to explore ever deeper layers of soil as growth progresses. Taproots undergo secondary growth and may become woody. Although these vertical taproots go through absorption phases early in the growth cycle and may absorb substances throughout their entire life cycle, their most important function is to transport substances absorbed by younger, smaller branch roots. Maximum length of taproot systems is longer than cereal root systems (Fageria, Baligar, and Wright, 1997). That is one reason why legumes can absorb nutrients and water from deeper soil layers compared to cereals. It is well documented that cereals and grasses are strong competitors for soil K. In grass-legume mixtures, low K and P availability may lead to the displacement of legumes by grasses. Overall, grasses develop larger root systems than legumes. Thus, superiority of grasses over legumes for many nutrients may be related to root morphology.

Root systems of several crop plants were noted to be fractural (irregular and fragmented patterns) with fractal dimensions of 1.48 to 1.58 for whole root systems (Tatsumi, Yamauchi, and Kono, 1989). Fractal analysis methods have been useful in describing both quantitative and qualitative morphological characteristics of maize root systems (Eghball et al., 1993).

SHOOT-ROOT RATIOS

The terms *shoot* and *root* are used here in a botanical sense and refer to the entire aerial and subterranean portions of higher seed plants, respectively (Aung, 1974). In the early part the twentieth century, shoot-root ratios were used rather extensively to characterize plant response to imposed nutritional changes. Root growth is closely related to whole-plant growth. This relationship is called *allometry* or relative growth. Root dry weight is related to total dry weight of a plant using the following equation (Yoshida, 1981):

$$W_R = HW_T^h \tag{2.1}$$

where W_R is root dry weight, W_T is total dry weight (shoot dry weight + root dry weight), and H and h are constants. Equation 2.1 can be converted into a logarithmic form to be

$$\log W_R = \log H + h \log W_T. \tag{2.2}$$

Thus, $\log W_R$ is a linear function of $\log W_T$. The relationship has been tested for different rice cultivars grown under various environmental conditions and can be expressed by the following equation (Yoshida, 1981):

$$W_R = 0.212\ W_T^{0.936}. \tag{2.3}$$

When plants are small (substitute 1 for W_T), W_R/W_T is ~0.2; W_R/W_T values approach 0.1 as plants grow larger (substitute 10^5 for W_T). In other words, ratios of root dry weight to total dry weight range from ~0.2 at the seedling stage to ~0.1 at the reproductive stage (heading) for rice (Yoshida, 1981).

This relationship between root and total dry weights gives an estimate of root mass that remains in soil if shoot weight is known. For example, when plants produce shoot dry weights of 3 t·ha^{-1} at heading, root dry weights remaining in soil should be ~330 kg·ha^{-1}. Partitioning of dry matter in roots relative to shoots is high during the seedling stages of growth and steadily declines throughout development (Evans and Wardlaw, 1976). Shoot-root ratios of common bean, rice, wheat, and cowpea increased as plants advanced in age (Fageria,

1992). Increases in shoot-root ratios indicated that shoots had a higher priority for photosynthate accumulation than roots. If shoot-root ratios decrease with time, roots have preferential utilization of photosynthates under the existing plant growth conditions.

Environmental stresses increase relative weights of roots compared to shoots (Eghball and Maranville, 1993). Decreases in availability of N, P, or water increased root-shoot ratios of perennial ryegrass (Davidson, 1969). Although deficiency of many mineral elements influences plant growth and root-shoot relationships, invariably water and N deficiency limit plant growth the most. Root-shoot ratios of 28-day-old maize plants were 0.27, 0.15, and 0.18 at volumetric soil moisture contents of 0.22, 0.27, and 0.32 $m^3 \cdot m^{-3}$, respectively (Mackay and Barber, 1985). In general, when low nutrient levels do not reduce maize grain yield by more than 20 percent, addition of N will reduce total root weights even though shoot weights increase (Barber, 1995). When plants are N deficient, relatively more photosynthate is used by roots as they develop greater length to aid the plant in obtaining more N. In an experiment with 18-day-old wheat seedlings deprived of N for seven days, soluble sugar contents in roots were higher than in corresponding roots of seedlings grown continuously with complete nutrient solutions (Talouizte et al., 1984). Champigny and Talouizte (1981) reported that under N deprivation, translocation of photoassimilates from shoots to roots increased because of increased sink strength of roots compared to shoot sinks. Root-shoot ratios of maize plants were higher when grown with low soil N compared to adequate N (Eghball and Maranville, 1993).

Soil salinity is another important soil chemical property that influences shoot-root ratios. The depressing effect of salinity on root growth is generally less severe than its effect on shoot growth. Shalhevet, Huck, and Schroeder (1995) summarized results of ten experiments relating shoot and root growth to salinity. The unifying feature was that both root and shoot responses were evaluated by measuring fresh or dry weights at the end of the experimental periods. Each of the experiments had either the same or stronger responses of shoots over roots because of the imposed osmotic potential. However, Slaton and Beyrouty (1992) observed shoot-root ratios of rice to generally be constant as a result of a functional equilibrium, which implied that shoot growth was proportional to root growth.

Partitioning of photosynthates and their effects on dry matter distribution was influenced by several environmental factors such as low temperature, drought, and mineral nutrient deficiency (Wardlaw, 1990). The mineral nutrients P and N exerted pronounced influences on photosynthate and dry matter partitioning between shoots and roots (Costa et al., 2002). Phosphorus- and N-deficient plants usually have more dry matter partitioned to roots than shoots, probably as a result of higher export rates of photosynthates to roots. It is understood that leaf expansion is highly sensitive to low P concentrations in tissue, leading to higher concentrations of sucrose and starch in P-deficient leaves because of reduced demand (Fredeen, Rao, and Terry, 1989). Thus, roots become more competitive for photosynthates than shoots, which leads to higher export of carbohydrates to roots with correspondingly lower shoot-root dry weights (Rufty et al., 1993). Cakmak, and Hengeler, and Marschner (1994) reported that dry matter partitioning between shoots and roots of common bean was affected differently by low supplies of P, K, and Mg. Although total dry matter production was somewhat similar in P-, K-, and Mg-deficient plants, K- and especially Mg-deficient plants had greater shoot than root dry matter, and P-deficient plants had smaller root dry matter than shoot dry matter (Cakmak, Hengeler, and Marschner, 1994). Shoot-root dry weight ratios were 1.8 in P-deficient, 4.9 in control, 6.9 in K-deficient, and 10.2 in Mg-deficient plants.

ROOT GROWTH PARAMETERS AND METHODS OF MEASUREMENT

Parameters commonly used to express root growth and distribution are number, weight, surface area, volume, diameter, length, and the number of tips (Bohm, 1979). In most plant research, weight, length, and density of roots are the most commonly measured parameters. Root weight is the most commonly used parameter for studies of root growth relative to environment. Washed roots are generally dried and weight determined. Fresh weight is frequently recorded in plant pathological studies investigating effects of nematodes and fungi on roots. To determine dry weight, washed and cleansed roots are dried in an oven at 105°C for about 10 to 20 hours, depending on amount of roots (Bohm, 1979). Drying can also take place at 60 to 75°C, which takes longer but has the advantage of preventing roots from being pulverized. Root weight is a good parameter for characterizing total mass of roots in soil. However, root weight is not a very good parameter for characterizing absorbing ability of roots in soil. Fine roots represent only a small fraction of total root weight but are the most active portion of the root system (Bohm, 1979).

Root length is one of the most important and widely used parameters for describing fine root systems and for predicting root response to changes in environment. The ratio of length to mass (specific root length) also has been widely used as an indicator for fine root morphology (Bauhus and Messier, 1999). Root length is a better parameter to relate root absorption of water and nutrients. According to Mengel (1985), the following properties represent the most important parameters for assessing P uptake: root length > P concentration in bulk soil solution > root radius > P buffer power > diffusion coefficient. This emphasizes the importance of root length for measuring root absorption of P from soil. Root length can be readily measured using a method described by Newman (1966), wherein root length is estimated by the equation

$$R = \pi AN/2H \qquad (2.4)$$

where R = the total length of roots; N = the number of intercepts between root fragments and the straight lines on a grid; A = the area of a square or rectangle of the grid; and H = the total length of the straight lines. Tennant (1975) proposed that the Newman formula could be simplified. He suggested that for a grid of indeterminate dimensions, the interaction counts can be converted to centimeter measurements using the equation

Root length (R) = 11/14 × number of intersections (N)
$\times$ grid unit. $\qquad (2.5)$

As proposed by Tennant (1975), the 11/14 of the equation can be combined with the grid unit, so a length conversion factor is obtained. The factors for the 1, 2, and 5 cm grid squares are 0.786, 1.57, and 3.93, respectively (Bohm, 1979).

Root density is another important parameter for root activity. Root length per unit soil volume has been defined as root density. This property can be calculated using the following equation:

Root density = (Total root length in cm)/(Soil volume
where roots have been collected in cm^3) = cm·cm^{-3}.   (2.6)

Mean root radii (r_0) can be calculated from the formula of Teo, Beyrouty, and Ghur (1995):

$$r_0 = (FW/L\pi)^{\frac{1}{2}}$$ (2.7)

where FW = fresh weight and L = root length in cm.

Most field methods used to study root growth are labor-intensive and require at least partial destruction of the experimental site (Bohm, 1979). To overcome these disadvantages, methods were developed to observe root growth in situ. McMichael and Taylor (1987) provided a historical overview of the development of various transparent-wall methods.

Some research institutes have developed large rhizotron facilities to examine plant root systems. The word "rhizotron" (coined from the Greek word *rhizos* for root and *tron* for instrument) can be defined as a facility or building for viewing and measuring underground plant parts through transparent surfaces (Klepper and Kaspar, 1994). These facilities allow simultaneous access to roots and shoots of plants growing in fieldlike environments. The first rhizotrons were built in the early 1960s and used to monitor seasonal and diurnal changes in root system growth and function, cultivar differences in root growth parameters, and effects of soil treatments on root growth and water uptake (Klepper and Kaspar, 1994).

Other similar techniques have been developed in recent years which include minirhizotrons and associated microvideo camera techniques to provide opportunities for quantifying in situ root systems of field-grown plants (Murphy et al., 1994). Linear regression of minirhizotron root counts on root length density (RLD) are often performed to determine the fit of sample data to theoretical models that have been proposed to convert minirhizotron root counts to root length densities. These models use the general equation

$$RLD = C \times (\text{root counts per cm}^2 \text{ of tube})$$ (2.8)

where the conversion factor C has a value of 1.0 (Upchurch and Ritchie, 1983), 2.0 (Melhuish and Lang, 1968), or 3.3 (Upchurch,

1985). Bland and Dugas (1988) reported that a value of $C = 2.0$ provided the best estimate of root length density from minirhizotron root counts of cotton, whereas no simple relationship for sorghum was evident. Studies with wheat have reported C values of 2.8 (Bragg, Govi, and Cannell, 1983), 3.5 (Meyer and Baris, 1985), and 20 (Belford and Henderson, 1985) with correlation coefficients *(r)* of 0.85, 0.90, and 0.95, respectively. Murphy et al. (1994) reported the empirically derived value of $C \geq 15$ for creeping bentgrass and annual bluegrass. Merrill (1992) presented detailed information about fabrication and usage of minirhizotron systems for observation of root growth in field experiments.

ROOT DISTRIBUTION IN SOIL

Because root distribution pattern in soil determines the zone of water and nutrient availability to plants, knowledge of root distribution is very important. Differences in root lengths, dry weight of roots at different soil depths, and extent of rooting at the seedling stage among wheat cultivars were related to differences in yield and ability to escape drought (Hurd, 1974). Upland rice cultivars, which are more drought tolerant than lowland cultivars, have deeper and more prolific rooting systems (Steponkus, Cutler, and O'Toole, 1980). Deep rooting of bean cultivars was positively associated with seed yield, crop growth, cooler canopy temperature, and soil water extraction, provided soil types did not restrict rooting potential (Sponchiado et al., 1989).

Because about 90 percent of the total NH_4^+, P, and K uptake and root length of flooded rice cultivars occurred within the surface 20 cm of soil, samples collected for routine soil tests should be taken from the top 20 cm (Teo, Beyrouty, and Gbur, 1995). Lowland rice plants develop surface matting of roots in the oxygenated zone near soil surfaces soon after application of flood waters (University of Arkansas Cooperative Extension Service Rice Committee, 1990). The presence of these roots in surface soil layers may contribute to high amounts of nutrients measured in the upper 20 cm of soil. Using the Claassen-Barber model to predict nutrient uptake by maize grown in silt loam soil, >90 percent of K and P uptake occurred in the top 20 cm of soil depth (Schenk and Barber, 1980). Silberbush and Barber (1984) reported that about 80 percent of P and 54 percent of K uptake by soybean was from 0 to 15 cm depths. Durieux et al. (1994) reported that more than half of the root length of maize was located in the surface 0 to 20 cm depth at all sampling times during a season. Roots of the peanut cultivar Florunner penetrated to depths up to 280 cm when grown in a sandy soil, and the most extensive root growth occurred in the top 30 cm (Boote et al., 1982). Sharratt and Cochran (1993) reported that 85 percent and 95 percent of the root mass of barley was located in interrows of the top 20 and 40 cm of soil, respectively. Welbank and Williams (1968) also found that nearly 80 percent of barley roots occupied the uppermost 15 cm of soil.

Maximum root densities of peanut were within the top 30 cm soil depth, which was located in the region above a tillage pan (Robertson et al., 1979). Even though the fibrous peanut roots below the usual harvesting depth of 30 cm constituted a small fraction of root weight, these roots may be important for water absorption (Boote et al., 1982).

A study conducted by Stone and Pereira (1994a,b) of four common bean cultivars and three upland rice cultivars to evaluate rooting depth grown in an Oxisol of central Brazil showed that 70 percent of roots were concentrated in the top 20 cm layer, and about 90 percent were concentrated in the top 40 cm soil depth of both crops (Figures 2.1 and 2.2). According to Beyrouty et al. (1988), nearly 100 percent of the root length of lowland rice grown on Crowley soil of Arkansas was measured in the upper 40 cm of soil. The same authors reported that 80 to 90 percent of the root length was measured between 0 to 20 cm soil depth at maximum tillering. As high as 80 percent of the total weight of roots of barley was recovered from the top 15 cm of soil, ~12 percent was between 15 and 30 cm deep, and ~8 percent was between 30 and 60 cm (Welbank et al., 1974).

ROOT DEVELOPMENT RELATIVE TO PLANT GROWTH STAGE

Root development varies with stages of plant growth and development. This information can guide researchers to the appropriate time for root growth observations during crop growth cycles. The most rapid development of maize roots occurs during the first eight weeks after planting (Anderson, 1987). As maize plants age, growth of roots

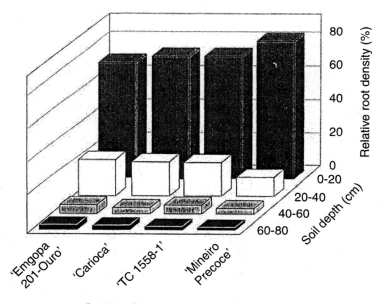

FIGURE 2.1. Distribution of relative root density for four common bean cultivars grown at different soil depths. *Source:* Reproduced from Stone and Pereira, 1994a.

generally increases at slower rates than shoots (Baligar, 1986). After silking, maize root length declines (Mengel and Barber, 1974). This decline in root length after silking is presumably due to the high C demand of grain, resulting in enhanced translocation of C and N to grain, including some C and N that roots would normally obtain (Wiesler and Horst, 1993).

Peanut root length density and root weight density increased at each soil depth increment from planting to 80 days after planting (Figure 2.3). Roots had penetrated to depths of 120 cm 40 to 45 days after planting, and spread laterally to 46 cm in midfurrow. The 0 to 15 cm soil depth increment had the highest mean root length density, which increased to a maximum at 2.1 cm·cm^{-3} 80 days after planting (Ketring and Reid, 1993). This meant that peanut roots were established both deeply and laterally in the soil profile early in the growing season. This would be advantageous in drought environments and helpful for water management.

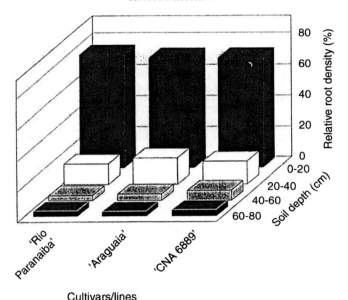

FIGURE 2.2. Distribution of relative root density of three upland rice cultivars grown at different soil depths. *Source:* Reproduced from Stone and Pereira, 1994b.

Sunflower (*Helianthus annuus* L.) rooting depth reached 1.88 m at the beginning of disk flowering and 2.02 m at the completion of disk flowering (Jaffar, Stone, and Goodwin, 1993). In a review of depth development of roots with time for 55 crop species (Borg and Grimes, 1986), it was shown that maximum rooting depth for most crop species was generally achieved at physiological maturity. Kaspar, Taylor, and Shibles (1984) noted that the rate of soybean root depth penetration reached a maximum during early flowering and declined during seed fill. However, some root growth was observed throughout the reproductive stage until physiological maturity.

Slaton et al. (1990) studied root growth dynamics of lowland rice and found that maximum root growth rates were reached between active tillering and panicle initiation, and maximum root length was reached by early booting. Beyrouty et al. (1987) noted that the most rapid rate of root and shoot growth in flooded rice occurred before panicle initiation, which corresponds to the plant transition between

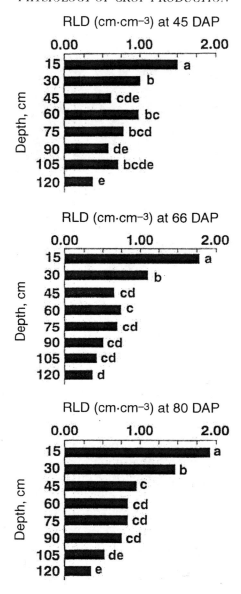

FIGURE 2.3. Root length density (RLD) by soil depth at 45, 66, and 80 days after planting (DAP). Different letters above each bar indicate significant differences between depths at P < 0.05. *Source:* Reproduced from Ketring and Reid, 1993, by permission from American Society of Agronomy, Madison, Wisconsin.

vegetative and reproductive growth. Approximately 77 percent and 81 percent of total shoot and root biomass, respectively, was achieved before panicle initiation. Following panicle initiation, elongation of roots and shoots increased only slightly until harvest (physiological maturity). Beyrouty et al. (1988) also reported that lowland rice root growth was most rapid during vegetative growth, with maximum root length occurring at panicle initiation. Root length either plateaued or declined during reproductive growth.

ROOT CHARACTERISTICS RELATED TO DROUGHT RESISTANCE

Drought is a major environmental stress, and it limits crop production on about 20 percent of lands in the world (Fageria and Baligar, 1993). It is a meteorological and hydrological event, involving precipitation, evaporation, and soil water storage (McWilliam, 1986). Impact of drought is a function of duration, crop growth stage, type of crop species or cultivar within species, type of soil, and adopted management practices. Flowering is the growth stage of crops most sensitive to drought relative to grain yield. A drought of about two weeks during flowering can result in a complete loss of grain yield. Under drought conditions, soil starts drying from the surface, but the deep soil horizons may remain moist and able to supply water to plant roots. Consequently, deep root portions may be more meaningful than shallow root portions when drought resistance of a cultivar is to be examined. For this reason, deep roots are considered a better measure for drought resistance in the field. Deep root-shoot ratios of rice cultivars have been defined as the weight (mg) of roots that are deeper than 30 cm·g^{-1} shoot. They range from <10 to >80 mg·g^{-1} (Yoshida, 1981).

SOIL ENVIRONMENT

Environment is a combination of the surrounding factors in which crops are grown (Fageria, 1992). Soil environment is dynamic in nature, and its influence on plant growth, and consequently on root growth, is rather complex. Root growth is influenced by soil physical, chemical, and biological properties. Detailed descriptions of these factors controlling root growth, and ultimately crop yield, are given.

Physical Factors

Important soil physical factors that influence root growth are temperature, moisture content, soil compaction, and flooding.

Soil Temperature

Soil temperature is important for determining rates and direction of soil physical processes which influence root growth and development. Each plant species has a minimum soil temperature requirement below which no root elongation occurs. Above that minimum temperature, root elongation rates increase almost linearly with temperature increases to a maximum temperature, which is also species dependent. With further increases in temperature, root elongation rates generally decrease rapidly. Since soil temperature varies diurnally to depths of ~50 cm under field conditions, root elongation would also be expected to vary diurnally (Taylor, Huck, and Klepper, 1972). Optimum temperature ranges for biochemical and metabolic activities of plants have been defined as the thermal kinetic windows (TKW) (Burke, Mahan, and Hontfield, 1988). Temperature either above or below the TKW results in stress that limits growth and yield. For example, the TKW for cotton growth is 23.5 to 32°C with an optimum temperature of 28°C, and biomass production is directly related to the amount of time that foliage temperatures remain within the TKW (Burke, Mahan, and Hontfield, 1988).

Miyasaka and Grunes (1990) reported that root extension rates for winter wheat seedlings grown in nutrient solution culture were significantly lower at 8 than at 16°C. Sharratt (1991) reported that root length density of barley grown in pots was reduced at root-zone temperatures of 5 and 10°C compared with 15°C. BassiriRad, Radin, and Matsuda (1991) observed that low temperatures limited root hydraulic conductivity and water uptake of barley and sorghum seedling detached roots within four hours following transfer of plants from 25 to 15 or 20°C conditions. Wraith and Hanks (1992) reported that soil water depletion increased with increases in soil temperature for winter wheat and maize, and that the depth of maximum water uptake was 0.2 to 0.4 m deeper for plants growing in warmer soils. Maximum rooting depth increased with enhanced soil temperature under low water applications, but this relationship was not consistent under high water applications. Thompson and Fick (1981) observed greater alfalfa sensitivity to flooding under high temperatures, and root growth ceased and shoot growth was reduced by 50 percent after two to eight days of flooding at growth temperatures of 32 and 16°C, respectively.

Mason et al. (1982) noted that root growth and elongation of field-grown soybean followed approximately the implementation of the 15 to 17°C isotherm. Bland (1993) also reported similar associations between maximum depth advancement of roots and soil temperature isotherms during part of the growth cycle for cotton and soybean grown in large-volume, controlled-temperature rhizotrons. Entz, Gross, and Fowler (1992) reported that preanthesis of spring and winter wheat root distribution was largely confined to soil profile regions with temperatures $\geq 15°C$.

Optimum soil temperature for root growth of barley was reported to be 15.5°C within 20 cm of the soil surface (Power et al., 1970). Sharratt (1991) reported root length densities of barley increased with increasing root zone temperatures from 5 to 15°C at 10 cm soil depths. Brar et al. (1990), studying root development of 12 forage legumes relative to soil temperature, noted that these plants differed in both main axis root length and lateral root development at various temperatures, with optimum temperature being in the range of 15 to 25°C for most species.

The response of legumes to varying root temperatures can also be influenced by strain of *Rhizobium,* shoot temperature, and availability of mineral N. At suboptimal temperatures, N_2 fixation was reduced because of lower enzyme activities, while degeneration of bacteroidal membranes and degradation of cytoplasm in cells decreased N_2 fixation at supra-optimal temperatures (Hernandez-Armenta, Wien, and Eaglesham, 1989). Root temperatures influenced early growth of arrowleaf clover through its influence on N_2 fixation, and root temperatures $>32°C$ were particularly detrimental because of combined effects of delayed nodulation and reduced N_2 fixation activity (Schomberg and Weaver, 1992).

Soil Moisture Content

Adequate soil moisture content is critical for optimal crop production. Water excesses, as well as deficits, lead to yield decreases. Limited plant-available water is often the greatest detriment to crop productivity. Supplies of water for crop biomass accumulation are provided from soil reservoirs by roots (Ketring and Reid, 1993). A common concept among farmers is that some early-season water stress encourages root growth and conditions crops to better withstand stress that may occur later in the season. Crops growing in relatively dry soils are thought to develop more extensive root systems than crops grown in moist soils (Jama and Ottman, 1993). Sharp and Davies (1985) reported that soil water deficits increase root growth and rooting depth, but may not affect total root length. Water stress can also reduce, instead of enhance, root growth in some cases (Saini and Chow, 1982; Cutforth, Shaykewich, Cho, 1986).

Enhanced root growth is one reason cited for the practice of delaying the first irrigation so plants can be conditioned to subsequent water stress. Various physiological mechanisms such as increased stomatal control, enhanced root hydraulic conductivity, and maintenance of photosynthesis at higher levels of water stress have also been proposed for water stress conditioning (Clemens and Jones, 1978). However, physiological conditions do not necessarily translate into improved plant performance or yield. Jama and Ottman (1993) reported that delaying the first irrigation for maize, which started the season with soils full of water in an arid region, did not encourage deep root growth or deep water uptake. Moreover, delaying the first irrigation did not result in water stress conditioning, and yields actually decreased. Water stress conditioning appeared to depend on the irrigation regime and the timing and severity of the stress imposed in sorghum (Garrity, Sullivan, and Watts, 1983).

Devries et al. (1989) concluded that root length densities increased for peanut grown in soil cores at soil depths <30 cm relative to imposed water stress. Under water stress, peanut roots reached greater depth compared to nonstressed plants (Narasimham, Rao, and Rao, 1977). Meisner and Karnok (1992) reported reduced root growth in the upper 40 cm soil depth during water stresses from 20 to 50 days after planting compared to root growth during the same period of well-watered plants. Severe water stress reduced maize root mass and length in greenhouse studies but increased root length for the same cultivar grown in moderate water stress in the field (Eghball and Maranville, 1993).

Soil Compaction

Recent use of heavy machinery combined with more intensive tilling and poor timing of field operations has led to reduced aggregate stability and creation of subsurface compacted layers in many irrigated soils around the world.

Soil compaction may affect nutrient uptake through modifications of root growth and changes in soil nutrient availability. Castillo et al. (1982) stressed the importance of soil volume explored by roots, as well as anatomical and morphological changes in roots caused by mechanical-impedance effects on water and nutrient uptake. Root growth and distribution were reduced when mechanical stresses were applied to soil surfaces for pea seedlings (Castillo et al., 1982). Uptake of Ca, K, Mg, and Mn was reduced, whereas uptake of B, Fe, Mg, and P was not affected. Oussible et al. (1993) reported root length densities of wheat to be reduced in compacted zones as measured by penetrometer load and bulk density. These authors also noted wheat yields were reduced by 12 to 23 percent in compacted soil compared to noncompacted soil. Both root growth and distribution were changed from subsurface compaction.

Compacted soil layers impede downward growth and distribution of roots (Yu et al., 1995). Mechanical impedance caused by compacted soil layers was important for influencing root elongation and proliferation and can result in physical and physiological constraints to overall plant growth (Tu and Tan, 1991). In limited-rainfall areas, compacted layers can compound these effects by reducing the ability of plants to exploit soil water reservoirs at deeper horizons. Thangraj, O'Toole, and De Datta (1990) suggested that water use is primarily determined by root system density and depth during periods of water stress. O'Toole and De Datta (1983) suggested that increased root density and depth may be responsible for drought avoidance in some rice genotypes. These studies indicated the importance of deep roots for avoiding periods of limited rainfall. Identification of genotypes with greater ability to penetrate compacted soil layers is important in developing superior drought-resistant cultivars. Yu et al. (1995)

reported that penetration ability of rice roots of cultivars ranged from 6 to 26 percent. Root thickness ranged from 0.9 to 1.5 mm and was positively correlated with root penetration ability. Rice cultivars from dryland hydrological origins had greater root penetration ability than did cultivars from wetland origins.

Flooding

Flooding induces many changes in plant roots, of which aerenchyma formation is an important adaptive mechanism (Laan et al., 1989). Aerenchymous roots are formed either by some cell wall separation and cell collapse (lysigeny) or by cell separation without collapse (schizogeny). Figure 2.4 shows cross sections of rice and wheat roots and lysigenous and schizogenous intercellular space. Both forms result in large longitudinal channels in root cortices, and such structures enhance diffusion of atmospheric or photosynthetic oxygen from shoots to roots so that aerobic respiration and growth can be maintained (Armstrong, 1979). Changes in root morphology occur after flooding for both wetland and non-wetland plant species. Flooding may increase branching of roots, development of new adventitious roots, and superficial rooting (Laan et al., 1989).

Cellular spaces exist in roots of various plants and may serve to some extent for oxygen passage. However, the amount of aeration varies greatly according to plant species. In the case of certain marsh plants such as rice, cortical cells of roots reveal columnar arrange-

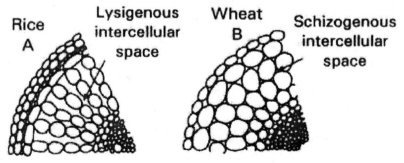

FIGURE 2.4. Cross section of a rice and wheat root showing lysigenous and schizogenous intercellular spaces. *Source:* Reproduced from Horiguchi, 1995, by permission from Food and Agriculture Policy Research Center, Tokyo, Japan.

ment and space is large. In the case of certain terrestrial plants, the arrangement is oblique and the space is small (Figure 2.5). There are marsh plants, however, which do not display columnar arrangement, while some terrestrial plants do (Horiguchi, 1995). In the case of rice, a large intercellular space develops schizogenously and lysigenously in root cortices even in normal growth environments. Horiguchi (1995) reported that in rice and other graminaceous plants, aerenchymous cells were well developed not only in nodes but also in internodes (Figure 2.6).

Inhibition of root growth relative to soil O_2 deficiency due to flooding has been reported for many vascular plants (Drew and Lynch, 1980). The incorporation of organic matter such as rice straw promotes the production and accumulation of organic acids in submerged soils. These acids are often toxic to rice at very low concentrations and may lead to reduced root growth and root decay (Kludze and DeLaune, 1995). Under conditions of high organic matter applications to flooded soils, ethylene gas produced by both microbial activity and plant roots may eventually inhibit root growth (Jackson, 1982).

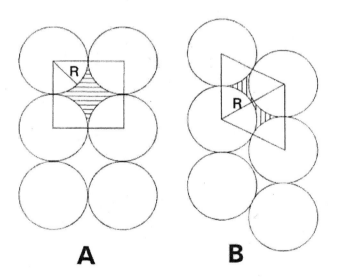

FIGURE 2.5. Comparison of intercellular space in plant roots when cells are arranged in column (A) and oblique (B) arrangements. *Source:* Reproduced from Horiguchi, 1995, by permission from Food and Agriculture Policy Research Center, Tokyo, Japan.

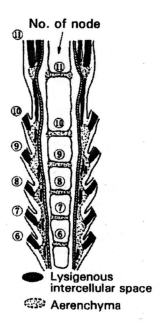

FIGURE 2.6. Distribution of aerenchymas and lysigenous intercelluar spaces for low internodes of rice plant. *Source:* Reproduced from Horiguchi, 1995, by permission from Food and Agriculture Policy Research Center, Tokyo, Japan.

Chemical Factors

Chemical factors that may affect root growth are soil acidity, soil salinity, toxicity of some chemical elements, and deficiency of essential nutrients. The soil chemical factors are interrelated, and separation of their interactions is impractical relative to plant growth, at least in soil as a growth medium. The situation is complicated by interactions of these factors with drought and plant genotype.

Soil pH

Soil pH is one of the most important chemical properties in determining plant growth. It determines nutrient uptake, deficiencies or toxicities of elements, and need to alleviate acidic conditions (liming). The pH

(H_2O) of nearly all agricultural soils is in the range of 4 to 9 (Fageria, Baligar, and Edwards, 1990). Nitrogen, P, K, S, Mg, B, and Cu have the greatest availability when the pH ranges from 6 to 7.5 (University of Kentucky, 1978). Solubility, concentration in soil solution, ionic form mobility, and availability of micronutrients for uptake by plant roots are strongly affected by the soil pH. As a rule, the availability of zinc (Zn) (desorption), manganese (Mn) (desorption and reduction), and, to a lesser degree, iron (Fe) (dissolution of Fe^{3+} compounds) increases, whereas molybdenum (Mo) availability decreases with decreasing soil pH (Marschner, 1991; Fageria, Baligar, and Clark, 2002).

The critical soil pH value, defined as the maximum pH at which lime increases crop yield, varies from crop to crop, soil to soil, and within cultivars of the same species. Fageria, Baligar, and Wright (1997) reported the influence of soil pH on root weight of rice, wheat, and common bean grown on an Oxisol of Central Brazil. Maximum root dry weight of rice was produced at the lowest pH (4.1), and decreases occurred at higher soil pH values (Fageria, Baligar, and Wright, 1997). Rice is considered a tolerant crop to soil acidity (Fageria, 2001). Root weights increased with increasing soil pH for wheat and common bean. Maximum root weight was obtained at pH 6.0 for wheat and at pH 5.9 for common bean. Critical pH values are also very dependant on method of pH measurement (water or $CaCl_2$). Bruce et al. (1988) noted that soil solution and soil pH measured in water were more sensitive indicators of soybean root growth than soil pH measured in 0.01 M $CaCl_2$.

Aluminum Toxicity

Aluminum toxicity is probably the most important growth-limiting factor for plants grown in most strongly acidic soils (Foy, 1992). The Al species Al^{3+}, $AlOH^{2+}$, and $AlOH_2^{+}$ which are responsible for phytotoxic effects appear to be the small fraction of total aluminum in soil solution. Alva et al. (1986) studied relationships between root length of four crops and different species of Al in solution. They concluded that nominal Al concentrations in solution are of little value as an indexes of Al toxicity for solutions. Where considerable quantities of polymeric Al species are present in solution, total Al concentration in solution is also of little value as index. When the effects of Al were studied in solution relative to differing ionic strength, concentrations of monomeric Al in solution was also a poor index of pH toxicity. The best index for assessing Al toxicity, as measured by root growth of soybean, subterranean clover, alfalfa, and sunflower, proved to be the sum of monomeric Al, which takes into account Al precipitation and polymerization as well as the effects of ionic strength.

Root growth is severely restricted by Al injury. Root tips and lateral roots become thickened, and fine branching and root hairs are generally reduced. Aluminum-damaged roots can explore only limited volumes of soil and are inefficient in absorbing nutrients and water (Wright, 1989). Among the soil acidity factors, Al concentration has been the most important in restricting root growth in subsoils. Plant growth at low pH is often limited not by H^+ activity but by toxicity of Al. Baligar et al. (1991) evaluated the influence of soil solution Al on root growth of a tolerant wheat cultivar, Yecorra Rojo, and an intolerant cultivar, Wampum. The longest root length of both cultivars was negatively correlated with soil and soil solution Al parameters (Figure 2.7) and was positively related to pH and Ca parameters. Bruce et al. (1988) studied diagnostic indices of Al as predictors of soybean root growth. These authors concluded that soil Al saturations were not good predictors of Al toxicity on root growth, but soil solution Al measurements were. Activities of Al^{3+} and $AlOH^{2+}$ gave the best association with relative root length (RRL), and values corresponding to 90 percent RRL were 4:M and 0.5:M, respectively. These results indicated that $Al(OH)_3^0$, $Al(OH)_3^+$, and $AlSO_4^+$ were not toxic Al species or forms.

Manganese Toxicity

Unlike Al, Mn seems to affect plant tops more directly than roots, but root damage follows when toxicity is severe (Suresh, Foy, and Weidner, 1989). The solubility and potential toxicity of Mn to any given crop depend upon many soil properties, including total Mn content, pH, organic matter content, aeration, and microbial activity. Manganese toxicity generally occurs at pH values <5.5, if soils contains sufficient concentrations of total Mn. However, Mn toxicity can also occur at high soil pH values in poorly drained or compacted soils where reducing conditions favor production of high concentrations of Mn^{2+} that plants absorb (Foy, 1992).

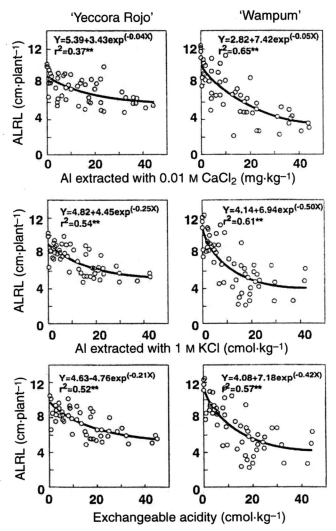

FIGURE 2.7. Average longest root length (ALRL) of 'Yeccora Rojo' and 'Wampum' wheat cultivars grown in 55 soil horizons relative to soil Al extracted by 0.01 M CaCl2, 1 M KCl, and exchangeable acidity. *Source:* Reproduced from Baligar, Wright, Ritchey, Ahlrichs, and Woolum (1991). Soil and solution property effects on root growth of aluminum tolerant and intolerant wheat cultivars. In: Wright, R. J., Baligar, V. C., & Murrmann, R. P. (Eds.), *Plant-Soil Interactions at Low pH*. Kluwer Academic Publishers, Dordrecht, the Netherlands, pp. 245-252. Reprinted with kind permission of Kluwer Academic Publishers.

Nitrogen

Nitrogen is usually the most important nutrient required for plant growth, development, and achievement of yield potential. Concern about N pollution of ground water has stimulated interest in fertilizer strategies that minimize N application rates. Nitrogen-use efficiency, defined as grain yield per unit N supplied, depends on the extent and effectiveness of roots for N uptake (Jackson et al., 1986). Nitrogen stimulated maize root length in the area of N application without affecting total root length (Durieux et al., 1994). Other studies also noted localized stimulation of root growth in response to N placement (Granato and Raper, 1989) and N source (Teyker and Hobbs, 1992). Anderson (1987) found that N additions appeared to stimulate root growth close to soil surfaces. Root weight was less affected by N than root length, and the effects may have depended on plant maturity (Durieux et al., 1994).

Fertilization with NH_4NO_3 at planting increased maize root weights in the soil surface 7 cm under field conditions (Anderson, 1987). Addition of N fertilizer also created longer roots without changing weight, which resulted in finer roots. Mackay and Barber (1986) concluded that field applications of N did not affect root length and root surface area of the maize hybrid 'Pioneer 3732', but increased root growth of the hybrid 'B73XM017'. In addition, the latter cultivar had higher grain yields than the former. Under high N and two moisture levels, Hatlitligil, Olson, and Compton (1984) noted greater root dry matter for 'B73XM017' compared to two other maize hybrids in greenhouse experiments. Baligar, Fageria, and Elrashidi (1998) reported that relative dry weights of roots due to absence of N in rice, common bean, maize, and soybean was 62, 44, 65, and 89 percent less than that of treatments where N, P, and K nutrients were applied at adequate levels.

Rooting of barley in the upper 10 cm of soil was more prolific where N fertilizer was banded rather than broadcast, but total N uptake was no different among treatments (Sharratt and Cochran, 1993). Under moderate N fertilization and water stress conditions, maize developed greater root masses and lengths than under normal nonstressful conditions (Eghball and Maranville, 1993). Maize root fractal dimensions were lower for zero N compared to no differences among root dimensions with applied N (Eghball et al., 1993). Under severe

N stress, maize roots appear to develop less branching than where plants receive adequate N. Root branching can be useful when nutrients are applied in localized zones, but maize roots seem to develop less branching when entire soils are N deficient (Eghball et al., 1993).

Other studies have reported that reductions in root growth may occur at high N supplies (Anderson, 1987; Comfort, Malzer, and Busch, 1988). High N rates may reduce deep root penetration and decrease potential use of deep soil nutrients and water. Bosemark (1954) concluded that with high N supplies, root growth stopped completely.

Cereal plants have been reported to respond to additional N nutrition through increased growth of the whole plant (Troughton, 1962). However, growth of shoots increased to a greater extent than roots. In herbage grasses, each increasing increment in level of N supply produced smaller increases in growth and resulted in retardation of growth at high levels (Troughton, 1957).

Phosphorus

Phosphorus is a key nutrient essential for root development in highly weathered tropical soils. Fageria, Baligar, and Wright (1997) reported that root dry weight was reduced 62 percent in rice, 74 percent in common bean, 50 percent in maize, and 21 percent in soybean without added soil P compared to adequate P in Brazilian Oxisol. In other studies conducted under greenhouse conditions, wheat, common bean, and cowpea were grown at varied levels of P (0 to 200 mg·kg^{-1}) on an Oxisols (Fageria, Baligar, and Wright, 1997). Maximum root dry weight for wheat was achieved at 152 mg P/kg, whereas maximum root dry weight for common bean and cowpea were achieved at 130 and 159 mg P/kg soil, respectively. These results indicate that increasing P levels increased root growth, but root growth was reduced at higher P levels, and crop species have different P requirements to achieve maximum growth potentials. Overall, root growth of cereals and legume crops was reduced if P was deficient. Most literature reports indicated that, within certain limits, both root and shoot growth vary similarly as P level increases. Above certain levels, further increases in P supply do not affect root or shoot growth (Troughton, 1962).

Potassium

Mullins et al. (1994) studied K placement effects on root growth of cotton grown on a fine sandy loam soil. Root density measurements taken in-row showed that root growth at depths >20 cm was improved with in-row subsoil additions of K. Tupper (1992) also observed increased cotton taproot length when K fertilizer was band-applied in subsoils of Mississippi having low soil-test K. On the other hand, Hallmark and Barber (1984) and Yibrin, Johnson, and Eckert (1993) reported that localized applications of K did not promote root growth. However, K has been shown to promote root growth of some vegetable crops (Zhao, Li, and Hung, 1991). Fageria (1992) determined root growth of rice grown in nutrient solution as well as in an Oxisol at deficient and adequate levels of K. At stress levels of K, rice root growth was reduced compared with nonstress levels. Baligar, Fageria, and Elrashidi (1998) reported that K deficiency reduced root growth by 23 percent in rice, 30 percent in common bean, 12 percent in maize, and 11 percent in soybean compared to adequate levels of K in an Inceptisol In other studies in which 13 maize genotypes were gown on an Oxisol, Baligar, Fageria, and Elrashidi (1998) reported 35 percent lower root dry weight at 0 mg K/kg compared to 200 mg K/kg soil. Fageria, Baligar, and Wright (1997) reported that root length and root weight of common bean grown at stress and nonstress K levels in an Oxisol of Central Brazil, and genotypes had different root lengths as well as root dry weights. On average, only slight decreases occurred in root lengths and root weights for plants grown at high K (200 mg·kg^{-1}) compared to control treatments.

Potassium nutrition can alter the health status of plants infected by phytophagous insects. Beneficial effects from improved K nutrition were noted for 59 percent of 231 insect and mite cases (Perrenoud, 1977). Increased K nutrition often offsets pest injury by increasing growth, but changes in plant structures also decreased suitability of host plants for pests in some cases. Alfalfa herbage yield, root weight, and root total nonstructural carbohydrates increased with increasing K fertilizer (Kitchen, Buchholz, and Nelson, 1990).

Calcium

Absolute Ca deficiency is difficult to identify on plants grown in acidic soils (Kamprath and Foy, 1985). Most acidic soils contain adequate total Ca for most plants, and Ca deficiency symptoms are rarely observed in the field. Only in highly leached, acidic, low-cation exchange soils (Oxisols and Ultisols) would absolute deficiencies likely occur (Garrity, Sullivan, and Watts, 1983). Levels of Ca required for essential growth functions are so low as to approach those of micronutrients. Hence, the majority of Ca in soils and in plants serves as excluders or detoxifiers of other elements such as Al, Mn, and heavy metals that might otherwise become toxic (Garrity, Sullivan, and Watts, 1983; Kamprath and Foy, 1985; Foy, 1992).

Gonzalez-Erico et al. (1979) evaluated maize response to deep incorporation of limestone on an Oxisol. They reported that incorporation of limestone to soil depths of 30 cm improved root growth, increased water utilization, and increased grain yield of maize. Similar results were obtained for maize and cotton when limestone was incorporated to depths up to 45 cm (Doss, Dumas, and Lund, 1979).

Suitable diagnostic indices for prediction of Ca limitations on root growth are either Ca saturation of the effective cation exchange capacity or Ca activity ratio of the soil solution, which has been defined as the ratio of Ca activity to the sum of the activities of Ca, Mg, K, and Na (Bruce et al., 1988). Values corresponding to 90 percent relative root length of soybean were 0.05 percent for the Ca activity ratio and 11 percent for Ca saturation. Calcium activity and Ca concentrations in soil solutions and exchangeable Ca were less useful for diagnostic indices (Bruce et al., 1988).

Magnesium

Magnesium is an essential macronutrient for plant growth. The most well-known and important role of Mg is its occurrence in chlorophyll molecules. In addition, Mg is required for many essential physiological reactions, especially phosphorylation reactions (Mengel and Kirkby, 1978).

Fageria and Souza (1991) determined the effects of Mg levels on root weights of rice, common bean, and cowpea grown in an Oxisol of Central Brazil (Figure 2.8). Dry weight of roots of rice was higher at the lowest Mg concentration compared with the highest soil Mg concentration. Initial exchangeable Mg levels of surface soils were 0.1 cmol·kg^{-1} and increased to 0.3 cmol·kg^{-1} within three days after liming and to 0.75 cmol·kg^{-1} at harvest time (33 days after sowing).

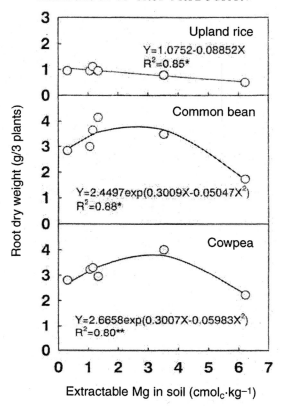

FIGURE 2.8. Root dry weights of upland rice, common bean, and cowpea grown with different Mg levels in an Oxisol. *Source:* Adapted from Fageria and Souza, 1991.

The lack of growth responses to applications of Mg indicated that this level of exchangeable Mg was adequate to meet Mg requirements of upland rice grown in this limed soil. Dry weight of roots of common bean increased by Mg application up to 3 cmol·kg^{-1} of soil based on quadratic regression equation. Similarly, significant quadratic responses were observed relative to root growth of cowpea and Mg levels in soil, and maximum root weight was achieved at 2.5 cmol Mg/kg of soil.

Micronutrients

Literature on the role of micronutrients on root growth of annual crops is limited. Baligar, Fageria, and Elrashidi (1998) have reviewed effects of deficiencies and toxicities of several elements on root morphology and growth of several crop species. Chairidchai and Ritchie (1993) reported that responses of wheat roots to Zn supply were pH dependent. At low pH (e.g., pH 4.0 and 4.6), when toxicity of Al^{3+} should be greatest, roots did not respond to increasing concentrations of Zn. On the other hand when Al^{3+} toxicity was absent at pH 5.2, increases in root growth were noted with increasing concentrations of Zn. Shoot growth appeared to be independent of root growth since wheat shoots responded to increases in Zn despite absence of root growth. Fageria (1992) studied root development of rice grown at deficient and adequate levels of Zn, Fe, Mo, Cu, B, and Mn. At deficient levels of these nutrients, root growth of rice was reduced, compared with adequate levels.

Application of B from 0 to 24 mg·kg^{-1} soil decreased root dry weight of upland rice and maize in greenhouse studies (Figure 2.9). However, application of B at lower levels to the same soil increased root dry weight in a quadratic fashion for common bean, soybean, and wheat (Fageria, 2000).

Salinity

Salinity results from the accumulation of salts in the rhizosphere at relatively high concentrations, which can impair crop root function and decrease yield. Salinity problems are more common in arid and semi-arid regions where evapotranspiration is higher than precipitation. When irrigation is practiced and soil permeability is low, soil salinization can occur. Subsoils with hydraulic conductivity that is 20 percent of the weighted hydraulic conductivity of overlying layers is considered a barrier to root growth (U.S. Department of Interior, Bureau of Reclamation, 1978). Sites are considered nonirrigable when they have salt barriers at depths of 1.8 m or less (Reichman and Trooien, 1993). Two potential effects of irrigation when salt barriers exist are development of perched water tables and accumulation of excess salts in root zones. Soil salinity may reduce root permeability and, consequently, reduce water and nutrient uptake (Fageria, 1992). Salts can limit nodule formation by reducing populations of *Rhizobium* in soils or by impairing *Rhizobium* ability to infect root hairs. A saline environment imposes osmotic and ionic stress on roots. Ionic stress in saline soil is due to high concentrations of Na^+, Mg^{2+}, Cl^-, and SO_4^{2-} and in some soils a toxic concentration of B. Salinity

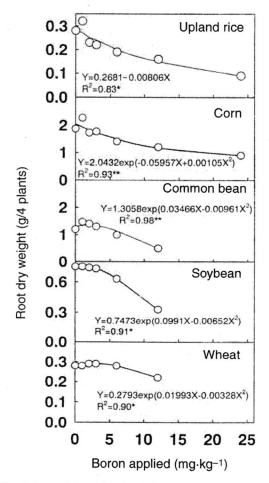

FIGURE 2.9. Root dry weights of upland rice, maize, common bean, soybean, and wheat grown with different B levels on an Oxisols of central Brazil. *Source:* Adapted from Fageria, 2000.

reduces root growth and permeability and consequently reduces water and nutrient uptake (Baligar, Fageria, and Elrashidi, 1998). Reduction of root mass and length, dieback of laterals, and reduced root systems are the common effects of soil salinity and alkalinity (Baligar, Fageria, and Elrashidi, 1998).

Biological Factors

Biological factors affecting crop root growth are differences among cultivars, diseases, insects, weeds, rhizobacteria, and mycorrhiza. Kaspar, Taylor, and Shibles (1984) reported differences in downward root growth among soybean cultivars. Root penetrating ability can vary among species and within species. Sorghum genotypes differ in rooting traits; those genotypes with greater root weights penetrated root systems more deeply, and higher root-shoot ratios were more drought tolerant (Jordan and Miller, 1980). Similarly, considerable diversity occurs in rooting traits of peanut genotypes (Ketring, 1984). Eghball and Maranville (1993) reported that root weight and length were different among maize genotypes.

Diseases, insects, and weeds have adverse effects on plant root growth. The impact of injury by pests on physiological and morphological processes of plants depends on plant structure injured, type of injury, timing of injury relative to plant phenology, and intensity of injury. Kitchen, Buchholz, and Nelson (1990) reported that potato leafhopper infestation of alfalfa reduced root weight. Similarly, Onstad, Shoemaker, and Hansen (1984) reported that retardation of alfalfa shoot yields by potato leafhoppers was associated with not only reduced root growth but also root carbohydrate accumulation. Yield reductions in nematode-attacked soybean plants were considered to be a result of suppression of root growth and disruption of root form and function (Radcliffe, Hussey, and McClendon, 1990). Termite attack on roots of upland rice grown in an Oxisol of Central Brazil is common, and plants have white heads and reduced yields.

The initial, and in some instances the only, effect of crop inoculation with plant growth-promoting rhizobacteria is often the promotion of root growth (Chanway and Nelson, 1991). Consequently, root growth enhancement is considered an important criterion in selection of microbial inoculants.

Rovira, Foster, and Martin (1983) reported that noninfecting rhizospheric microorganisms may affect mineral nutrition of plants through their influence on growth and morphology of roots, physiology and development of plants, and nutrient uptake and availability processes.

Microbial activity in the rhizosphere may increase or decrease availability of nutrients, especially micronutrients. Organic acids and/or chelating compounds that are present in root exudates may be responsible for mobilizing Mn or Fe and detoxifying Al, but utilization of these organic acids by rhizosphere microorganisms reduces uptake of these micronutrients by plants (Marschner, 1991). Phytosiderophores released by grass roots under Fe or Zn deficiency are readily degraded by microorganisms (Marschner, 1991). If sugars in root exudates are utilized by microorganisms with effective organic acids producers or metal chelating compounds (e.g., siderophores), positive effects of rhizosphere microorganisms on micronutrient availability might be expected.

Siderophores produced by *Pseudomonas* spp. increase solubility and mobility of Fe in soil solutions, and increase Fe supply to roots. Similarly, noninfecting rhizosphere microorganisms are important in Mn availability to plants. In well-aerated calcareous soils, Mn availability to plant roots can be decreased by Mn-oxidizing bacteria in the rhizosphere. Plant roots infected with mycorrhizae improve availability of P, Fe, and Cu to host plants (Marschner, 1991).

Arbuscular mycorrhizal fungi (AMF) benefit plants by allowing them to grow and produce in relatively harsh mineral toxic environments (Clark and Zeto, 2000). This has been attributed extensively to the ability of AMF to expand the volume of soil for which mineral contents are made available to plants compared to what roots themselves would contact. The nutrients enhanced most in host plants grown in many soils (e.g., high and low soil pH) are N, P, K, Ca, Mg, Zn, and Cu, which are enhanced when plants are grown in acidic soils. Many AMF have also the ability to ameliorate Al and Mn toxicities for plants grown in acidic soil (Clark and Zeto, 2000).

MANAGEMENT STRATEGIES FOR MAXIMIZING ROOT SYSTEMS

Several management strategies that might improve root systems of crop plants are suggested.

1. Improve soil conditions for root growth by deep plowing to break compacted layers that roots cannot readily penetrate. If water tables are near soil surfaces, drainage would be a good alternative. When depths to rooting-restrictive hardpans are relatively shallow (<0.25 m), chisel plowing can be effective for

disrupting compacted layers. However, such disruption does not last unless organic matter is incorporated to hold channels open.

2. Root growth may be manipulated through cultural practices such as mulching that can affect soil temperature. Warmer soil temperatures generally produce larger root systems. In addition, crop residue on soil surfaces decreases soil evaporation and improves water-use efficiency.

3. Greater sowing depth may decrease seedling emergence rate, possibly due to lack of aeration, which could increase survival of emergent seedlings, as well as increase both depth of rooting and water availability.

4. Larger seeds generally produce seedlings with more extensive root systems. Sowing good quality seeds could improve root systems.

5. Add and/or maintain adequate soil organic matter to improve internal drainage, aeration, water-holding capacity, and create adequate or improve physical, chemical, and biological environments for ample and deep root proliferation. For example incorporation of organic matter reduces metal toxicities and increases availability of P. This practice has been an appropriate and effective measure to improve root environments of annual crops.

6. To prevent harmful accumulation of salts in the root zone, leaching may be required. Steady-state leaching requirements (L_r) may be estimated (Reichman and Trooien, 1993) as

$$L_r = (D_d/D_a) = (EC_a/EC_d) \qquad (2.9)$$

where D_d and D_a are depths of drainage water and applied water, respectively, and EC_a and EC_d are electrical conductivity of applied water and drainage water, respectively.

7. Use of adequate lime on acidic soils is still a dominant practice for improving crop yields and improving root development. Results indicate that adequate levels of lime application and good mixing in soils can ameliorate subsoil acidity (Fageria et al., 1991; Fageria and Zimmermann, 1996).

8. Use of adequate fertilization to improve nutrient status of crops can improve root growth in low-fertility soils such as Oxisols of the Cerrado region in Brazil. Phosphorus deficiency is one of

the most important yield-limiting factors limiting root proliferation in these soils.

9. Gypsum ($CaSO_4 \cdot 2H_2O$) or phosphogypsum (e.g., by-products of
   phosphoric acid manufacturing processes) applications to
   leach Ca into deeper soil profiles where Ca can replace Al on
   cation exchange complexes. Much of the Al displaced by Ca
   can be leached from the root zones. This practice works well
   in sandy soils or Oxisols with clay loam aggregates which behave like sands (Foy, 1992).

10. Integrated management systems may be used to control diseases,
    insects, and weeds. Use of appropriate crop rotations has been
    an important practice. Reduced tillage, N side-dressing, and
    early planting can be included in integrated pest management
    programs with no risk of increasing potential for root damage
    from western maize rootworm (Roth, Calvin, and Lweloff,
    1995). Miltner, Karnok, and Hussey (1991) reported that cyst
    nematodes suppressed soybean root growth on susceptible
    cultivars, whereas root growth of tolerant cultivars was
    stimulated by the presence of soybean cyst nematodes. Use of
    tolerant or nutrient-use-efficient cultivars may be an impor-
    tant practice to improve root growth in stress environments.
    Stress-tolerant genotypes are being identified and bred world-
    wide to solve some of the most difficult problems of soil fertil-
    ity such as subsoil acidity, salinity, low plant availability of Fe
    in calcareous soils, and low P availability in acidic soils (Foy,
    1992). Rhizobial species/strains differ markedly in tolerance
    to low pH, and toxicities of Al and Mn in tropical soils have
    been identified. Use of improved strains can improve N_2 fixa-
    tion and root growth of legumes grown in acidic soils.

11. Identify and use adapted and improved strains of soil fungi
    and bacteria to improve uptake of essential and toxic elements
    and water by plants (Kucey, Janzen, and Leggett, 1989).

12. Ridge-till tillage systems at growth stage V7 tend to result in
    shorter internodes and more functional nodal roots which may
    be important to maize for lodging resistance and recovery
    from maize rootworm (Thomas and Kaspar, 1997).

13. Tailor plants to fit soils having physical and chemical problems
    that are not economically correctable with current technology.

CONCLUSION

Besides anchoring plants, roots absorb water and nutrients from soil. Root architecture can alter water and nutrient uptake and consequently affect plant physiology. Roots also synthesize numerous plant hormones such as cytokinins and gibberellins that influence plant physiological processes. Altered hormone levels may affect not only root systems but also leaf photosynthesis and plant development. Many studies have been conducted to understand the various aspects of root growth and function. Well-developed root systems are important in determining plant yield. Several environmental factors also affect root growth and development and consequently root architecture. These factors have been divided into physical, chemical, and biological categories. Important features of root distribution that need to be evaluated are (1) rooting density which influences thoroughness with which a unit volume of soil is explored; (2) duration and rate of root extension which governs ability of roots to explore new soil; and (3) depth of penetration, a measure of root ability to explore subsoil for water and nutrients. Although many root characteristics can be measured, root length is one of the most meaningful parameters relative to water and nutrient absorption by crop plants. Management needed for maximizing root growth requires that we improve our understanding of root dynamics in differing crop ecosystems and environments. Even though some roots may penetrate to considerable depths, the major portion of roots occurs in the upper layers of soil. Depth of root penetration is influenced by several environmental factors. Concentration of roots in upper layers of soil does not imply that roots growing in deeper layers are not important. Indeed, they may be vital when available water has been extracted from surface soil layers. Growth of roots seldom continues throughout the entire plant life cycle, and many studies point to cessation of growth at about the time of heading in cereals. In legume crops, root development may reach its maximum at about pod-setting growth stage. However, root growth can continue well into grain development stages of growth under favorable nutrient and moisture conditions. Level of nutrition can affect growth of roots and shoots differently, root growth generally being reduced under high levels of nutrition.

Chapter 3

Physiology of Growth and Yield Components

INTRODUCTION

Crop yield is determined by plant growth and partitioning of biomass to marketable parts of plants. Both growth and partitioning are affected by developmental stages of crops. Knowledge of physiological processes of growth, development, and partitioning into yield components is necessary for a basic understanding of maximizing crop yield. It is important first to define growth.

Growth refers to biomass accumulation and can be measured by leaf area, shoot, root, and total weights, or plant height, and these can be used to compare within cultivars or between cultivar means (McCauley, 1990). The term growth also applies to quantitative changes that occur during development and may be defined as an irreversible change in the size of cells, organs, or whole organisms (Wareing and Phillips, 1981). Wilhelm and McMaster (1995) defined growth simply as irreversible increases in physical dimension of individuals or organs with time. Therefore, examples of growth are irreversible lengthening of leaf blade tissue or increases in leaf area. In agriculture, where the purpose of most enterprises is to convert solar energy into dry matter, an equally useful definition of growth is increased dry weight. Growth is affected by environmental factors such as climate, soil, and plants themselves. Growth is sometimes confused with development. These two terms are interrelated but different.

Plant development may be defined as the sequence of ontogenetic events, involving both growth and differentiation, leading to changes in function and morphology (Landsberg, 1977). Development is generally measured over time between various physiological stages. Development includes processes of organ initiation (morphogenesis) but extends to differentiation and ultimately includes processes of senescence. The process by which plants, organs, or cells pass through various identifiable stages during their life cycles can be considered a functional definition of development. Phyllochrons (interval between similar growth stages of successive leaves on the same culm) have been used extensively to describe and understand development of grasses (Wilhelm and McMaster, 1995). Development may be affected slightly by factors influencing growth but is more closely related to heat unit accumulation (Stansel, 1975). Growth and development in crop plants do not proceed at constant or fixed rates through time. They are modified by environmental factors such as temperature, light intensity and duration, nutrition, and cultural practices. Therefore, calendar date is not suitable as quantitative descriptions for developmental stages of plants. Many attempts have been made to define precise and easily applicable methods for describing each important period and stage during cereal development. The scales used to assess development of cereals have been described by Landes and Porter (1989). These authors identified 23 scales, which are separated into those that codify external development of plants and those that describe events at apical meristems. Similarly, growth stages of legumes such as soybean and common bean have also been described (Fageria, 1992).

THEORY OF GROWTH AND YIELD
COMPONENTS ANALYSIS

Techniques used to quantify components of crop growth are collectively known as growth analysis. Growth analysis techniques have made substantial contributions to current concepts of crop yield physiology. Functional growth analysis, based on experiments in which plants are grown under standard conditions and harvested at regular intervals, can provide the first clues toward an understanding of variation in growth rates among genotypes or species (Lambers, 1987). Plant growth analysis can also be used to evaluate effects of cultural management practices such as fertilization, plant spacing and density, irrigation treatments, and disease, insect, and weed control. Plant growth analysis, demographic analysis, and yield component analysis are three procedures used to study relationships associated with plant growth and development (Jolliffe, Eaton, and Doust, 1982).

Plant growth analysis is generally expressed as indexes of growth such as crop growth rate, relative growth rate, net assimilation rate, leaf area ratio, and leaf area index. Discussion of these growth indexes follows.

Crop Growth Rate

Dry matter accumulation rate per unit land area is referred to as crop growth rate (CGR). It has been calculated as follows:

$$CGR = (W_2 - W_1) / SA (t_2 - t_1) \qquad (3.1)$$

where CGR is crop growth rate expressed in $g \cdot m^{-2}$ per day or $kg \cdot ha^{-1}$ per day, W_1 and W_2 are crop dry weights at the beginning and end of intervals, t_1 and t_2 are corresponding days, and SA is the land area occupied by plants at each sampling. Crop growth patterns can be defined accurately by taking plant samples at different time intervals during the growing season. Values of CGR are normally low during early growth stages and increase with time, reaching maximum values at about the time of flowering. Analysis of CGR is important for evaluating treatment differences among crop species or cultivars within species in relation to yield. Some CGR values of some important annual crops are presented in Table 3.1. Lowland rice had maximum CGR values and common bean had minimum.

Relative Growth Rate

Relative growth rates (RGR) of plants at specific instants in time (t) are defined as increases in plant dry matter per unit plant material per unit time (Radford, 1967). These RGR values can be calculated using the following formula and are expressed as gram per gram dry weight per day ($g \cdot g^{-1}$ dry wt. per day) or $kg \cdot ha^{-1}$ dry wt. per day.

$$RGR = (1/W) \times (dW/dt) = (d/dt) (\log_e W) \qquad (3.2)$$

where W is the dry weight and dW/dt is the change in dry weight per unit time. Values of RGR are generally higher during early growth stages of crop growth and decrease with advancement in age. Curves

TABLE 3.1. Maximum growth rates during growth cycles of upland and lowland rice, common bean, maize, and soybean grown on an Oxisol and Inceptisol in central Brazil.

Crop species[a]	CGR (kg·ha^{-1}/day)
Upland rice (90-99 days after sowing)	229
Lowland rice (98-111 days after sowing)	385
Common bean (43-62 days after sowing)	66
Maize (35-53 days after sowing)	311
Soybean (102-120 days after sowing)	138

Source: Fageria, 1998.

[a]Lowland rice was grown on an Inceptisol and the other crops were grown on an Oxisol. Growth cycle of upland rice was 130 days; lowland rice was 141 days; common bean was 96 days; maize was 119 days; and soybean was 158 days.

for RGR of crops are opposite to dry matter accumulation during life cycles of crops.

Net Assimilation Rate

Dry matter accumulation per unit of leaf area is termed net assimilation rate (NAR) and is expressed as grams per square meter of leaf area per day (g·m^{-2} leaf area/day), and can be calculated using the following formula (Brown, 1984):

$$NAR = (1/A) (dW/dt) \tag{3.3}$$

where A is leaf area and dW/dt is the change in plant dry matter per unit time. The objective of measuring NAR is to determine efficiency of plant leaves for dry matter production. NAR values decrease with crop growth due to mutual shading of leaves and reduced photosynthetic efficiency of older leaves (Fageria, 1992).

Leaf Area Ratio

Leaf area ratios (LAR) of plants at specific instants in time *(t)* are defined as ratios of assimilatory plant material per unit of plant material. This growth parameter is calculated as follows:

$$LAR = (A/W) \tag{3.4}$$

where A is leaf area and W is plant dry weight.

Leaf Area Index

Leaf area index (LAI) is defined as leaf area per unit soil area $(cm^2 \cdot m^{-2})$. This growth index can be calculated as follows (Watson, 1958):

$$LAI = (A \times N)/10,000 \tag{3.5}$$

where A is leaf area (cm^2) and N is number of tillers (cereals), branches (legumes), or plants per m^2.

Watson (1958) pointed out that crop growth rates (CGR) could be calculated from LAI and NAR as follows:

$$CGR = LAI \times NAR \tag{3.6}$$

Demographic Analysis

Bazzaz and Harper (1977) first applied demographic analysis to study plant growth. Demography follows births and deaths of individual components, and demographic analyses study dynamics of leaf populations and flower populations (Abul-Faith and Bazzaz, 1980). Hunt (1978) pointed out that demography can be used in addition to traditional plant growth analysis.

YIELD COMPONENTS ANALYSIS

Yield components analysis subdivides plant development into stages of growth. Indexes called yield components represent those stages of growth, and plant yield or productivity is the mathematical value of yield components (Jolliffe, Eaton, and Doust, 1982). Hardwick (1976) studied origins of yield components analysis to improve agricultural productivity. For example, yield components analysis provided key physiological information during the selection of some modern high-yielding rice cultivars (Yoshida, 1972). The interpretation of quantitative relationships among yield components has been advanced in recent years through the use of correlation analysis and partial regression analysis. Yield components are important in many crop research programs. Plant breeders often seek to improve yield by selecting components of yield such as seeds per plant or thousand-kernel weight. A complex character of yield may be defined as a character for which variation is determined by variations in numbers of component traits (Bos and Sparnaaij, 1993; Piepho, 1995). Perry and D'Antuono (1989) concluded that 80 percent of overall increases of some Australian spring wheat cultivar yields were due to increased harvest index values. Grains per ear and grains per m^2 were strongly and positively correlated with grain yield, but weak negative correlations were noted between 1000-grain weight and yield. Cultivars with semi-dwarf backgrounds had equal biomass, but higher yields, harvest index values, ear numbers per m^2, and grains as compared to modern tall cultivars. Perry and D'Antuono (1989) also concluded that genetic improvement substantially increased yield potentials of wheat, which was achieved through substantial increases in grain numbers per m^2 and with improved harvest indexes. Mean harvest indexes were 0.29 for older wheat cultivars and 0.36 for more recent semi-dwarf cultivars.

Among components for grain yield of cereal crops, numbers of spikelets appear to be important in developing high-yielding cultivars (Feil, 1992). However, increasing numbers of spikelets per panicle in rice may cause overproduction of spikelets on secondary branches (Yamamoto et al., 1991). In contrast to spikelets on primary branches, grain spikelets on secondary branches generally do not fill sufficiently (Kato, 1993). Thus, increasing numbers of spikelets per panicle does not always result in higher yields. Breeders frequently face conflicting requirements for improving yield, that is, increasing both numbers of spikelets per panicle as well as filled grain percentages (Kato, 1997).

Grain yield in field crops is determined by various yield components. Important yield components in cereals are panicles or ear per unit area, numbers of spikelets per panicle or ear, and spikelet weights. Similarly, yield components in legumes are determined by numbers of pods per unit area, grains per pod, and dry weights of grain. In cereals, relationships between yield and yield components can be expressed as follows (Fageria, 1992):

Grain yield = numbers of panicles or ears m^{-2}
　　　　　× numbers of spikelets per panicle or ear
　　　　　× 1,000 grain weight (g)
　　　　　× filled spikelets (percent) × 10 $^{-5}$　　　　(3.7)

For example, to obtain a rice crop of 6 t·ha^{-1} grain yield, it would be necessary to have a combination of the following yield-attributing characters or components:

1. 400 panicles per m^2,
2. 80 spikelets per panicle,
3. 85 percent filled spikelets, and
4. 22 g weight 1,000 grains.

By putting these values into the above equation,

$$\text{Grain yield} = 400 \times 80 \times 0.85 \times 22 \times 10^{-5} = 6 \text{ t·ha}^{-1}. \quad (3.8)$$

This yield equation is based on the fact that combinations of grain numbers, grain weights, and spikelet numbers are essential to raise yield levels in most grain crops. Several studies have shown that these components are negatively correlated and difficult to combine (Sinha and Swaminathan, 1984). Nevertheless, we know that despite having negative correlations, yields have been improved in many crops, especially for wheat and rice. An understanding of these crop development properties can be useful in breaking these correlations. Appropriate agronomic practices such as seeding rates to obtain optimum numbers of spikelets per unit area can also help offset negative correlations observed.

Among yield components, numbers of panicles have usually been the most variable for rice even though it has been related to grain yield (Figures 3.1 and 3.2). Grain yield increased quadratically with increasing numbers of panicles, and maximum yield was achieved at about 583 panicles per m^2 (calculated from regression analysis). Varying plant density and tillering performance can vary numbers of panicles per unit area.

Other yield components have also been associated with crop yield, but many have not. For example, harvest index (grain yield/grain plus

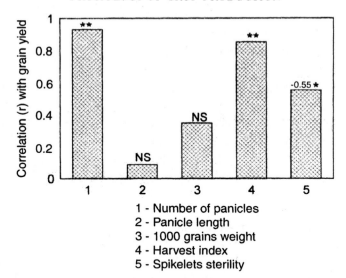

1 - Number of panicles
2 - Panicle length
3 - 1000 grains weight
4 - Harvest index
5 - Spikelets sterility

FIGURE 3.1. Correlation between grain yields and yield components of lowland rice. *Source:* Reproduced from Fageria, 1998.

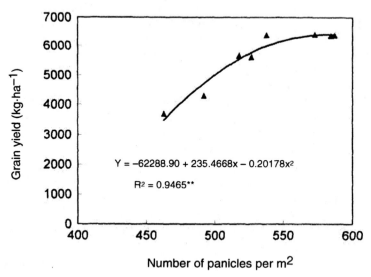

$$Y = -62288.90 + 235.4668x - 0.20178x^2$$

$$R^2 = 0.9465**$$

Number of panicles per m^2

FIGURE 3.2. Relationship between numbers of panicles and grain yield of lowland rice. *Source:* Reproduced from Fageria, 1998.

straw yield) has been positively correlated with grain yield, but spikelet sterility has been negatively correlated. Beyond certain plant densities, negative correlations exist between components of yield capacity so that yields tend to become constant under given sets of conditions. Such correlations appear to be developmental rather than genetic (Yoshida, 1972). Dofing and Knight (1994) evaluated performance and yield component contributions for yield of five uniculm barley genotypes. Results from path analysis demonstrated that spikes per square meter was the primary determinant of grain yield, followed by kernels per spike, with kernel weight of only minor importance. Increasing spikelets per square meter caused relatively large reductions in kernels per spikelet and kernel weights, while increasing kernels per spikelet caused only minor reductions in kernel weights.

Yield components do not influence crop yields independently but are closely related to one another. In many cases, increases of one component cause declines of other components. For example, increased numbers of panicles were associated with decreased numbers of grains per panicle. It has also been a general tendency that increases in number of grains per unit area cause decreases in percentage of ripened grains (Matsushima, 1976). This means it is possible to increase yield components independently with some management practices, but grain yields generally have not increased as expected.

Examples of using the yield equation for grain yields of legumes are given as follows:

$$\text{Yield (t·ha}^{-1}) = \text{pod number m}^{-2} \times \text{grain number per pod}$$
$$\times \text{1,000 grain weight (g)} \times 10^{-5} \qquad (3.9)$$

or

$$\text{grain number m}^{-2} \times \text{1,000 grain weight (g)} \times 10^{-5} \qquad (3.10)$$

To obtain yields of 1.5 t·ha^{-1} cowpea, the following combination of yield components would be needed: 155 pods per m^2, seven grains per pod, and 140 g for 1,000 grains. By putting these values into the above equation, the result would be

$$\text{Grain yield} = 155 \times 7 \times 140 \times 10^{-5} = 1.5 \text{ t·ha}^{-1} \qquad (3.11)$$

In addition, Akhter and Sneller (1996) used the following yield component model to determine soybean grain yields:

Plot yield = plants per plot × [(main stem nodes/plant
  × pods/node × seeds/pod × grams/seed)
  + (branches/plant × pods/branch
  × seeds/pod × grams/seed)]   (3.12)

Numbers of main stem nodes, main stem pods, branches, pods on branches, and seeds were counted for each plant. Pods per main stem node were calculated by dividing numbers of pods on main stems by numbers of main stem nodes. Numbers of pods per branch were calculated by dividing total numbers of pods on branches by numbers of branches. Numbers of seeds per pod for main stems and branches were calculated by dividing total numbers of seeds per plant by total numbers of pods per plant. Seed weights were determined by dividing total weights of seeds per plant by numbers of seeds per plant. These yield components could be classified into primary yield components affecting final yield (seed numbers and seed sizes), secondary yield components affecting seed numbers (seeds per pod and pod numbers), and tertiary yield components affecting pod numbers (reproductive node numbers, pods per reproductive node, percent nodes being reproductive, and node numbers).

Greater understanding of how yield components influence yield formation on phenotypic and genotypic levels can be obtained by applying path analyses to determine direct and indirect effects of primary, secondary, and tertiary yield components on their respective response variables (yield, seed numbers, and pod numbers, respectively). Such analyses will help clarify which yield components are most important for mediating environmental (especially crop growth rate and light interception) and genetic effects on yield (Board, Kang, and Harville, 1999).

Analysis of 12 soybean cultivars grown in narrow and wide row spacing revealed that yield was positively correlated with pod numbers and total dry matter at the R_5 growth stage, but not seeds per pod or seed size (Board, Zhang, and Harville, 1996). These results led to speculation that genetic improvements of late-planted soybean yields may result from selection for pod numbers or yield components influencing pod numbers. Because pod numbers are largely determined by crop growth rates during emergence to the R_5 growth stage (Board, Kamal, and Harville, 1992), genetic differences in pod numbers could be caused by genetic differences in crop growth rate. However, path analysis revealed that seed or pod numbers are poor predictors for genetic effects on yield (Board, Kang, and Harville, 1999).

Number of pods in legumes such as soybeans has been a critical yield component trait (Fageria, Baligar, and Jones, 1997) even though some cultivars known to initiate large numbers of pods did not produce high yields. In fact, shedding of growing pods, shell enlargement, seed abortion, or nonoptimal seed growth may contribute to reductions of yield potential. Environmental stresses due to high temperature (Egli and Wardlaw, 1980) and water deficits (Meckel et al., 1984) may cause pod and seed losses. Thus, it is important to simulate the dynamics of pod and seed setting and to specify effects of genetic and climate factors on each yield component (Colson, Bouniols, and Jones, 1995). For example, Board, Kang, and Harville (1999) reported that although several yield components of soybean were being formed simultaneously, changes in source strength affected some traits. That is, rapid seed filling commenced near the R_6 growth stage, and seed growth rate commenced during rapid seed filling and was influenced by numbers of cotyledonary cells, which formed between the R_3 (Peterson et al., 1992) and R_6 (Egli, 1994) stages of growth. Other studies indicated that changes in assimilatory capacity during this period of growth could reduce cotyledonary cell numbers, resulting in slower seed growth rates and decreased seed size (Guldan and Brun, 1985). Seeds per pod are determined starting from shortly after the R_1 (Peterson et al., 1992) to near the R_6 growth periods when pod extension is completed (Pigeaire, Duthion, and Turc, 1986). Major factors affecting pod numbers (pods per reproductive node and reproductive node numbers) are also determined during the R_1 to R_6 growth periods (Board and Tan, 1995). Board, Kang, and Harville (1999) concluded that the best genetic approach to improve yield at late plantings of soybeans is to select for cultivars or genotypes with high pods per reproductive node but which at the same time do not show compensatory interactions between seed numbers and seed size.

Higher crop yields are achieved only when appropriate combinations of yield component traits occur. Similarly, higher dry matter yields are related to higher grain yields. Understanding the physiology of dry matter production and yield component traits can help in manipulation of these yield-determining factors to favor higher yields.

DRY MATTER PRODUCTION AND GRAIN YIELD

Dry matter does not mean any single substance but refers to the entire organic dry matter that is produced from essential activities of photosynthesis and protein metabolism. Organic dry matter includes glucose, starch, cellulose, amino acids, protein, and other organic components (Akita, 1995). Dry matter of plants refers to total biological yield which includes leaves, stalks, straw, roots, and other plant parts that may not be used commercially as well as economically useful parts such as grains, tubers, and fruits. For agricultural purposes, dry matter yields of many crops refer to aboveground (soil) grown plant parts but may contain belowground plant parts for some crops such as potatoes, yams, taro, cassava, and some vegetable crops. Whole-plant aboveground biomass is the economically useful part of forage plants, while grain is the main useful part of cereals and soybeans. For cereals and legumes, grain yields are related to total biomass yields as well as other yield component traits. For example, grain yield of cereals and soybeans has been related to harvest index as follows (Donald and Hamblin, 1976):

$$\text{Grain yield} = \text{biological yield (straw + grain)} \times \text{harvest index} \qquad (3.13)$$

According to the above equation, crop production can be measured as total biomass or economically useful plant parts. Total yield of plant material is known as biological yield, and ratio of grain yield to biological yield is harvest index (Huhn, 1990). Grain yields correlated positively with biological yields of pigeon pea, mung bean, and urd bean (Snyder and Carlson, 1984). These authors concluded that biological yield did not seem to be related to limited high grain yield, but harvest index was related to limited yield of these three plant species. Therefore, greater emphasis could be placed on breeding for high harvest indexes for cultivars of these plant species.

Efficiency of grain production in crop plants has also been frequently related to harvest index. Sinclair (1998) and Hay (1995) stated that harvest index is an important trait associated with the dramatic increases in crop yields during the twentieth century. Harvest index reflects partitioning of photosynthate between grain and vegetative plant parts. Improvements in harvest index emphasize the importance of C allocation in grain production. By definition, harvest index values are <1, but some researchers prefer to express values as percentages.

Grain yield has also been related directly to harvest index even though biological yield and harvest index may not be related. Snyder and Carlson (1984) reviewed relationships of harvest index to yield and reported that harvest index correlated positively with grain yield and negatively with biological yield of barley, oats, rye, wheat, soybean, pigeon pea, and mung bean. Thus, selecting for higher harvest indexes should increase grain yield in most cases, particularly when biological yield is relatively stable. Increased harvest indexes have contributed to increasing yields of rice, barley, and wheat; values of more than 50 percent have been achieved for these cereals. However, limits of how far harvest index values can be improved is unknown. It appears that these values will not be able to rise much above 60 percent in cereals, although they may go further in root and tuber crops (Evans, 1980). Further increases in yield potential will depend on improvements in rates of photosynthesis and growth. Increases in harvest index of maize have been modest (from 45 percent in the 1930s to 50 percent in the 1980s), and much of maize yield increases have resulted from increases in total crop mass (Russell, 1985). Accumulation of high levels of N is essential for high maize grain yields, and high levels of N are commonly associated with crops having high harvest indexes (Sinclair, 1998). Under conditions where N has been limited, low harvest indexes of crops have been advantageous. Limited N can be partitioned into vegetative tissue of low N concentration, which results in higher total production of plant mass than in tissues of high N concentration. However, increasing grain yield and crop harvest index with high N grain requires concomitant increases in crop N accumulation (Sinclair, 1998).

Ranges in harvest index values among crop plants are considerable. Snyder and Carlson (1984) noted that harvest index for a semidwarf wheat was 47 percent, compared to 40 percent for standard cultivars, and varied from 23 to 50 percent for 21 late-duration, high-yielding rice cultivars. Harvest index values for peanut ranged from 20 to 47 percent for bunch types, 3 to 31 percent for semi-spreading types, and 10 to 22 percent for spreading types. Harvest index values for 23 cultivars of common bean varied from 39 to 58 percent. Snyder and Carlson (1984) also reported a cassava cultivar (M Col 22) at nine months of age having 80 percent of its total weight in root material. This has been one of the highest percentages of economic yield reported for any root crop. These large genetically controlled ranges in harvest indexes among various crops, coupled with significant correlations between harvest index and economic yield, have contributed to plant breeder success for using harvest index as a selection criterion to increase economic yields of crops (Snyder and Carlson, 1984). For several important crop species, harvest indexes (Table 3.2) and biological yield and economic yield based on glucose (Table 3.3) have been reported.

TABLE 3.2. Harvest index (percent) of five cereal crops.

Crop	Minimum	Maximum	Average
Millet	16	40	26
Sorghum	25	56	27
Maize	25	56	42
Rice	34	55	44
Wheat	35	49	41

Source: Van Duivenbooden, De Wit, and Van Keulen, 1996; Fageria, Baligar, and Jones, 1997.

TABLE 3.3. Biological and economic yield and harvest indexes of some important field crops.

Crop	Biological yield (g $CH_2O \cdot m^{-2}$)	Economic yield (g $CH_2O \cdot m^{-2}$)	Harvest index
Rice	1324	657	0.50
Wheat	1957	1125	0.57
Maize	2240	1157	0.52
Soybean	816	392	0.49
Sugar beet	2174	1436	0.66

Source: Adapted from Osaki, Fujisaki et al., 1993.

Differences in translocation of photosynthate to grain may account for differences in harvest indexes (Gent and Kiyomoto, 1989). The ability of plants to retain and efficiently remobilize photosynthate to grain through maturity may also influence harvest index and yield. Almost half of plant photosynthate before anthesis is lost from wheat plants at maturity (Austin et al., 1977), and similar fractions may be lost during grain filling. Photosynthate is lost by respiration and by abscission of plant material during senescence. Wheat plants that efficiently remobilize photosynthate to grain during maturation may retain more photosynthate and provide higher harvest indexes (Gent and Kiyomoto, 1989).

Plants with any combination of morphophysiological traits that offer high harvest indexes and high grain yield deserve consideration in plant breeding programs. Limits to which harvest index can be increased are considered to be around 60 percent (Austin et al., 1980). Hence, cultivars with low harvest indexes would indicate that further improvements in partitioning of biomass are likely. On the other hand, cultivars with harvest indexes between 50 and 60 percent would probably not be improved further (Sharma and Smith, 1986).

Harvest index has been shown to be positively related to grain yield in wheat (Kulshrestha and Jain, 1982; Singh and Stoskopf, 1971). Donald and Hamblin (1976) suggested that plant breeders should use biomass and harvest index traits as early-generation selection criteria. Harvest indexes along with grain yields have been selection criteria for improving yields of cereals by several researchers (Austin et al., 1980; Riggs et al., 1981; Sedgley, 1991; Peng et al., 2000).

It was noted that 72 percent of grain yield variability in field studies of 49 spring wheat cultivars were associated with harvest index values from single plants grown under greenhouse conditions (Syme, 1972). Indirect selection for oat grain yields using harvest index was 43 percent as efficient as direct selection (Rosielle and Frey, 1975). Harvest indexes of spaced plants were superior to grain yields for prediction of wheat grain yields in large field plots (Fischer and Kertesz, 1976). Bhatt (1977) reported that harvest index was a useful selection criterion for improving grain yields of two wheat crosses in segregating generations. Harvest index was found to have merit as a selection criterion for grain yields in two crosses of spring wheat and was considered more reliable at high plant populations than at low population densities (Nass, 1980).

Genetic improvement in yield potentials of many crops has resulted from increases in harvest indexes (Austin et al., 1980; Riggs et al., 1981; Peng et al., 2000), which have been associated with ideotypic characters such as short stature in wheat and uniculms in maize and sunflower (Sedgley, 1991). However, many researchers have also reported improvement in yield potentials of many crops to be associated with increased biomass yields in wheat (Waddington et al., 1986), maize (Tollenaar, 1989), and soybean (Cregan and Yaklich, 1986). Improvements in grain yields have been related to both dry matter accumulation and harvest indexes in wheat, barley, and oat (McEwan and Cross, 1979; Wych and Rasmusson, 1983; Wych and Stuthman, 1983). Peng et al. (2000) reported that increasing trends for yields for rice cultivars released before 1980 were mainly due to improvements in harvest indexes, while increases in total biomass were associated with yield trends for cultivars released after 1980. Hybrid rice had about 15 percent higher yields than inbreds mainly because of increases in biomass production rather than in harvest indexes (Song, Agata, and Kawamitsu, 1990; Yamauchi, 1994). This would indicate that further improvement in rice yield potentials might come from increased biomass production rather than increased harvest indexes.

Grain yields are related to dry matter production as well as harvest indexes (Fageria, 1992). Leaf area index is one of the most important plant growth indexes for determining dry matter yields, and consequently yield. Increasing LAI values increases dry matter production. Obtaining improved or optimum LAI values for plant species or cultivars within species may become advantageous traits to use in plant breeding programs.

Dry matter production at different growth stages for lowland and upland rice grown on an Inceptisol and Oxisol in central Brazil is presented in Figures 3.3 and 3.4. In the case of lowland rice (Figure 3.3), vegetative dry matter increased relatively slowly during the first 40 to 50 days but increased greatly from 50 to 120 days. Increases in dry matter during growth of this plant were attributed mainly to high photosynthesis from increased leaf area (Fageria, Santos, and Baligar 1997). Vegetative dry matter decreased 24 percent during grain filling, as active translocation of assimilates occurred from leaves/stalks

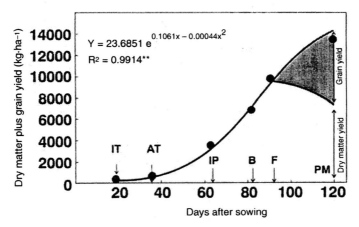

FIGURE 3.3. Dry matter accumulation and grain yield of lowland rice grown in central Brazil on an Inceptisol. *Source:* Reproduced from Fageria, 1998. *Note:* IT = initiation of tillering; AT = active tillering; IP = initiation of panicle; B = booting; F = flowering; and PM = physiological maturity

to grain (Guindo, Wells, and Norman, 1994). Grain yield was about 6600 kg·ha^{-1} or 49 percent of total dry matter production for this plant. Total dry matter production of upland rice grown on an Oxisol in Brazil increased nearly linearly as plants aged (Figure 3.4). Maximum vegetative dry matter was produced at about flowering and decreased by 21 percent due to assimilate translocation from leaves/ stalks to grains. Grain yield was about 4900 kg·ha^{-1} or 42 percent of total dry matter production for this plant. Protein metabolism dominates during vegetative growth phases, while carbohydrate metabolism dominates during reproductive growth phases (Murayama, 1995). Rates of protein, wall substances (cellulose and lignin), and carbohydrate accumulation in rice plants with growth stages are presented in Figure 3.5.

Maize vegetative dry matter consistently increased as plants aged (Figure 3.6). Dry matter accumulation for maize was different from that noted for lowland and upland rice (Figures 3.3 and 3.4), common bean (Figure 3.7), and soybean (Figure 3.8). Vegetative dry matter decreased after flowering for lowland and upland rice, common bean, and soybean, but total dry matter increased for maize. Both curves for dry matter of common bean and soybean were exponentially quadratic, and dry matter production increased up to 80 days for common bean and up to 120 days for soybean, Thereafter, dry matter production declined in both of these crops. Decreases in dry matter yield were likely related to translocation of assimilates to grain as well as loss of older leaves senescencing. For both common bean and soybean, grain dry matter was not included. These results indicated that dry matter production differs among crop species as well as during crop growth stages. Data for vegetative dry matter production at harvest and grain yield concerned with Figures 3.3 to 3.8 are provided in Table 3.4. Among cereals, dry matter production was related to grain yield. For example, maximum dry matter was produced by maize, and this crop also produced higher grain yields than the other crops. Lowland rice was second in dry matter production as well as in grain yield. However, such trends were not observed among legumes. For example, soybean produced higher dry matter than common bean but produced less grain and had a lower harvest index than common bean.

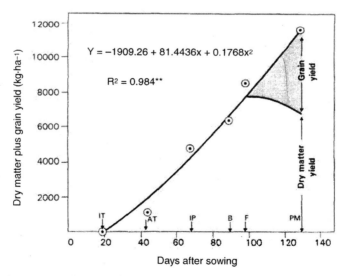

FIGURE 3.4. Dry matter accumulation and grain yield of upland rice grown in central Brazil on an Oxisols. *Source:* Reproduced from Fageria, 1998. *Note:* IT = initiation of tillering; AT = active tillering; IP = initiation of panicle; B = booting; F = flowering; and PM = physiological maturity

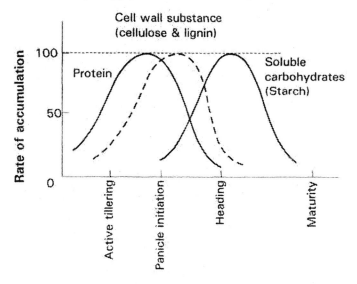

FIGURE 3.5. Accumulation rates of major organic substances in rice. *Source:* Reproduced from Murayama, 1995, with permission from Food and Agriculture Policy Research Center, Tokyo, Japan.

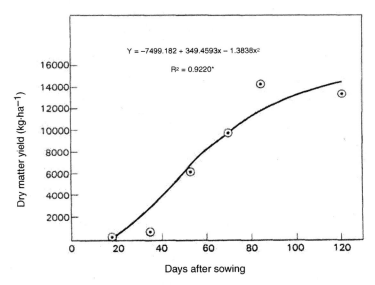

FIGURE 3.6. Dry matter production of maize grown in central Brazil on an Oxisol. *Source:* Reproduced from Fageria, 1998.

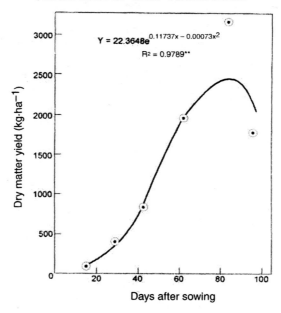

FIGURE 3.7. Dry matter production of common bean grown in central Brazil on an Oxisol. *Source:* Reproduced from Fageria, 1998.

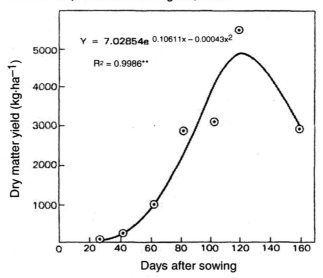

FIGURE 3.8. Dry matter production of soybean grown in central Brazil on an Oxisol. *Source:* Reproduced from Fageria, 1998.

TABLE 3.4. Shoot (straw) dry weight, grain yield, and harvest indexes of five annual crops grown on an Oxisol in central Brazil.

Crop species	Shoot dry weight $(kg \cdot ha^{-1})$	Grain yield $(kg \cdot ha^{-1})$	Harvest index
Upland rice	6642	4794	0.42
Lowland rice	9611	6862	0.42
Maize	13670	8148	0.37
Common bean	1773	1674	0.49
Soybean	2901	1323	0.31

Source: Fageria, 1998.

Unlike self-pollinated cereals in which improved partitioning of dry matter resulted in higher grain yields, similar improvements have not been obtained for grain legumes. Cereals are generally more responsive to increased soil inputs such as N fertilization, which accounts for improved cultivar performance through selection for responsive genotypes (Kelly, Kolkman, and Schneider, 1998). It has been widely reported that dry matter production of cereals such as rice, wheat, and maize is twice that for legumes such as soybean, common bean, and adzuki bean (Osaki, Shinano, and Tadano, 1992; Shinano, Osaki, and Tadano, 1994). It has been postulated that the low productivity of soybean was generally due to high contents of protein and lipids in seeds, because production of these compounds from photosynthates requires large amounts of energy (Yamaguchi, 1978). Based on studies of growth efficiency in harvesting organs and in whole plants during vegetative growth stages, low productivity of Leguminosae was mainly ascribed to high respiratory losses of carbohydrates from leaves and stems and not from harvested organs (Shinano et al., 1991; Shinano, Osaki, and Tadono, 1993), in contrast to findings reported by Yamaguchi (1978).

Tanaka and Osaki (1983) fed $^{14}CO_2$ to whole rice, wheat, maize, soybean, and common bean plants before analyzing for release of $^{14}CO_2$ from plants. They reported that larger amounts of $^{14}CO_2$ were released from soybean and common bean plants than from rice, wheat, and maize. Soybean also respired larger amounts of $^{14}CO_2$ from storage substances than did rice (Shinano et al., 1991). These results indicated that legumes consume larger amounts of current photosynthates and/or temporary storage substances for respiration in leaves and stems than Graminaea plants. Further, dry matter production efficiency per unit N absorbed was reported to be lower in legumes than in cereals (Osaki, Shinano, and Tadano, 1992). This was ascribed to low growth efficiency of shoots but not to nodulation and low growth efficiency of grains (Osaki, Shinano, and Tadano, 1992; Shinano, Osaki, and Tadano, 1995). Shinano, Osaki, and Tadano (1994) suggested that when $^{14}CO_2$ was fed to leaves of rice and soybean, the amount of $^{14}CO_2$ released from soybean leaves was larger than that for rice under light conditions, because of higher photorespiration of soybean. In addition, photosynthesized C in leaves of soybean was more distributed into organic and amino acids than for rice (Shinano, Osaki, and Tadano, 1994). Accordingly, it has been hypothesized that photosynthesized C distribution mechanisms differ between cereal and legume crops (Nakamura et al., 1997). That is, photosynthesized C was actively distributed into tricarboxylic acid (TCA) and amino acid pools in legume crops grown under light, which resulted in higher respiratory rates compared to cereal crops. Higher proportions of photosynthesized C were distributed into carbohydrate pools, and photorespiratory activity was lower in cereals than in legumes.

Table 3.5 relates lowland rice dry matter production, grain yield, total biological yield, and harvest index to different levels of N when grown on an Inceptisol in central Brazil. Dry matter production, grain yield, and total biological yield increased as level of N increased. Harvest index also increased quadratically with increases of N, but responses were not significantly different. That is, grain yield was related to dry matter production and harvest index in a quadratic fashion, and application of adequate N is one management practice that is needed to improve dry matter and grain yields in annual crops.

DURATION OF REPRODUCTIVE GROWTH PERIOD

Duration of the reproductive growth period is very important because numbers of grains for plants are determined during this period. It has been reported that numbers of spikelets per ear in cereal crops can be increased by increasing length of the reproductive growth period (Yoshida, 1972). In grain crops (cereals), inflorescences and

TABLE 3.5. Shoot (straw) dry weight, grain yield, total biological yield, and harvest index of lowland rice grown on an Inceptisol in central Brazil.

N Rate (kg·ha⁻¹)	Shoot dry weight (kg·ha⁻¹)	Grain yield (kg·ha⁻¹)	Total biological yield (kg·ha⁻¹)	Harvest index
0	5278	3678	8956	0.41
30	6764	4281	11045	0.39
60	7294	5602	12896	0.43
90	7303	5647	12949	0.44
120	8215	6344	14559	0.44
150	8623	6362	14986	0.42
180	9060	6372	15433	0.41
210	9423	6389	15812	0.40
R^2	0.96[a]	0.97[a]	0.98[a]	0.58[b]

Source: Fageria, 1998.

[a]Significant at the 1 percent probability level

[b]Nonsignificant

leaves grow at the same time. Therefore, distribution of assimilates between inflorescences and leaves will also determine sizes of inflorescences. Most high-yielding rice cultivars are known to have small flag leaves. This is because flag leaves compete with developing inflorescences for assimilates.

Temperature and photoperiod are two major environmental variables that affect length of the preflowering period. In general, high temperature accelerates and low temperature delays flowering, but extremely high temperatures can also delay flowering (Yin, Kropff, and Goudriaan, 1997). Longer vegetative time periods (to heading) resulted in higher barley grain yield, but longer heading to maturity periods had no effect on grain yield (Dofing, 1997). Longer pre-heading time periods appear necessary to develop adequate numbers of kernels per spike and sufficient numbers of leaves to provide photoassimilate during grain fill.

DURATION OF GRAIN-FILLING PERIOD

The grain-filling period is the duration from anthesis to physiological maturity and beyond. After physiological maturity, no significant increases in wheat grain dry matter occurred (Mou and Kronstad, 1994). However, harvest index generally increases over time after anthesis in soybean (Speath and Sinclair, 1985), cotton (Sadras, Bange, and Milroy, 1997), sorghum and maize (Muchow, 1990a,b), barley (Goyne et al., 1996), and sunflower (Bange, Hammer, and Rickert, 1998). Grain filling generally results from translocation of photosynthate from source to sink. For example, 80 to 90 percent of carbohydrate for grain growth of many plants is derived from photosynthetic activity after anthesis, and in the case of wheat, only 10 to 20 percent of carbohydrate comes from available reserves in vegetative plant parts (Spiertz and Vos, 1985). Kernel weight was also correlated with grain-filling rate in wheat (Bruckner and Frohberg, 1987). Wheat cultivars differ in kernel weight and rate and duration of grain filling (Darroch and Baker, 1990). Grain-filling period is also influenced by temperature and light (Grabau, Van Sanford, and Meng, 1990).

Sayre et al. (1998) reported grain-filling rates for 15 wheat genotypes to average from 145 kg·ha^{-1} per day for plots treated by fungicide (rust disease control) to 115 kg·ha^{-1} per day for plots without fungicide treatment. Figure 3.9 shows relationships between average daily temperatures on grain-filling period, as well as kernel weight and grain yield, of oat. For each of these traits, values decreased linearly as average daily temperature increased, but decreases in grain-filling period were less than for the other traits. Fageria (1998) reported relationships between N application rate and grain-filling rate in lowland rice grown on an Inceptisol in central Brazil. Nitrogen application increased grain-filling rate from 129 kg·ha^{-1} per day with no added N to 231 kg·ha^{-1} per day with 180 kg N·ha^{-1}.

Longer grain-filling periods commonly result in higher yields for many plant species grown under optimum environmental conditions. Because rice grain size is physically limited, yield capacity is largely determined by number of grains per unit land area. Therefore, increasing grain filling period would be meaningful if number of grains per unit area does not limit rice grain yield. Hartung, Poneleit, and Cornelius (1989) reported that grain-filling duration and dry matter

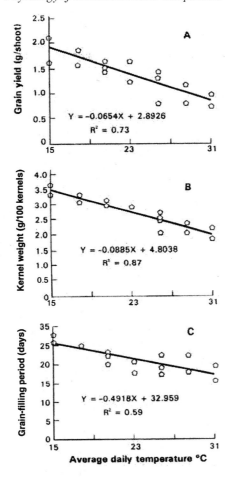

FIGURE 3.9. Linear regression of oat grain yield (A), kernel weight (B), and grain-filling period (C) as related to average daily temperature. *Source:* Reproduced from Hellewell et al., 1996, with permission from American Society of Crop Science, Madison, WI.

accumulation rates are associated with maize yield potential and might be used for indirect yield selection of this crop. Daynard, Tanner, and Duncan (1971) found that 70 to 80 percent of the yield differences among three maize hybrids were due to differences in the grain-filling period. A significant phenotypic correlation of 0.8 was reported between grain yield and grain-filling period (Cross, 1975). Egli (1981) reported similar associations for seed growth characteristics of several plant species, particularly for those plants with seed sizes in excess of 250 mg per kernel. Yield improvements for maize achieved in long-term recurrent selection procedures have been associated with increased grain-filling period duration (Crosbie, 1982). Spiertz, Tem, and Kupers (1971) and Daynard and Kannenberg (1976) reported positive correlations between length of grain-filling period and grain yield in spring wheat and maize hybrids. However, Dofing (1997) reported that longer vegetative periods (time from sowing to heading) resulted in higher grain yields for barley, but longer heading to maturity periods had no effect on grain yields. In addition, faster grain-fill rates resulted in higher grain yields and shorter grain-filling periods (Dofing and Knight, 1994). Longer preheading periods appear necessary to develop adequate numbers of kernels per spikelet and sufficient numbers of leaves to provide photoassimilate during grain fill. Improved productivity of new oat cultivars has been due to short grain-fill periods and increased panicle filling rates (Peltonen-Sainio, 1991). Board, Wier, and Boethel (1994) reported that soybean grain yield reductions were caused by reduced seed size from shorter effective grain-filling periods and slower seed-filling rates. Effective seed-filling periods are that part of the seed-filling period whereby seed dry matter accumulates at linear rates to increase final seed sizes (Gbikpi and Crookston, 1981).

Seed-filling rate is the dry matter accumulation rate during the period of linear seed filling (Board, Wier, and Boethel, 1997). Further, C:N balances during grain filling or ripening periods are important in determining final yields of field crops. Generally, C accumulation in harvested organs occurs mainly during ripening, while N accumulation in harvested organs occurs largely from leaves that accumulate N before ripening. A small part of N in harvested organs also accumulates directly from N absorbed by roots during the ripening period (Osaki, Morikawa et al., 1993).

For cereals such as rice and wheat, grain N originates from both remobilization of vegetative tissue N and from active uptake and assimilation of N during grain fill. Remobilized N is usually the predominant N source for cereals grown under field conditions (Simmons, 1997). As N is remobilized from vegetative tissue, photosynthetic capacity of tissue decreases. The total loss of leaf areas and termination of assimilate production should result in termination of grain fill (Frederick, 1997). Therefore, increasing plant N uptake during grain fill should theoretically delay leaf senescence and sustain leaf photosynthesis, resulting in longer grain-filling periods and heavier kernels.

Translocation of photosynthates from leaves to roots during grain filling decreases with plant age because sinks (organs to be harvested) require large amounts of photosynthates, and photosynthetic activity of lower leaves which commonly supplies photosynthates to roots decreases with mutual shading. As a result, root activity is assumed to limit N absorption during grain filling or maturation. In crops with high biological yields (e.g., maize, sugar beet, and sunflower), N absorption during maturation remains high (Osaki et al., 1995). This means that root ability to absorb N after flowering or grain filling is important for controlling productivity of some crops.

Gebeyehou, Knott, and Baker (1982) reported that the length of grain filling had significant positive correlations with kernels per spikelet (r = 0.39*), kernel weight (r = 0.56**), and grain yield (r = 0.39*) for 16 durum wheat cultivars. Knott and Gebeyehou (1987) noted relationships between grain yield and length of grain filling in three durum wheat cultivars, and significant correlations occurred between grain yield and length of grain filling.

Grain sink capacity (potential of grains to achieve maximum mass) is genetically determined for plants grown under optimum growth conditions. However, actual grain capacity for plants grown under field conditions can be mediated by environmental perturbations such as water deficiency and heat stress. Delayed sowings in temperate climates caused slower rates of grain filling, shorter duration of grain filling, and decreased final weights of maize kernels (Cirilo and Andrade, 1996). Decreased irradiation and temperature 19 days after silking to physiological maturity reduced amounts of assimilates available for grain filling sufficient to decrease kernel weight (Cirilo and Andrade, 1996). Grain dry matter accumulation depends on rate and duration of crop growth, and availability of assimilate reserves from other plant parts. Potential sink strength for assimilates of grain is established during the lag phases of grain development when endosperm cell division is occurring (Reddy and Daynard, 1983). A shortage of assimilate supply or unfavorable environmental conditions during grain filling affects potential kernel size in maize (Jones et al.,

1996). Increased temperatures during grain filling also increase metabolic rates and sink strengths of maize kernels, and rates of grain filling (Jones, Ouattar, and Crookston, 1984). Moreover, several authors have suggested that reduction in assimilate supply to maize kernels induces early black layer formation to decrease grain filling duration and kernel size (Tollenaar and Daynard, 1978; Afuakwa, Crookston, and Jones, 1984; Cirilo and Andrade, 1996).

Temperature has large effects on grain filling and yield components. Hellewell et al. (1996) studied the effects of temperature treatments during grain filling on oat grain yield and yield components by growing three genotypes with varying maturity (early, mid, and late season) in controlled environmental chambers with varying planting dates to synchronize heading date among cultivars. At heading, each cultivar divided into nine treatments of day-night temperature combinations of 31, 23, and 15°C. Temperature treatments were imposed until plants were fully mature. Grain yield was 87 percent, kernel weight 51 percent, grain-fill period 27 percent, and grain-filling rate 45 percent greater for plants grown at 15°C day temperature than at 31°C day temperature with night temperature of 15°C. For night temperature, grain yield was 24 percent, kernel weight 12 percent, and grain-filling period 27 percent greater at 15°C night temperature than at 31°C night temperature with day temperature of 15°C. Figure 3.9 shows relationships between temperature and grain yield, kernel weight, and grain filling for oat.

MANAGEMENT STRATEGIES
FOR IDEAL YIELD COMPONENTS

Environment, management, and genotypes interact to determine overall yields of crops. This means ideal yield components can be achieved through ideal combinations of favorable plant growth environments, adoption of adequate management practices, and use of appropriate genotypes. Among environmental factors, adequate moisture supply and optimum temperature and solar radiation are important in determining yield components and consequently yield. From the vast amounts of rainfall, temperature, and solar radiation data, it should be possible to adjust sowing dates and adopt appropriate plant species to determined environmental variables. The key aim to optimize crop productivity by matching ontogeny (sequence of development stages) to weather resources of the environment (duration of favorable temperature, water supply, and solar radiation), and to minimize unfavorable extremes during vulnerable stages of growth (Summerfield et al., 1997). For annual crops, it is well understood that water deficiency and extreme temperatures during flowering reduces grain yield more than any other stage of growth. Hence, sowing date may be adjusted so that the most sensitive growth stage does not coincide with potential unfavorable weather conditions. Other options may be to use cultivars with different growth cycles, in which the most sensitive growth stage does not occur during anticipated stress periods. Planting drought-resistant and extreme-temperature-resistant plant species or cultivars is another strategy that can be used to reduce risk of crop yield losses due to prevailing stresses.

Early-flowering wheat cultivars are favored in regions that experience late-season heat stress. Such cultivars are better able to avoid reductions in kernel weight due to high temperature stresses during grain filling (Bruckner and Frohberg, 1987). Genotypic adaptation is especially important for short-season growth regions, in which sufficient heat units may not be available for crops to reach maturity. In such environments, inclement weather during harvest may result in complete losses of crops. However, early maturity in such environments is associated with reduced grain yield (Dofing, 1995). Therefore, reasonable breeding objectives would be to increase grain yield while keeping time to maturity constant.

Use of adequate levels of fertilizers, plant density, and plant spacing is important in producing ideal yield components. For example, N and P are important nutrients that determine numbers of panicles or grains produced per unit area. A typical example for increased numbers of panicles with applications of N of lowland rice grown on an Inceptisol in central Brazil is presented in Figure 3.10. Maximum numbers of panicles were obtained with applications of about 200 kg N/ha. Plants grown with low levels of N had higher spikelet sterility compared to higher levels of N (Figure 3.11). Spikelet sterility was about 35 percent for plants grown with zero added N, and was reduced to 28 percent with 210 kg N/ha. Small percentages (<15 percent) of floret sterility should be considered normal, as all florets in an ear do not set. This means that adequate N can improve rice grain yield by increasing numbers of panicles per unit area and reducing spikelet sterility. Nitrogen accumulation in dry matter and grain

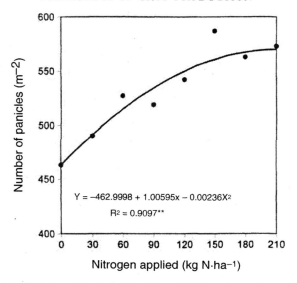

FIGURE 3.10. Relationship between N applied and number of panicles in lowland rice grown on an Inceptisols in central Brazil. *Source:* Reproduced from Fageria, 1998.

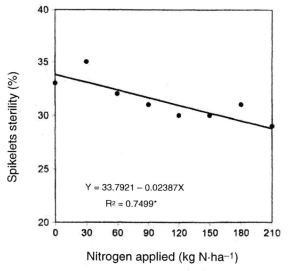

FIGURE 3.11. Relationship between N applied and spikelet sterility in lowland rice. *Source:* Reproduced from Fageria, 1998.

increased dry matter as well as grain yields of lowland rice grown on an Inceptsol in central Brazil (Figures 3.12 and 3.13). Among micronutrients, B is important in reproductive development of higher plants and reduces spikelet sterility. Subedi, Budhathoki, and Subedi (1997) reported strong wheat genotypic responses to B applications, and genotypic variations were evident. Boron at 1 kg·ha^{-1} reduced numbers of late ears, increased numbers of grains per ear, and increased grain yields in B-responsive wheat genotypes. However, added B did not influence B-tolerant genotypes. This means that addition of adequate amounts of B in B-deficient soils and use of B-tolerant genotypes are two attractive strategies for improving crop yields of annual crops.

The special distribution of plants in crop communities is an important determinant of yield, and many experiments have been conducted to determine spacing between rows and between plants within rows to maximize yield (Duncan, 1986; Egli, 1988). Two general concepts are frequently used to explain relationships between row spacing, plant density, and plant yield. First, maximum yield can be obtained only if plant communities produce sufficient leaf area to

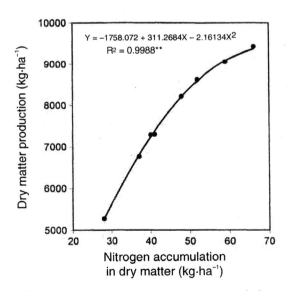

FIGURE 3.12. Nitrogen accumulation in dry matter and dry matter yield of lowland rice grown on an Inceptisol in central Brazil. *Source:* Reproduced from Fageria, 1998.

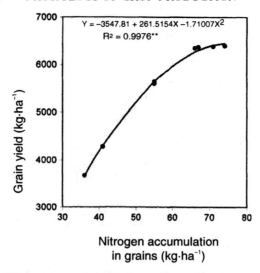

FIGURE 3.13. Nitrogen accumulation in grains and grain yield of lowland rice grown on an Inceptisol in central Brazil. *Source:* Reproduced from Fageria, 1998.

provide maximum light interception during reproductive growth (Johnson, Green, and Jordan, 1982). Second, equidistant spacing between plants will maximize yield because it minimizes interplant competition. Duncan (1986) defined three phases of soybean yield responses to increased plant density. Phase I covered ranges of plant densities where no competition among plants occurs, and yield is directly proportional to plant density (e.g., yield per plant is constant). Phase II begins at plant densities sufficiently large to intercept essentially all light at full canopy and is terminated at densities where further increases in density cause no increases in yield. Phase III includes all plant densities where increased competition among plants exists.

CONCLUSION

For maximum or highest potential yield of crop species or cultivars, it is important that crop species or cultivars sown have adaptability to growing environments for which the plants will be grown and biotic and abiotic stresses can be effectively controlled. Grain yield represents the summation of numerous physiological processes and morphological developments, and these processes should be understood to measure or achieve highest yield potentials. Annual crop growth cycles are divided into three growth stages: vegetative, reproductive, and grain filling. In the vegetative growth stage, canopy architecture needs to be defined. This includes numbers of tillers and leaf area index. During vegetative growth, most photosynthetic capacities and yield capacities such as numbers of panicles or heads should be determined. Under optimum growth conditions, the contribution of preanthesis photosynthate reserves to final grain weight was 5 to 10 percent in wheat and 20 percent in barley (Evans and Wardlaw, 1976). Bingham (1969) noted that long vegetative growth periods often resulted in high grain yields of wheat. However, Sharma (1992) reported that combinations of short vegetative growth periods and long grain-filling periods might produce high grain yields for wheat. The vegetative phase duration has been positively correlated with leaf area and increased source size. Photosynthetic capacity as well as yield capacity is influenced by crop and environmental conditions. In the reproductive growth stage, numbers of grains need to be estimated (i.e., sink size) so that potential storage capacity for individual crops may be determined. Nearly all stem growth in cereal crops occurs during the reproductive growth stage, which becomes competitive with developing inflorescences (Evans and Wardlaw, 1976). Numbers of grains are very sensitive to N deficiency, drought, and extreme low and high temperatures. During the reproductive growth stage, division growth stage (about one week before flowering) is very sensitive to adverse environmental conditions, and numbers of grains are reduced. Longer vegetative growth periods or more rapid leaf area development during this phase increase source size, while longer grain-filling periods increase yield if sufficient sources of assimilate are present (Corke and Kannenberg, 1989). During grain filling, weights of grains are determined from up to one week to ten days after flowering. This is a very sensitive period of time for high potential yield. During this phase, potential grain storage capacity is determined by final yield. Any adverse environmental factors such as low radiation, N deficiency, low and high temperatures, and drought lower grain weights as well as increase grain sterility to reduce yield. This means management strategies such as using adequate levels of nutrients, especially N, adequate moisture, adjusting planting dates to avoid extreme low and high temperatures, and using drought-resistant and nutrient-efficient cultivars can improve annual crop grain yields.

Economic yield of crops is determined by two parameters, namely, biological yield and harvest index. Biological yield is a function of growth duration and crop growth rate at successive growth stages. Harvest index is controlled by partitioning of photosynthates between harvested and nonharvested organs during plant growth cycles (Tanaka and Osaki, 1983). Significant changes in partitioning of photosynthate in crop plants (expressed as harvest index) have been important for changing grain yields. An important feature of changes in harvest index is the N budget for plants that must accompany changes in harvest index. N availability and the relative priority given to grain production help determine desired harvest indexes (Sinclair, 1998). Thus, economic yield is closely related to growth processes.

Breeding for physiological efficiency has been proposed as an approach to improve crop yield (Wallace et al., 1993). Days to maturity is the most important physiological trait affecting that outcome. All crop plants have specific growth rates. Through genetic manipulation, growth rates can be increased. However, increasing growth rates do not necessarily translate into higher seed yields unless greater amounts of photosynthate from vegetative biomass are partitioned to seeds. Indirect selection for the three major physiological traits affecting yield, namely, biomass, harvest index, and days to maturity, should result in improved yields (Wallace et al., 1993). Simultaneous selection of many traits is required because genetically established interrelationships occur among these physiological traits. Increases in maturity result in increased biomass, whereas increases in harvest index result in decreases in maturity and decreased biomass (Kelly, Kolkman, and Schneider, 1998; Kelly, Schneider, and Kolkman, 1999). Each of these physiological traits and their correlations among themselves can be quantified by yield system analyses of yield trial data (Wallace and Yan, 1998). Long-term direct selection for yield will gradually exploit the most useful genetic variability for harvest index and maturity.

Chapter 4

Photosynthesis and Crop Yield

INTRODUCTION

Photosynthesis is the basic process underlying plant growth and production of food, fuel, and fiber required to sustain life (Tolbert, 1997). An understanding of photosynthesis is therefore necessary to appreciate processes that determine yield in agriculture, forestry, ecology, and many other fields. This most important biochemical process in green plants, which literally means building by light, probably evolved 3500 million years ago (Shopf, 1993). In general, photosynthesis is the process by which plants synthesize organic compounds from inorganic substances using light. During photosynthesis, C from atmospheric carbon dioxide (CO_2) is fixed to become part of many organic molecules that constitute plant tissues. Because of this, total dry matter production of crop plants is correlated with photosynthetic rates integrated over plant growth cycles. Yield is related to total dry matter for many plants, especially associations with harvest index (see Chapter 3).

Light energy is converted into chemical potential energy when processes of photosynthesis occur. Chemical products of photosynthesis are translocated to sites of utilization and are incorporated into plant parts of economic interest such as grain in annual crops. Efficiency of light energy conversion to economic products depends on various factors such as CO_2 supply, light intensity, light interception, soil fertility level, water availability, temperature, and genetic factors related to the plant itself (Berry and Bjorkman, 1980; Eastin and Sullivan, 1984). Detailed discussion of these factors has been provided by Moss (1984). In photochemical reactions, carbohydrates are produced, and O_2 and water are released according to the following equation:

$$6CO_2 + 12H_2O \xrightarrow[\text{chloroplasts}]{\text{sunlight}} C_6H_{12}O_6 + 6O_2 + 6H_2O \qquad (4.1)$$

The process of photosynthesis consists of three partial processes, which are (1) diffusion of CO_2 to chloroplasts, (2) photochemical reactions, and (3) dark reactions (Yoshida, 1981). In processes of diffusion of CO_2 to chloroplasts, CO_2 in air (normally about 300 mg·kg^{-1} or 0.03 percent v/v) is transported by turbulence due to wind and diffusion due to concentration gradients, through the leaf stomates and intercellular air spaces to the chloroplasts. In photochemical reactions, light energy is used to split water to produce molecular O_2, reduced nicotinamide adenine dinucleotide phosphate (NADPH), and adenosine triphosphate (ATP) according to the following equation (Yoshida, 1981):

$$2H_2O + 2ADP + 4NADP + 2P_i \xrightarrow[\text{chloroplasts}]{\text{sunlight}} O_2 + 2ATP + 4NADPH \qquad (4.2)$$

In dark reactions, NADPH and ATP produced in sunlight reactions are used to reduce CO_2 to carbohydrates and other compounds according to the following equation:

$$CO_2 + 2ATP + 4NADPH \longrightarrow CH_2O + H_2O + 4NADP + 2P_i + 2ADP \qquad (4.3)$$

This process does not require sunlight and takes place in darkness once energy is captured by light-harvesting molecular networks and stored into appropriate compounds. The objective of this chapter is to compile the latest developments in this field. This information should be helpful to agricultural scientists of different disciplines as reference material for research and educational purposes.

CROP PHOTOSYNTHESIS

Seedlings begin as heterotrophic plants depending totally on food mobilized from seed endosperm, and pass through a transition phase when photosynthesis commences even while endosperm mobilization reactions continue. Finally, seedlings depend entirely on photosynthesis and become autotrophic. Nelson and Larson (1984) generalized that duration of the preautotrophic phase to be 12 days (heterotrophic phase being six days and transition phase being an additional six days). Duration of preautotrophic phases of maize has been reported to range between 14 and 26 days (Cooper and MacDonald, 1970); variations occur because of temperature differences. De Datta (1981) reported that rice seedlings become autotrophic at the three- to four-leaf stage, somewhere in the range of 14 to 22 days after seedling emergence.

When photosynthesis occurs, dark respiration is assumed to take place simultaneously. Hence, measured photosynthesis consists of differences between true photosynthesis and dark respiration. On the basis of this concept, gross photosynthesis refers to the sum of net photosynthesis and dark respiration (Yoshida, 1981):

$$\text{Gross photosynthesis} = \text{Net photosynthesis} + \text{Dark respiration.} \tag{4.4}$$

Net photosynthesis is normally measured by CO_2 intake or by O_2 output. Dry weight increases are also often used to estimate net photosynthesis after appropriate corrections have been made for mineral content. Rates of dry weight changes are expressed as

$$dW/dt = D - N, \tag{4.5}$$

where D = daytime net total CO_2 taken into plants and N = night total CO_2 evolved by plants (McCree, 1974). The term dW/dt is expressed per unit ground area (g dry matter/m^2 per day or week), and is called crop growth rate (CGR). Crop growth rates are used to measure primary productivity of crops in the field (Yoshida, 1981).

Crop photosynthesis values collected from the field are primarily determined by photosynthetic capacity per unit leaf area or whole plant canopy, leaf area index, and light interception efficiency of leaf canopies (Yoshida, 1981; Mae, 1997). It has been concluded that leaf area index and amount of light interception for many crops are closely associated with grain yields (Yoshida, 1981).

CANOPY PHOTOSYNTHESIS

Canopy photosynthesis describes photosynthetic activity per unit ground area and combines genotype efficiency, leaf morphology, and canopy architecture (Wells, Meredith, and Williford, 1986). Once exponential phases of canopy leaf area development are complete, individual leaves begin to senesce, and if not balanced by production of new leaves, contribute to declines in canopy photosynthesis (Kelly and Davies, 1988). Contributions of leaf senescence to declining canopy photosynthesis and to C availability for yield have been identified as potential limitations in crop production (Wullschleger and Oosterhuis, 1992).

Net photosynthetic rates of crop plants vary with leaf position, nutritional status, water status, plant species, cultivar within species, and growth stage. Crop species are generally classed to have "C_3" or "C_4" metabolism based on whether their initial sugars produced in photosynthesis consist of three or four C molecules. Table 4.1 provides examples of C_3 and C_4 crops. The two types of photosynthetic pathways differ in chloroplast arrangement, primary photosynthetic enzymes, temperature response, water-use efficiency, light saturation, and response to CO_2 concentrations (Table 4.2). Tanaka and Osaki (1983) determined photosynthetic rates per unit leaf area for some important field crops (Table 4.3). No clear differences in photosynthetic

TABLE 4.1. Examples of C_3 and C_4 field crops.

C_3 crops	C_4 crops
Barley	Maize
Bean (common)	Millet (pearl)
Cotton	Sorghum
Cowpea	Sugarcane
Oat	
Peanut	
Rice	
Soybean	
Sugarbeet	
Wheat	

Source: Fageria, 1992; Fageria, Baligar, and Jones, 1997.

TABLE 4.2. Characteristics of C_3 and C_4 plants.

Characteristics	C_3	C_4
Photosynthetic efficiency	Low	High
Photorespiration	High	Low
Water utilization efficiency	Low	High
Optimum temperature for photo-synthesis	10-25°C	30-45°C
Response to light intensity	Low	High
Response to CO_2 concentration	Low	High
Response to O_2 concentration	Low	High
Major pathway of photosynthetic CO_2 fixation	Reductive pentose phosphate cycle	C_4-dicarboxylic acid and reductive pentose phosphate cycle
Transpiration ratios	High	Low
Leaf chlorophyll a to b ratio	Low	High

Source: Fageria, 1992; Fageria, Baligar, and Jones, 1997.

rates exist among crop species, except for maize and alfalfa. Maize has about twice the photosynthetic rate of many other crops, and alfalfa had an intermediate photosynthetic rate. Variation in photosynthetic rates is often correlated with concentrations of N compounds in leaves (Evans, 1983; Hirose and Werger, 1987). This correlation is explained by the fact that some 75 percent of all N in mesophyll cells of C_3 plants is associated with photosynthesis (Lambers, 1987). Further, larger fractions of N (about 25 percent) become components of the enzyme Rubisco (ribulose-bisphosphate carboxylase). Rates of photosynthesis are therefore closely correlated with Rubisco activity in leaves (Evans, 1983). Net photosynthesis of individual rice leaves reaches maximum values of about 40 to 60 kilolux (800 to 1200: moles·m^{-2}·sec^{-1}) near half-full sunlight (Yoshida, 1981). However, photosynthesis of well-developed canopies increases with increasing light intensity of up to full sunlight, and no indication of light saturation appears to be reached (Murata, 1961). In early stages of crop growth, the main determinant of photosynthesis is the extent of leaf area development. As leaf area indexes (LAI) increase, so do the ex-

TABLE 4.3. Photosynthetic rate per unit leaf area in various crop plants.

Crop plant	Photosynthetic rate ($mg\ CO_2 \cdot dm^{-2} \cdot h^{-1}$)
Alfalfa	50-55
Bean (common)	30-40
Maize	60-80
Pea (field)	25-30
Potato	25-35
Rice (lowland)	40-50
Soybean	30-35
Sugarbeet	30-35
Sunflower	40-42
Wheat	28-39

Source: Mahon, 1983; Tanaka and Osaki, 1983.

tents of light interception increase, and often exceed 95 percent for many cereal crops with LAI values of about 4. Once canopies close from leaf density, further increases in LAI have little effect on plant photosynthesis, which may be influenced by incident radiation and structures of canopies (Evans and Wardlaw, 1976).

Net photosynthesis rates of active, healthy single rice (C_3 plant) leaves are about 40 to 50 mg $CO_2 \cdot dm^{-2} \cdot h^{-1}$ under light saturation conditions. Ishii (1988) noted no significant differences in photosynthetic rates of single rice leaves at heading stage of growth for 32 Japanese cultivars bred from 1880 to 1976. However, 25 percent increases in canopy photosynthesis were observed in new rice cultivars (bred between 1949 and 1976) compared to old cultivars (bred from 1880 to 1913). Improvement in canopy photosynthesis was attributed to higher efficiency of light interception by canopies because of changes in leaf angles. Old cultivars had droopy leaves with high mutual shading, while new cultivars had erect leaves with less mutual shading (Ishii, 1988). Similarly, Conocono, Egdane, and Setter (1998) concluded that much of the large increases in rice yield over the past three decades could be attributed to improvements in canopy structure that enhanced canopy light interception and photosynthesis.

Canopy photosynthesis can be described as the product of canopy light interception and radiation use efficiency (Loomis and Connor, 1992). Positive linear relationships have been noted between crop growth rates and canopy light interception for most field crops. In cotton, close relationships have been described between leaf area index and canopy light interception (Heitholt, 1994) and between light interception and lint yield (Heitholt, Pettigrew, and Meredith, 1992). Leaf photosynthetic rates depend on incident radiation and leaf absorbency. Absorbency is affected by leaf external and internal reflectance and by leaf pigment contents, especially chlorophyll (Maas and Dunlap, 1989).

C_3 AND C_4 PHOTOSYNTHESIS

Carbon dioxide is converted into carbohydrates during photosynthesis by two biochemical processes known as Calvin pathways C_3 and C_4 (Hatch and Slack, 1970). The first stable product from fixation of CO_2 in the Calvin cycle is 3-phosphoglyceric acid. In recent years, however, studies revealed that certain species have additional photosynthetic reactions in which the first detectable product resulting from CO_2 fixation is not the three-carbon compound, 3-phosphoglyceric acid, but rather the four-carbon compound, oxaloacetic acid, which is quickly transformed into either malic or aspartic acid. To distinguish between plants whose leaves have this adaptation and those that do not, it is customary to refer to species for which the Calvin cycle alone functions in photosynthesis as C_3 species. Likewise, the symbol C_4 is applied to species in which four carbon acids are the first stable product of CO_2 reduction (Moss, 1984).

The C_4 photosynthetic pathway is present in both monocotyledonous (monocot) and dicotyledonous (dicot) plants, and has reduced energy wasting processes of photorespiration that appear in the photosynthetic pathway of C_3 plants. In C_3 plants, CO_2 is fixed through ribulose bisphosphate (RuBP) carboxylase (RuBPCase) so that RuBPCase-mediated carboxylation of RuBP produces the compound 3-phosphoglycerate. This reaction is inhibited nearly 50 percent when atmospheric O_2 competes with CO_2 at active sites of RuBPCase. The C_4 photosynthetic pathway eliminates photorespiration by splitting photosynthetic reactions between two morphologically distinct cell types: bundle sheaths (BS) and mesophylls (M). RuBPCase is physically separated from atmospheric O_2 by being compartmentalized in internal BS cells. This scheme is advantageous in hot, dry conditions but can be energetically wasteful (Nelson and Langdale, 1992). Some studies have indicated that certain C_4 plants function as C_3 plants when the use of the C_3 pathway is energetically favorable (Khanna and Sinha, 1973). Plants in the C_4 group have higher photosynthetic efficiency compared with plants in the C_3 group. Characteristics distinguishing these two groups of higher plants are presented in Table 4.2. It should be noted that C_4 plants represent an adaptation to habitats with relatively high temperatures, high irradiance, and limited water supply. The C_4 plants tend to be at a disadvantage in cool climates. Important field crops in these two groups are presented in Table 4.1. Detailed information about C_3 and C_4 plants is provided by Downton (1971, 1975) and Black (1971). Al-Khatib and Paulsen (1999) compared photosynthetic responses among C_3 (wheat and rice) and C_4 (millet) plants to note their reactions at high temperature and sensitivity to light reactions. These studies were conducted to note differences between plants with good adaptation to high temperatures (millet) and those not as readily adapted to high temperatures. Leaf photosynthesis of millet and rice increased at temperatures from 22 to 32°C before decreasing at 42°C, whereas wheat had highest leaf photosynthesis at 22°C and decreased as temperature increased. These results indicated that leaf photosynthetic response differences at high temperature were associated with light reactions. The extreme sensitivity of wheat to high temperature was attributed to injury of photosystem II.

RADIATION USE EFFICIENCY

One of the most important physiological factors associated with crop production is utilization of solar radiation, which is influenced by canopy structure (Daughtry, Gallo, and Bauer, 1983). Radiation is transmitted through and between leaves to cause radiant flux densities, and spectral composition changes rapidly with canopy depth (Gardner, Pearce, and Mitchell, 1985). Radiation use efficiency (RUE) represents crop canopy ability to convert intercepted solar energy into dry matter. Total amounts of aboveground dry matter produced by crops grown in nonstressed environments is directly related to amounts of intercepted photosynthetically active radiation (IPAR) (Monteith, 1977; Kiniry et al., 1989). Gallagher and Biscoe (1978) demonstrated that this relationship is linear and its slope equals RUE. Crop RUE values vary with crop type and environment.

Temperature and crop ontogeny are important factors affecting RUE (Andrade, Uhart, and Cirilo, 1993; Otegui et al., 1995). Mendham, Shipway, and Scott (1981) reported that RUE for winter rape was 1.2 g/MJ or 2.67 g/MJ PAR (assuming 45 percent of total solar radiation to be PAR). The RUE value for summer rape grown in Australia was 1.5 g/MJ solar radiation or 3.33 g/MJ PAR (Mendham, Russell, and Jarosz, 1990). These values were determined for plants at preflowering, and therefore did not include photosynthetic capacity of pods. Published values of seasonal RUE values for grain sorghum varied from 2.3 to 4.0 g/MJ (Kiniry et al., 1989; Muchow, 1989; Rosenthal, Gerik, and Wade, 1993).

Morrison and Stewart (1995) examined the effects of varying row widths and seeding rates on amount of IPAR and RUE values of summer rape. After flowering, plants intercepted more PAR when grown in 15 cm wide rows compared to 30 cm wide rows. For both row widths, IPAR values increased as seeding rates increased from 1.5 to 12.0 kg·ha^{-1}. The RUE values were higher for plants grown in 15 cm wide rows than for plants grown in 30 cm rows, but decreased with increasing seeding rates. The mean RUE value was 2.83 g/MJ PAR. When pod areas were added to leaf areas, IPAR values increased and RUE values decreased. Similarly, Morrison, McVetty, and Scarth (1990) reported highest grain yields from plants grown in 15 cm wide rows at 1.5 kg·ha^{-1} seeding rates.

Kiniry et al. (1989) noted that RUE values of rice, sorghum, sunflower, maize, and wheat did not decrease as plants approached anthesis. Therefore, it may be possible to use RUE values over the entire life cycle of plants. Sinclair and Horie (1989) attempted to explain mechanisms about the effects of N on crop biomass accumulation by understanding quantitative relationships among leaf N content, CO_2 assimilation rates, and crop NUE values for rice, maize, and soybean. These authors noted that RUE values within each plant species were nearly constant at high leaf CO_2 assimilation rates but decreased appreciably at low CO_2 assimilation rates. At leaf CO_2 assimilation rates typical for these species, RUE values were predicted to be near 1.2 g/MJ for soybean, 1.4 g/MJ for rice, and 1.7 g/MJ for maize.

Competitive use of N for construction of either large leaf areas or high leaf N contents was examined for these plants during early growth. Maize plants had highest biomass accumulation because these plants had low leaf N contents and greatest leaf growth, and at the same time had high RUE values. For each rate of N supply to leaves, optimum leaf N contents existed to provide maximum biomass accumulation. Muchow and Sinclair (1994) obtained field data on CO_2 assimilation rates and RUE values for maize and sorghum relative to differing leaf N contents induced by varying amounts of applied N fertilizer. The CO_2 assimilation rates were most responsive to leaf N content when low values of leaf N existed, and no differences were noted between plant species for relationships between CO_2 assimilation and leaf N content. Similarly, RUE values increased with leaf N content per unit leaf area, but maximum RUE and leaf N values were higher in maize compared to sorghum. The maximum RUE value was 1.7 g/MJ and the maximum canopy leaf N content was about 1.8 g N·m^{-2} for maize, while the maximum RUE value was less than 1.3 g/MJ and the maximum canopy leaf N content was less than 1.3 g N·m^{-2} for sorghum.

Intercepted photosynthetically active radiation light by canopies for many crop models is calculated from Beer's Law equation:

$$IPAR = PAR \times [1-\exp(-k \times LAI)] \tag{4.6}$$

where PAR is photosynthetically active radiation, k is the light extension coefficient, and LAI is leaf area index (Thornley, 1976). Plants are more evenly distributed when sown in narrow row spacing and efficiency of light interception is improved by plants grown under these conditions (Flenet et al., 1996). Increases in light interception when row spacing was reduced have been reported for maize (Egharevba, 1975), sorghum (Muchow et al., 1982), soybean (Board, Harville, and Saxton, 1990), and sunflower (Zaffaroni and Schneiter, 1989). In addition, greater light interception often increases crop yields (Parvez, Gardner, and Boote, 1989).

The RUE values change with crop ontogeny (Orgaz et al., 1992; Trapani et al., 1992) and with LAI. In sunflower, RUE values have been reported to increase exponentially from 1.4 at LAI values near zero to 3 at LAI values of 4. Shortly after anthesis, RUE values decline sharply (Steer, Milroy, and Kamona, 1993), as do LAI values.

Values of RUE near 1.3 are assumed for plants beginning anthesis or grain filling (Trapani et al., 1992). Senescence and remobilization of leaf N to grain are the most probable causes of decreases in RUE values during grain filling because these reactions do not commence until after anthesis. When kernel growth begins in sunflower, RUE values of 1.3 have been used to make allowances for effects of remobilization (Hall, Whitfield, and Connor, 1990).

Light interception during reproductive crop growth is important for determining crop yields. Maximum light interception during reproductive growth was needed for maximum soybean yields (Wells, 1991; Board and Harville, 1993). However, yields can also be increased by changing planting patterns without increases in light interception (Egli, 1988; Wells, 1991). Yield increases were the result of more efficient utilization of assimilates to produce pods and seeds (Egli, 1994). This apparently occurred because similar amounts of assimilate were distributed over many flowering sites (nodes). When consideration was given to soybean yield responses to changes in plant density, ranges in plant population occurred with yield increases even though no changes in light interception occurred (Duncan, 1986). Increases in yield were attributed to increases in plant size, but no clearly defined plant size (i.e., height, mass, numbers of branches) was provided. Board, Harville, and Saxton (1990) and Parvez, Gardner, and Boote (1989) reported increased yields for plants grown in narrow rows to be associated with increases in nodes per unit area. Larger plants were also noted to have higher yields (Duncan, 1986), which might indicate that these larger plants may have been related to greater numbers of nodes as noted by Board, Harville, and Saxton (1990) and Parvez, Gardner, and Boote (1989). Yield responses from plants grown in narrow rows occur more frequently when plants are grown in high-yielding environments (Johnson, 1987). It appears that the benefit of higher numbers of nodes is associated with high rates of canopy photosynthesis for plants grown in optimum environmental conditions (Egli, 1994), and plants grown under these conditions tend to have lower incidence of environmental stresses.

Proportions of incident radiation and diffuse radiation that are diffused affect RUE values for crops (Bange, Hammer, and Rickert, 1997). Decreasing levels of incident radiation and increasing proportions of diffuse radiation have been noted to theoretically increase RUE values for maize, soybean, and peanut (Sinclair, Shiraiwa, and

Hammer, 1992; Hammer and Wright, 1994). These authors hypothesized that increases in RUE values resulted from more efficient use of light by sunlit leaves at low light levels, and that increases in the diffuse component of light was being spread over greater areas of sunlit and shaded leaves. The fraction of diffuse radiation in a field on clear days can be 10 to 15 percent of incident radiation, and this may increase to 100 percent with heavy cloud cover (Milthorpe and Moorby, 1988). During cloudy weather, more diffusive radiation is received from all angles and penetrates further into crop canopies, resulting in more efficient light utilization (Russell, Jarvis, and Monteith, 1989). In addition, RUE improve because incident radiation flux densities are lower in cloudy weather and are likely to be below saturation levels for photosynthesis (Sheehy and Chapas, 1976).

LEAF AREA INDEX

Leaf area index is an important yield-determining factor for field-grown crops because LAI is a major determinant of light interception and transpiration. Therefore, LAI is an important parameter to estimate yields for many crop growth models that use net photosynthesis, assimilate partitioning, canopy mass, and energy exchange (Fortin, Pierce, and Edwards, 1994). Today, leaf area meters are available to accurately and rapidly determine leaf areas of crop plants. In addition, individual leaf area (L_a) can be calculated using simple formulas based on leaf lengths (L_l), maximum widths (L_w), and a correction factor (k) (McKee, 1964):

$$L_a = L_l \times L_w \times k. \tag{4.7}$$

A correction factor value of 0.75 can be used for maize (Bollero, Bullock, and Hollinger, 1996) and rice (Yoshida et al., 1976) at most growth stages, with exceptions being for seedlings and plants at maturity. At these two stages of growth, the *k* value is slightly lower. For rice, this value is normally 0.67 (Yoshida et al., 1976).

Canopy light interception and photosynthesis are closely related to LAI critical values, which are required to intercept 95 percent incident irradiance (Pearce, Brown, and Blaser, 1965). Eastin (1969) suggested that optimum leaf arrangements for each genotype-row spacing-population combinations were needed.

Values for LAI increase until maximum values are attained around flowering, and then subsequently decrease (see Chapter 3). Maximum observed LAI values are greater than 5.5 for rice, maize, potato, and sugarbeet, 4.5 to 5.0 for soybean and common bean, and about 3.5 for spring wheat (Tanaka and Osaki, 1983). Leaf area is influenced by climatic, soil, and plant factors. One plant leaf area trait is specific leaf area (SLA) or mass per area, which is influenced by many environmental and genetic factors. For many plant species, evidence indicates that maximum rates of photosynthesis per unit leaf area under light-saturated, ambient air conditions are nearly proportional to SLA. This appears to be because large portions of leaf N are invested in cells, namely chloroplasts, where photosynthetic reactions occur. In mature rice leaves, about 80 percent of total leaf N is allocated to chloroplasts (Morita, 1980), as is the case with other C_3 plants (Makino and Osmond, 1991). Plant densities and spacing are other major factors influencing leaf area of plants grown under field conditions. Tetio-Kagho and Gardner (1988) reported that maximum LAI values for maize (species with low plasticity) were 1.7, 2.6, and 4.0 at tasseling, respectively, for densities of 1.7, 2.6, and 6.3 plants per m^{-2}. There is no doubt that LAI is determined genetically in crop plants and can be manipulated through breeding.

Increasing LAI increases dry matter production, but net canopy photosynthesis cannot increase indefinitely because of increased mutual shading of leaves. Irradiance is lower for leaves within a canopy, which leads to decreased photosynthesis rates per unit leaf area (Yoshida, 1972). Similarly, optimum LAI values exist beyond which yields no longer increase. Figure 4.1 shows relationships between leaf area index and grain yield of soybean. Regression of grain yield with LAI resulted in quadratic relationships. Maximum grain yields were achieved at a LAI value of 5.0. This yield response was related to light interception (Figure 4.2), which was maximal at 93 percent for a LAI value of 4.0. As indicated in Figure 4.2, light interception does not increase as LAI increases above 4.0.

Plant population influences the amount of radiation intercepted by canopies (Tetio-Kagho and Gardner, 1989a) as well as RUE values (Rochette et al., 1995). Higher plant populations increase LAI and vegetative dry mass. Fractions of light intercepted by upper canopies are also higher with high plant populations but decrease total dry mass per plant (Tetio-Kagho and Gardner, 1989b). Dale, Coelho, and

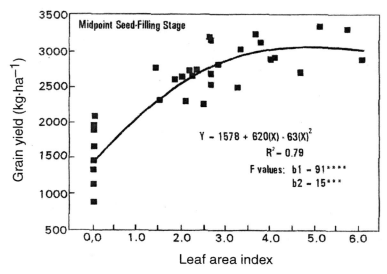

FIGURE 4.1. Relationship between soybean grain yield and leaf area index. *Source:* Reproduced from Board, Wier, and Boethel, 1997, with permission from American Society of Agronomy, Madison, WI.

Gallo (1980) successfully incorporated plant population into an LAI prediction equation and noted that population was important for determining LAI values of maize during two of three phenological periods. Major, Beasley, and Hamilton (1991) reported that maize RUE values were also a function of LAI. Compared to older hybrids, modern maize hybrids responded more favorably to high population densities because of higher leaf area indexes at silking, which resulted in more interception of PAR and more dry matter accumulation during vegetative plant development (Tollenaar and Aguilera, 1992). Modern hybrids at high plant densities also have higher leaf photosynthesis rates, compared with older hybrids, despite their higher LAI values (Dwyer, Tollenaar, and Stewart, 1991). Modern hybrids also have higher RUE values during grain filling, which further contributes to higher dry matter yields (Tollenaar and Aguilera, 1992). Modern hybrids grown at high plant densities also have higher crop growth rates from one week before to three weeks after silking, which contributes to more kernels per plant and higher grain yields (Tollenaar, Dwyer, and Stewart, 1992). Cox (1996) reported that low plant densities (4.5

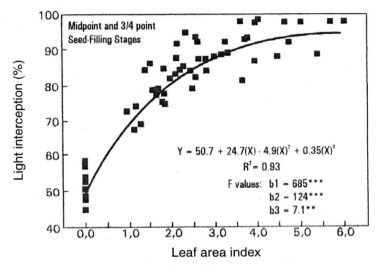

FIGURE 4.2. Relationship between leaf area index and light interception by soybean canopies. *Source:* Reproduced from Board, Wier, and Boethel, 1997, with permission from American Society of Agronomy, Madison, WI.

plants/m^2) compared to high plant densities (9 plants/m^2) had 40 percent lower LAI values from mid-vegetative to early grain filling, which offset higher photosynthetic efficiency and resulted in lower crop growth rates during vegetative development and 25 percent less dry matter accumulation at silking. Tollenaar (1991) suggested that modern maize hybrids compared to older hybrids responded more positively to high plant densities because of increased stress tolerance.

PARTITIONING OF ASSIMILATES

Photosynthesis rates and assimilate distribution within vegetative and reproductive organs are physiological processes that determine crop yield. Synthesis, translocation, partitioning, and accumulation of photosynthetic products within plants are controlled genetically and are influenced by environment (Snyder and Carlson, 1984). Unlike self-pollinated cereals, where improved partitioning has resulted in higher grain yields, similar improvements have not resulted in grain legumes. In general, cereals are more responsive to increased inputs such as N fertilization, which accounts for improved cultivar performance through selection for responsive genotypes (Kelly, Kolkman, and Schneider, 1998). However, neither improved partitioning nor responsive cultivars have been useful selection criteria for improving yield of grain legumes. In North America, soybean is the major grain legume and has been bred almost exclusively for yield by both private and public sectors. Steady plant yield gains of 0.65 percent per year have been attributed solely to genetic improvement (Kisha, Sneller, and Diers, 1997). One constraint recognized by soybean breeders is the narrow genetic base available within cultivated soybean cultivars. Similarly, Morrison, Voldeng, and Cober (1999) reported that with 58 years of soybean genetic improvement in Canada (1934-1992), 0.5 percent increases per year were observed for seed yield, harvest index, and photosynthetic rate, while 0.4 percent per year decreases were noted for LAI. Increases in seed yield with year of release was highly correlated with increases in harvest index, photosynthesis, and stomatal conductance and decreases in leaf area index. According to Morrison, Voldeng, and Cober (1999), recently released soybean cultivars have been more efficient at producing and allocating C resources to seeds than were their predecessors.

Changes in concentration of assimilates occur simultaneously with plant partitioning during seed filling in annual legumes (Cabellero et al., 1998). This leads to net changes in yield of assimilates derived from processes of accumulation and redistribution. Caballero et al. (1998) reported concentration of assimilates in common vetch plant parts changed over the entire seed-filling period. During rapid seed growth (from 200-250 to 450-550 g dry matter/kg per seed), assimilation was dominant for starch (net gain of 1086 kg·ha^{-1}), redistribution for total soluble sugars (net loss of 104 kg·ha^{-1}), and both processes were associated with distribution of total N (net gain of 193 kg·ha^{-1}). During seed ripening (from 450 to 550 to >800 g dry matter/kg per seed), most chemical components had net losses, except acid-detergent lignin. Maximum total accumulation of the most mobile chemical components occurred during the period when seeds were 450 to 550 g dry matter/kg per seed, except total soluble sugars. According to Caballero et al. (1998), three patterns of distribution were noted for individual assimilate constituents in legumes between any two maturity stages. Redistribution from vegetative structures to seed accumulation indicated little photosynthetic activity. Ratios of redistribution to seed accumulation of <1 indicated that redistribution alone could not supply seed needs, and that photosynthetic activity would have to supply the remainder needed. Ratios >1 indicated that low-mobility compounds appeared with no net gain from photosynthetic activity

Botella, Cerda, and Lips (1993) reported that types of N affected sink priority determination in plants. Ammonium-fed plants allocated assimilates preferentially to developing young tillers in wheat (vegetative growth), while NO_3-fed plants had preferential assimilate allocation to grains (reproductive growth). Nitrate-fed plants retained most of their photosynthetic assimilates in the main stem. Seed heads (spikelets) of NO_3-fed plants attracted two to three times more assimilate compounds than did spikelets of NH_4-fed plants. This indicated that spikelets were more effective sinks for NO_3-fed than for NH_4-fed plants. This was because of the more effective competition of young tillers for assimilates in NH_4-fed compared to NO_3-fed plants. These allocation patterns explained why NO_3-fed wheat plants were more efficient in producing higher percentages of tillers with mature spikelets and having larger numbers of grains than NH_4-fed plants. Spikelets of NO_3-fed plants had high priority for assimilate allocation compared to NH_4-fed plants (Botella, Cerda, and Lips, 1993).

RESPIRATION DURING PHOTOSYNTHESIS

Phyto-mass production relates to balances between rates at which plants fix C by photosynthesis and rates at which they lose C by respiration. Atmospheric CO_2 is the source or sink for both processes. Measurement of CO_2 exchange between the atmosphere and vegetation is a direct method to quantify short-term crop productivity. This exchange also contributes to an understanding of developmental and functional relationships between factors controlling both photosynthesis and respiration (Rochette et al., 1996).

Respiratory activity of plants in the light, measured as CO_2 release through the tricarbonic acid (TCA) cycle or O_2 consumption by the respiratory chain, varies between 25 and 100 percent of dark respiratory activity (Kromer, 1995). Photosynthesis and dark respiration are metabolic pathways that produce redox equivalents and ATP to meet cell energy demands for growth and maintenance. In this way, plant growth is closely related to respiration. There is no growth without respiration (Yoshida, 1981). During respiration, O_2 is consumed and CO_2 is evolved through metabolism of carbohydrates and release of energy as ATP. Thus, respiration is essentially opposite of photosynthesis whereby building and storing energy occurs. The respiration equation is (Stoskopf, 1981)

$$C_6H_{12}O_6 + 6O_2 \longrightarrow 6CO_2 + 6H_2O + energy. \qquad (4.8)$$

Mitchell (1970) estimated 25 to 50 percent losses of CO_2 from respiration for plants grown in the field and suggested average loss values of 33 percent. During dark periods, respiration continues even though photosynthesis is no longer active. In addition to dark respiration, light respiration or photorespiration also occurs.

Some photosynthesized C is rapidly respired within a few days after assimilation begins (Shinano, Osaki, and Tadano, 1991). This type of respiration, designated as current photosynthate respiration by Osaki and Tanaka (1979), generally corresponds to growth respiration described by McCree (1974). Since some C compounds that were once composition of shoots are respired slowly, this type of respiration was designated as storage substance respiration (Osaki and Tanaka, 1979) and generally corresponds to maintenance respiration (McCree, 1974). Both types of respiration were reported to be more active in leguminosae than in gramineae plants (Tanaka and Osaki, 1983).

MANAGEMENT STRATEGIES
FOR MAXIMIZING PHOTOSYNTHESIS

Basic research in photosynthesis, especially during the past two decades, has revealed many possible approaches toward increasing efficiency of CO_2 assimilation by crops. Efficiency is defined here as rates of net CO_2 uptake per unit leaf area or per unit ground area (Zelitch, 1975). Total irradiation available for photosynthesis and duration of photosynthesis are undoubtedly important factors determining final plant productivity. Hence, productivity can be increased by planting cultivars with rapid rates of leaf area expansion, using closer plant spacing to capture more sunlight, and breeding plants with leaves that have more erect angles of elevation to absorb sunlight more effectively (Zelitch, 1975).

Genotypic differences in components of photosynthesis have been noted and used as selection criteria for plant improvement (Gifford and Evans, 1981). This means using appropriate genotypes for given environmental conditions, which can improve photosynthetic rates and consequently crop yields. Harvest index (discussed in Chapter 3) is a useful measure of how products of photosynthesis are partitioned. Improving harvest index will continue to be worthwhile for plant breeders and physiologists, especially as net photosynthetic CO_2 assimilation is improved.

Nitrogen supply can affect plant growth and productivity by altering both leaf area and photosynthetic capacity (Novoa and Loomis, 1981). Strong relationships between leaf N concentration and single leaf photosynthetic rates occur (Bondada et al., 1996), which appears to be associated with large fractions of leaf N composed in photosynthetic enzymes (Shiraiwa and Sinclair, 1993). High correlation between light-saturated leaf CO_2 assimilation and leaf N per unit leaf area have been reported, and hyperbolic relationships between CO_2 assimilation rates and N per unit leaf area differed markedly among species (Sinclair and Horie, 1989). Thus, use of adequate N is an important strategy for improving photosynthesis in crop plants. High percentages of N contribute to maintaining photosynthetic integrity. For example, 50 to 70 percent of total N in leaves was directly associated with chloroplasts in maize (Hageman, 1986). Strong correlations have been demonstrated between net photosynthetic rates and leaf N or protein contents (Edwards, 1986; Sinclair and Horie, 1989). Photosynthetic rates in maize increase linearly with leaf N contents until critical leaf N values have been reached, after which photosynthetic rates remain constant above critical N contents despite further increases in N (Wong, Cowan, and Farquhar, 1985). Since N can limit crop growth in many environments, high correlations of photosynthetic rates with leaf-soluble protein indicates that leaf N may be a good criterion for estimating photosynthetic capacity in some environmental conditions (Edwards, 1986). Plants deficient in N will have lower photosynthetic rates, accumulate less dry matter, and produce lower yields (Dwyer et al., 1995).

Numerous physiological processes associated with wheat growth are influenced by N fertility. In many wheat-producing areas, increasing rates of N fertility results in maximum LAI (Frederick and Camberato, 1994), vegetative dry weights (Blacklow and Incoll,

1981), and leaf N concentration (Frederick and Camberato, 1995). Flag leaf CO_2 exchange rates during grain fill are usually positively correlated with N concentration (Hunt and Poorten, 1985), except when drought-induced stomatal closure occurs (Rawson, Gifford, and Bremner, 1976). Maintenance of green leaf areas late into a growing season often results in longer grain filling and higher wheat yields (Frederick and Camberato, 1995). Gwathmey and Howard (1998) reported that soil applications of K may also increase canopy photosynthetic photon flux density in cotton.

Maize growth is affected by planting date, and planting before or after optimal dates results in reduced LAI, leaf area duration, total dry matter production, and grain yield (Swanson and Wilhelm, 1996). In temperate regions, maize potential productivity appears to be more limited by amount of solar radiation that is available around silking (determines grain set) than during grain filling (determines grain weight). Early and intermediate plantings tend to better utilize solar radiation for grain production (Otegui et al., 1995). Planting at appropriate dates is important for improving LAI, solar RUE, and consequently plant photosynthesis and yield.

Appropriate row spacing and plant densities are other management strategies for improving photosynthetic efficiency of crop plants, and consequently yield. For example, soybean yield responses were greater to narrow (50 cm) compared to wide (100 cm) row spacing in the southeastern United States (Board and Harville, 1994). Greater soybean yields grown in narrow rows (Board, Kamal, and Harville, 1992) resulted because of increased pod numbers created by greater light interception and higher crop growth rates during early reproductive development (R_1 through R_5 stages) using growth stage definitions of Fehr and Caviness (1977). Much of the increased pod numbers resulted from branch nodes (Board, Harville, and Saxton, 1990). Because crop growth rates reflect canopy apparent photosynthesis (Imsande, 1989), higher pod numbers are direct responses to greater assimilation capacity.

Development of crop cultivars that produce leaf areas quickly and tolerate high plant densities could enhance yields. A breeding strategy for increasing leaf areas per plant is to incorporate leafy traits into inbred lines. Plants bearing leafy traits are characterized by extra leaves above ears, low ear placement, highly lignified stalks and leaf parts, early maturities, and high yield potentials (Shaver, 1983).

Increasing plant population is a management tool for increasing capture of solar radiation within canopies. De Wit (1967) reported that crop canopies convert only 5 percent of incident solar radiation into chemical energy during crop growth cycles. Pepper (1974) reported that increased plant densities promote better utilization of solar radiation by corn canopies. However, efficiency of conversion of intercepted light into economic maize yields can decrease with high plant population densities because of mutual shading of plants.

Growing crop plants under optimum temperatures is another strategy for improving photosynthetic efficiency. Optimum temperature for photosynthesis and reproductive growth of wheat, for example, is near 20°C and higher temperatures impair photosynthetic activity (Camp et al., 1982; Al-Khatib and Paulsen, 1984). Thus, the supply of photosynthate to grain is diminished (Wardlaw, Sofield, and Cartwright, 1980). Farmers cannot of course influence weather, but they can use planting date and maturity length to minimize crop exposure to supra-optimal temperatures.

CONCLUSION

Photosynthesis is a biochemical process by which water and CO_2 react to form carbohydrates. This process is intimately linked to plant productivity that occurs with chlorophyll (green pigment in plant leaves) and requires light energy to drive photosynthetic processes. It has been reported that 90 to 95 percent of dry weight of plants is derived from photosynthetically fixed CO_2. Photosynthethic measurements can be made at the cellular level, on single leaves, on individual plants, or even on whole crop canopies. On a crop level, photosynthetic capacity can be expressed as RUE, which is calculated as the slope of linear relationships between accumulated crop biomass and intercepted solar radiation (Monteith, 1977). Monteith (1977) also defined efficiency of conversion of radiation to dry matter as RUE (g/MJ) and noted that RUE values are dependent on radiation extinction coefficients, biochemical conversion efficiency, and CO_2 exchange coefficients. Each plant species has characteristic RUE values, with C_4 plants having higher values than C_3 plants (Kiniry et al., 1989), except at low temperatures. Management practices such as high fertility increased RUE due to increased photosynthetic activity. Other practices such as using appropriate cultivars and row spacing also affect RUE values. Crop plants grown at different planting dates will be exposed to different levels of solar radiation during a season, and this may influence RUE values through effects on radiation transmission coefficients.

Plant breeding has been important in the past and hopefully will continue to be important in the future to improve photosynthetic efficiency through appropriate manipulation of plant architecture or crop canopies. Some management strategies such as cultivars adapted to given environmental conditions, appropriate crop canopies, appropriate plant spacing and densities, adequate levels of fertilizer, supplying adequate soil moisture during crop growth, and adjusting time of sowing for optimum temperature during crop growth can improve photosynthetic efficiency in annual crops and consequently improve yields. In terms of total dry matter production by crop communities, LAI, canopy photosynthetic rates, and RUE are major determinants of crop growth rates. Of these parameters, LAI is the most variable and can be widely changed by manipulating plant densities and applying fertilizers, especially N and P. Indeed, a major objective of agronomic practice should be to attain sufficiently large LAI values for crop plants to obtain maximum production (Yoshida, 1972).

Chapter 5

Source-Sink Relationships
and Crop Yield

INTRODUCTION

High crop yields are determined by ability of plants to produce high levels of photoassimilate and/or to partition large proportions of carbohydrate efficiently into harvested organs (Daie, 1985; Faville et al., 1999). Assimilate-producing plant parts such as leaves are known as the source, and plant parts to which assimilate is translocated such as grains and fruits are known as the sink. Thus, source-sink relationships may be defined as relationships between plant parts in which one part serves as the producer of materials (source) translocated to other plant parts where materials accumulate or are consumed (sink) (CSSA, 1992). Limitations to grain yields of cereals that cannot be explained by unfavorable factors such as diseases, deficiencies or toxicities of nutrients and elements, or water disorders may be assessed for potential assimilate translocation to developing grains (source capacity) and potential to accumulate assimilates (sink capacity) (Uhart and Andrade, 1995). Source-sink relationships are important for determining quantity of crop yields. Several studies have shown maize grain yield to be limited by sink capacity (Goldsworthy and Coleogrove, 1974; Allison, Wilson, and Williams, 1975; Barnett and Pearce, 1983; Jones and Simmons, 1983). In contrast, Tollenaar and Daynard (1978) noted source limitations for yield potentials at high latitudes, and Uhart and Andrade (1991) reported limitations by both source and sink capacities at cool temperatures. Changes in source/sink ratios affect C partitioning during grain filling so that carbohydrate accumulates in stems when sinks are limiting and carbohydrates are remobilized when sources are limiting (Jones and Simmons, 1983; Uhart and Andrade, 1991). Nitrogen partitioning can also be affected by source/sink ratios during grain filling. Greater leaf and stem N concentrations occur when sinks are limiting (Christensen, Below, and Hageman, 1981; Reed et al., 1988), and greater vegetative N remobilization generally occurs when sources are limiting (Reed et al., 1988).

Leaves are the dominant primary source for producing assimilates in crop plants, although green stems and floral organs can sometimes contribute substantially (Gifford and Evans, 1981). Wheat and barley panicles, stems, leaf sheaths, and other plant parts have fairly high photosynthetic activities during grain filling, which contribute substantially to grain yield (Kumura, 1995). On the other hand, photosynthetic activities of plant parts other than leaf blades are very small for rice (Imaizumi, Kiyota, and Ishihara, 1988).

Yields of economically important crops have increased significantly in the past few decades. Plant breeders have enhanced source and sink capacities of plants, and together with improved management practices, significantly higher yields have been achieved. Yield potentials of important food crops such as wheat, soybean, maize, and peanut have been improved by much as 40 to 100 percent within the twentieth century through plant breeding (Gifford et al., 1984; Ho, 1988). Potato tuber yields (dry weight) as a proportion of total plant weight (harvest index = tuber weight/tuber plus tops weight) have increased from 75 to 81 percent (Inoue and Tanaka, 1978). Similarly, harvest index has improved from 35 to 50 percent in lowland rice over the past four decades (Fageria, 1992).

Higher photosynthetic capacities of modern crop cultivars have been achieved through modification of plant canopies, which intercept more solar radiation (see Chapters 1 and 4). For example, increased leaf numbers and more erect leaf postures have been developed for maize (Tanaka and Yamaguchi, 1972) and rice (Yoshida, 1981), and larger individual leaves have been developed for wheat (Evans and Dunstone, 1970). Higher dry matter accumulation capacities of harvestable organs have been accomplished mainly by increasing either numbers of grains as for rice (Yoshida, 1981) or increasing sizes of individual grains as for wheat (Evans and Dunstone, 1970). Capacity for dry matter production in leaves may either be higher or lower than capacity for dry matter accumulation in other plant parts. Therefore, either source- or sink-limiting conditions may exist with crops. That is, better yields may be achieved by successful regulation of source-sink relationships for production and utilization of photo-assimilate within plants. However, neither source or sink manipulation alone can improve crop yields indefinitely (Ho, 1988). The objective of this chapter is to review source-sink relationships from a practical point of view and suggest practical measures for improvement of physiological aspects of crop plants and consequently to improve crop yields.

SOURCE-SINK TRANSITIONS

Photoassimilate production and distribution change during plant development. Young leaves are heterotrophic to begin with and depend in part on carbohydrate imported from other portions of the plant during vegetative growth stages and afterward may serve as sinks as seed heads form. Leaves usually produce excess photoassimilates and act as both source and sink, but later export photoassimilates to seed endosperms that serve as sinks (Turgeon, 1989). In dicotyledonous plants, transition photoassimilate from source to sink begins shortly after leaves begin to unfold. At this stage of development, the major morphogenetic events that determine leaf shape are completed. Leaves of dicotyledonous plants stop importing and begin exporting photoassimilates when they are 30 to 60 percent fully expanded (Turgeon, 1989). However, developing leaves continue to import photoassimilate from source leaves for a time after they begin to export their own photosynthetic products (Anderson and Dale, 1983; Turgeon, 1989).

Essentially all plant organs at some stage of plant development will act as sinks to receive photoassimilates. Ability of sink organs to import photoassimilates is termed sink strength (Ho, 1988). Proportions of imported assimilates used in respiration reactions by sink organs can be substantial (Farrar, 1985). Thus, sink strengths of sink organs, measured as absolute growth rate or net accumulation rate of dry matter, fail to assess true ability of sink organs to receive assimilates, and is a measure of apparent sink strength. Import rates of assimilate, measured as the sum of net C gain and respiratory C loss by sink organs, should provide more appropriate estimates of actual sink strength (Walker and Ho, 1977). Although actual sink strengths will be affected by availability of assimilate supplies and proximity of sinks to sources, the most critical determinant is the intrinsic ability of sinks to receive or attract assimilates (Cook and Evans, 1983). The intrinsic ability of sinks is the true measure of potential sink strength. Potential sink strength is genetically determined and can be expressed when supplies of assimilate are sufficient to meet demands. In addition, environmental conditions for metabolic activities of sink organs need to be optimal (Ho, 1988).

The intrinsic ability of seed endosperms to attract assimilates (i.e., kernel sink capacity) is one of the most important physiological determinants of grain yield of cereal crops, and may be one of the major limitations to yield potential (Jones, Schreiber, and Roessler, 1996). Critical formative periods following anthesis and preceding linear grain filling in which kernel sink capacities and subsequent potentials for rate and duration of growth are determined are final kernel sizes formalized to determine grain yields. It is during the formative period (may last approximately two weeks) that endosperm cells develop (Kowles and Phillips, 1988; Jones, Schreiber, and Roessler, 1996). Proplastids, which will eventually differentiate into amyloplasts and accumulate starch during linear fill, are also initiated during this formative period (Boyer, Daniels, and Shannon, 1976). Hence, kernel sink capacities are established during early stages of kernel development and are largely a function of numbers of cells and amyloplasts formed in the endosperm (Jones, Roessler, and Ouattar, 1985). Collectively, these variables determine numbers of potential sites for starch deposition, and thus capacities for kernel dry matter accumulation and grain yield. Both numbers of endosperm cells and numbers of starch granules are correlated with kernel mass at maturity (Capitanio, Gentinetta, and Motto, 1983; Jones, Roessler, and Ouattar, 1985).

PHYSIOLOGICAL ASPECTS OF SOURCE-SINK RELATIONSHIPS IN ANNUAL CROP PLANTS

Limiting factors to dry matter production or photosynthetic rates per unit leaf area can be related to source or sink capacities. When source capacity is beyond particular sink capacity, sinks control rates of dry matter production. Conversely, when sink capacity is beyond source capacity, dry matter production is controlled by source capacity. Several sinks can compete for assimilate from similar sources.

Three fundamental growth phases occur with cereal crops: vegetative, reproductive, and grain filling. Major sinks during vegetative growth are leaves, roots, and tillers. During the reproductive phase, major sinks include developing panicles, internodes, and several leaves at the top of each culm. Grains constitute the major sink during grain filling. Figure 5.1 depicts dry matter yields of vegetative and harvested organs of rice, common bean, and potato through their growth cycles. Harvested organs of rice develop later in the growth cycle than bean or potato. Based on source-sink concepts of grain yield formation, rice grains are the major sink during plant maturation, while potato has both leaves and tubers as major sinks throughout growth (Figure 5.2) (Tanaka, 1972; Tanaka and Yamaguchi, 1972). Based on this concept, improvement of cereal grain yields may be obtained by improving either sinks or sources during the ripening phase of growth. Which is most practical depends upon several conditions.

Potential capacities of sinks during grain filling may be expressed in terms of yield components: numbers of panicles per unit area, numbers of spikelets per panicle, size of each hull, and numbers of filled grains. These yield components are determined at panicle initiation or flowering, with the exception of numbers of filled grains. On the other hand, potential capacities of sources may be expressed by LAI, leaf longevity, leaf extinction coefficient to light, and potential photosynthetic rates of leaves. Of these, leaf longevity and potential photosynthetic rates of leaves may be altered considerably by crop

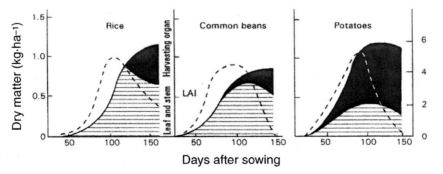

FIGURE 5.1. Dry matter production (leaves and stems—solid lines), weight of harvested organs (shaded areas), and LAI (dashed lines) of rice, common bean, and potato at successive growth stages. *Source:* Reproduced from Tanaka, 1980.

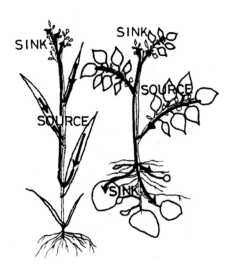

FIGURE 5.2. Hypothetical plants for different source-sink relationships of rice and potato. *Source:* Reproduced from Tanaka, 1980.

management practices after flowering (Tanaka, 1980). Thus, the more active sources are during vegetative and reproductive phases, the larger sinks will be during reproductive and grain fill stages. In this manner, source-sink relationships show sequential development with crop growth.

Harvest organs of root crops such as potato start growing during early growth stages (Figure 5.1). After commencement of tuber growth, both tubers and new leaves of upper stems or branches become sinks for photosynthesizing leaves (source) for long periods (Figure 5.2). Tubers and new leaves may compete with each other if source capacity cannot fulfill demands that sinks develop (Inoue and Tanaka, 1978). New leaves may compete with harvested organs as sinks during some growth stages and will become sources for harvested organs during later growth stages. For this reason, analyses of yield based on source-sink relationships are more complicated for root crops than for cereal crops (Tanaka, 1980).

Vegetative, reproductive, and grain-filling growth phases overlap each other for grain legumes. For example, after commencement of flowering of common bean, pods start to grow and both new leaves and flower primordia continue to differentiate on growing stems for fairly long periods. Young pods constitute the major sink during flowering. Pods (sink) and leaves (source) grow simultaneously, and these organs may compete with each other when source capacity is insufficient to meet demands of the sinks. Sink capacities during ripening (pod numbers) are decided by competitive conditions during short periods at or after flowering. Thus, the nature of source-sink relationships and their effects on yields of grain legumes are intermediate between cereal crops and root crops.

AN EXAMPLE OF YIELD MANIPULATION USING SOURCE-SINK CONCEPTS RELATIVE TO NITROGEN AND CROP DEVELOPMENT

Nitrogen is very important in determining source-sink capacities of crop plants during much of their growth cycle. Nitrogen determines both source (leaf canopy) and sink (inflorescence) size. The effects of N on sink sizes during the growth cycles of 45 lowland rice cultivars were grouped by days to maturity: very short (<110 days), short (111 to 120 days), medium (121 to 130 days), and long (>130 days) grown with N at 0 and 90 kg N/ha (60 kg basal + 30 kg top dressing at panicle initiation) were determined (IRRI, 1987). Cultivar differences for N absorption ability were observed up to maximum tillering but not thereafter. Concentrations of N in plants at flowering for both 0 and 90 kg/N ha rates were highly correlated with growth duration ($r = 0.736**$ and $r = 0.734,**$ respectively, for 0 and 90 kg N/ha). Concentrations of N at flowering was also correlated with the sink size and yield (IRRI, 1987). Optimum growth duration for sink size and yield was observed at 125 days (IRRI, 1987). Likewise, highest contributions of plant N to sinks at flowering and to grain yield were observed at 125 days growth duration (IRRI, 1987).

Yields decreased when growth duration was shorter than optimum because of small sinks (low panicle numbers) caused by low amounts of plant N during late stages of spikelet initiation. Lower yields of short-duration cultivars also caused by low contribution of plant N to sink formation at flowering when panicle initiation occurred and before when maximum numbers of tillers formed. Nitrogen absorbed after panicle initiation contributed little to sink differentiation or panicle formation, and did not affect numbers of spikelets, although N contributed to grain size (IRRI, 1987). On the other hand, when growth duration was longer than optimum, yields decreased because of sink shortages caused by high numbers of degenerated or nonviable spikelets. In this IRRI (1987) study, percentages of degenerated spikelets were correlated with growth duration ($r = 0.86**$). This occurred despite the fact that plants had large amounts of N at flowering.

Water drought conditions were also unfavorable to sink sizes in rice. For example, when drought occurred commonly during flowering, rice with shorter growth cycles yielded more than those with longer growth cycles. Even though short-growth-cycle cultivars had lower potential sink sizes, they had less damage from drought than long-growth-cycle cultivars. Longer-growth-cycle cultivars with their potentially greater sink sizes had more drastically reduced sink sizes than short-growth-cycle cultivars because of drought-induced reduced flower or spikelet formation (Fageria, 1992).

The uptake of N in plants at critical growth stages was correlated with growth duration in the dry season ($r = 0.74**$) and the wet season ($r = 0.90**$) when plants were grown under similar cultural practices. The uptake of N in plants at flowering was also highly correlated with the amount of sink and yield in the dry season ($r = 0.93**$) and the wet season ($r = 0.89**$). Yield was governed by sink strength (i.e., spikelet numbers) despite large differences in percentages of ripened grains between the dry and wet seasons (IRRI, 1988). Sink contributed more to yield in the dry season than in the wet season.

Studies were conducted at IRRI (1988) to determine relationships between growth duration and grain yield in lowland rice. Optimum growth duration for yield and sink size was about 125 days in both the dry and wet seasons, but optimum growth duration was slightly shorter for enhanced yields than for increased sink size. Growth duration was negatively correlated with the percentages of ripened grains during both the dry and wet seasons at $r = -0.54**$ and $r = -0.60,**$ respectively, for dry and wet season (IRRI, 1988). These results were in contrast to those of lowland rice, which generally is grown under favorable environmental conditions.

Lower yields of short-duration cultivars can be explained by shortages of sinks because of low N in plants during stages of growth when spikelets were initiated and low contributions of N to amount of sink in plants at flowering. High positive correlations between plant

N uptake ability and yield, sink, N harvest index (NHI = N in the grain/N in the grain plus straw). Contributions of N to sink formation were also noted in short-duration rice cultivars (IRRI, 1988, 1989). Thus, N absorption ability of plants during early growth stages was more important for short-duration cultivars than for medium- and long-duration cultivars. Lower yields of long-duration cultivars were attributed to degenerated sinks caused by longer vegetative lags. Hence, optimum growth duration can be explained primarily by duration of vegetative lags. Nitrogen was primarily responsible for increases in maize grain yields observed during the past 50 years (Olson and Sander, 1999). Nitrogen is essential to C flow and protein synthesis of higher plants (Yamazaki, Watanabe, and Sugiyama, 1986; Sugiharto et al., 1990). Bruns and Abel (2003) reported that increasing N concentration in maize plant tissue was positively associated with grain yield.

RELATIONSHIP BETWEEN SINK AND RESPIRATION

Tanaka and Yamaguchi (1968) evaluated quantitative roles of respiration in crop production and defined growth efficiency (GE) by the equation

$$GE = W/(W + R), \qquad (5.1)$$

where W = dry matter production (g) and R = respiration expressed as CH_2O utilized.

Growth efficiency values for rice and maize were 0.70 (Yamaguchi, Kawachi, and Tanaka, 1975) and 0.45 for soybean (Yamaguchi, Kawachi, and Tanaka, 1975; Yamaguchi, Watanabe, and Tanaka, 1975). These differences were ascribed to differences in composition of chemical components of organs, especially to greater source of protein and oil (Yamaguchi, 1978; Penning de Varies et al., 1989; Shinano et al., 1993). Figure 5.3 shows C distribution during the growth cycle of rice. Figure 5.3 also illustrates that when rice plants are grown under favorable conditions, major portions (more than 80 percent) of new photosynthetic assimilates are translocated to sinks and utilized as new biomass or as substrate for growth respiration. Only limited amounts of photosynthetic assimilates are stored

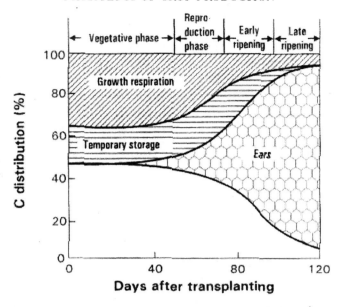

FIGURE 5.3. Distribution of ^{14}C assimilated in rice plants during different growth stages. *Source:* Reproduced from Tanaka, 1980.

temporarily in sources or old organs. In the latter case, photosynthetic assimilates will be consumed gradually during maintenance respiration.

FORMATION OF YIELD SINKS

Some flowers of gramineous and leguminous plants remain sterile after anthesis. In addition, some young fruits and seeds formed after fertilization fail to develop normally. Only those flowers that have been fertilized and kernels that have set will accumulate sufficient dry matter to fill grains. Therefore, grain growth is often divided into set and filling stages for gramineous crops (Kumura, 1995) and pod or seed set for leguminous plants. Filling is the process by which dry matter is accumulated into grains, fruits, or seeds.

Whether young grains and seeds are successfully set is usually determined at relatively early stages of grain/seed growth. After passing this critical stage of grain/seed development, failure of grain/seed growth rarely occurs under ordinary conditions. In most crops, capacity of grain/seed sinks adjust to suit source capacity and to supply of photosynthetic assimilates. If source capacity is high, rates of setting and yield sinks are usually both high (Kumura, 1995).

Several studies have indicated that soybean yield is source restricted during early reproductive growth periods. For example, Kokubun and Watanable (1983) noted that during this early growth stage (i.e., from R1 to R5 growth stages) the most sensitive time may alter source strength. Reduced source capacity during this growth period affected yield mainly through pod numbers. Other studies where late planting dates were used showed that crop growth rates during early reproductive periods were responsible for pod numbers through regulation of numbers of reproductive nodes on branches and numbers of pods per reproductive node on whole plants (Board and Harville, 1993; Board and Tan, 1995). In addition, Board and Harville (1998) noted that reduced crop growth rates during early reproductive periods caused by late plantings had from 147 percent to 63 percent greater effects on yield compared to depressed crop growth rates during late reproductive periods. This indicated that source restriction was greater during early reproductive periods than during late reproductive periods. These authors concluded that soybean yield was affected most by stresses that reduced crop growth rates during early reproductive periods.

MODIFICATION IN SOURCE-SINK RELATIONSHIPS WITH CULTIVAR IMPROVEMENT

Crop production has been significantly improved by modification of source-sink relationships during the past few decades. Major improvements have been made for rice, wheat, maize, and potato, and to some extent for legume crops. Figure 5.4 shows development trends in modern high-yielding rice cultivars compared to old cultivars and future ideotypes. From this figure, it can be noted that source strength has improved, partially through modification of plant architecture to intercept more solar radiation (see Chapter 1). In addition, sink size increased by increasing harvest index and grain size (see Chapter 3). Harvest index of traditional tall cultivars increased about 0.3 and that of semidwarf cultivars 0.5. In principle, harvest index could be increased further (up to about 0.65), which would improve sink capacity.

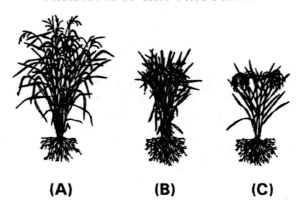

(A)   (B)   (C)

FIGURE 5.4. Traditional old (A), modern high yielding (B), and future ideotype (C) plants of lowland rice. *Source:* Reproduced from Khush, 1995.

For example, grain yields of 16.6 t·ha^{-1} (rough rice) was recorded by a Japonica F_1 hybrid rice in China (Amano et al., 1993), where harvest index of this hybrid was 0.67. In Australia and California, rice yields of 13 to 15 t·ha^{-1} have been achieved (Kropff et al., 1994). In comparison, average rice yields for many countries varied from 0.9 to 8.1 t·ha^{-1} depending on climate, cultivar, fertilizer, pesticide usage, and other cultural practices (FAO, 1994).

MANAGEMENT STRATEGIES FOR MAXIMIZING SOURCE-SINK RELATIONSHIPS

Yield potentials of cereal and seed legume crops have been increased through plant breeding, primarily by improving sink capacity (Gifford et al., 1984). Modern cotton cultivars also partition greater proportions of dry matter into fiber and seeds than obsolete older cultivars (Wells and Meredith, 1984; Pace et al., 1999). Moreover, modern cultivars tend to mature earlier than older cultivars.

Source/sink ratios depend on interactions between genotype and environment, which can be influenced by crop management factors such as planting date, population density, nutrient supply, adequate soil moisture, and control of biotic factors such as diseases, insects, and weeds. Timing and amount of nitrogen application can particularly improve source-sink relationships. Studies have reported that adequate N in plant leaves at critical growth stages can improve both source and sink capacities. When proper environmental conditions prevail (e.g., soil moisture availability and N topdressings during reproductive phases), increases in source and sink capacities usually occur during ripening from increases of spikelet numbers per panicle. Therefore, grain yield can be increased if source capacity during ripening is sufficiently large to support increased sink capacity (Tanaka, 1980). Other management strategies to maximize yields include breeding crop cultivars possessing optimum source-sink balances for responses to N and other management factors.

Manipulation of source/sink ratios by artificial reduction in grain numbers per inflorescence has been used to estimate potential kernel weight and study grain-filling processes of several cereals (Simmons, Crookston, and Kurle, 1982; Blum et al., 1983; Peterson, 1983; Bruckner and Frohberg, 1991). Actual kernel weight is usually less than potential kernel weight because of plant competition for light, water, and nutrients (Peterson, 1983).

CONCLUSION

Crop yields are the end result of growth and development during different phases of crop life cycles. Plants have variously located organs that are closely connected in source-sink relationships. Relative strengths of sources and sinks are decided by sequential development of each organ. To maintain high specific absorption rates of N (mg N/day per g root dry weight) during maturation, photosynthetic assimilate distribution into roots is important. Cereal and legume crops form strong sinks in aboveground parts, and root-shoot interactions become weaker as growth progresses. In annual crops, plant type and canopy structure are important for maintenance of high root activities. The major supply of photosynthetic assimilates for roots comes from lower leaves of annual crop plants. Root crops benefit more than cereal and legume crops from carbohydrate supplies transported to roots because roots are directly attached to sink organs (Osaki et al., 1997). Penning de Varies et al. (1989) reported that about 14 percent of dry matter produced by rice from emergence to half anthesis was allocated to roots, with the remainder going to shoots. Dry matter of shoots was equally divided between culm and leaves during this growth period. Grain yields of major cereal crops during later growth stages are largely determined by source-sink relationships, in which florets are the primary photosynthetic sink and the top three leaves on stems are the primary source. In rice, over 80 percent of carbohydrates accumulated in grains is produced by the top two leaves (Li et al., 1998). Similarly, these authors investigated genetics underlying relationships between source leaves (top two leaves) and sink capacity of rice, and concluded that 50 percent of the phenotypic variation of primary sink capacity (grain weight per panicle) was attributed to variation of flag leaf area. To achieve high yields, N should be actively absorbed through roots even during maturation (Osaki et al., 1995, 1996). When large amounts of N are supplied to leaves from roots, photosynthesis should remain high during maturation, which assures a supply of carbohydrates to both roots and aboveground plant parts. Thus, activities of roots and shoots were assumed to be mutually regulated, and called root-shoot interactions. Improving potential crop yields through breeding been largely by selecting plants for high yield of the sink of economic interest.

Chapter 6

Carbon Dioxide and Crop Yield

INTRODUCTION

Elevated atmospheric CO_2 concentrations and associated climate changes may affect crop production in coming decades, as well as world food supplies (Rosenzweig and Hillel, 1998). During the latter part of the twenty-first century, crops are expected to be grown in environments with twice the present atmospheric CO_2 concentrations, with average temperatures being about 2 to 5°C warmer than at present (Houghton et al., 1996). In addition, unexpected late springs and early frosts and periodic episodes of heat and drought stresses are predicted to occur more frequently under new weather environments, and changes could exacerbate climate effects on many aspects of crop growth and development to decrease crop yields and quality (Reddy et al., 1999).

Current knowledge about interactions of CO_2 with key environmental factors is insufficient to show definitive conclusions regarding the magnitude and even the direction of potential future changes in crop yields. Although elevated CO_2 alone tends to increase crop production, predicted warmer temperatures could negatively affect plant growth and development, possibly counterbalancing positive CO_2 effects (Rawson, 1995; Tubiello et al., 1999). Management practices may also modify crop responses to climate. Crops grown with low N may respond to elevated CO_2 less than those grown with adequate fertilizer (Sionit et al., 1981). Rain-fed crops might exhibit greater relative growth responses to elevated CO_2 than those grown under irrigated conditions due to reduced stomatal conductance and water use (Morison, 1985; Chaudhuri, Kirkham, and Kanemasu, 1990a,b; Tubiello et al., 1999).

Natural environmental gases include CO_2, carbon monoxide (CO), methane (CH_4), ozone (O_3), [nitrous oxide (N_2O), nitrogen oxide (NO), and nitrogen dioxide (NO_2)] called NO_x. Some synthetic gas compounds possessing attributes similar to those mentioned have been introduced since the 1930s, and include several chlorofluorocarbons and chlorofluorohydrocarbons (collectively called CFCs). These latter gases trap solar radiation within the earth's atmosphere similar to those in greenhouses, leading to their being called, collectively, "greenhouse gases" (GHGs) or active radiative gases (Lal et al., 1998).

Global warming and its potential effects on increased temperatures and raising sea levels have been of worldwide concern. For instance, it has been estimated that temperature may increase between 1.1 and 1.9°C and sea levels rise between 0.14 and 0.24 m by 2030 (Figure 6.1). Increases in concentrations of greenhouse gases in the atmosphere are predicted to raise mean temperatures by 2 to 3°C by 2050, together with more frequent episodes of water deficit and high temperature events (Wilks and Riha, 1996; Ferris et al., 1999). Atmospheric CO_2 is important not only to plant growth but also to global energy balances (Baker, Spaans, and Reece, 1996). Further, CO_2 is the most important GHG, because increases in its concentration causes about 50 percent of total radiative forcing (Rodhe, 1990). Concentrations of atmospheric CO_2 have risen from about 280 $\mu mol \cdot mol^{-1}$ to 362 $\mu mol \cdot mol^{-1}$ in the past nine decades and are predicted to continue to rise by an average of 1.5 $\mu mol \cdot mol^{-1}$ per year (Siegenthaler, 1990; Grayston et al., 1998). In addition, Amthor (1998) reported that the earth's atmospheric CO_2 concentration increased about 30 percent during the past 200 years, from near 280 to more than 360 $\mu mol \cdot mol^{-1}$ (Figure 6.2). Seneweera and Conroy (1997) also reported that atmospheric CO_2 concentrations have risen from 315 $\mu mol \cdot mol^{-1}$ in 1958 to 360 $\mu mol \cdot mol^{-1}$ in 1996 with current rates of increase of 1.9 $\mu mol \cdot mol^{-1}$ per year. Given the reluctance of developed countries to reduce use of fossil fuels, it is inevitable that CO_2 concentrations will reach between 510 and 760 $\mu mol \cdot mol^{-1}$ during the next 50 years.

Goudriaan (1995) reported that during the past 50 years CO_2 rates of increase have grown exponentially about 2.4 percent per year. A single descriptive equation for atmospheric concentrations of CO_2 as a function of a year number (t) is (Goudriaan, 1995)

$$CO_2 = 285 + 52\exp [0.024(t - 1980)] = \mu mol \times mol^{-1} \quad (6.1)$$

Observed values fit well with this formula. This formula assumes preindustrial levels of CO_2 at 285 $\mu mol \cdot mol^{-1}$ and CO_2 levels of 337 $\mu mol \cdot mol^{-1}$ for the reference year 1980. Each $\mu mol \cdot mol^{-1}$ CO_2 in the atmosphere correspond to about 2.1 Pg (1Pg or petagram = 10^{15} g = 1 Gt or gigaton). Goudriaan (1995) used the following equation to calculate approximate rates of C emission to the atmosphere with fossil fuel combustion (F):

$$F = 4.8 \exp [0.024 (t - 1980)] = Pg/year. \quad (6.2)$$

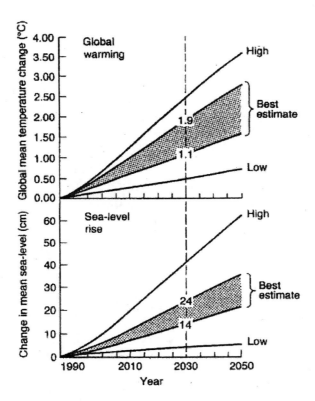

FIGURE 6.1. Projected global warming and sea level rise from 1990 to 2050. *Source:* Reproduced from Warrick, R. A. and E. M. Barrow (1990). "Climate and sea level change: A perspective." *Outlook on Agriculture* 19:5-8. Used by permission.

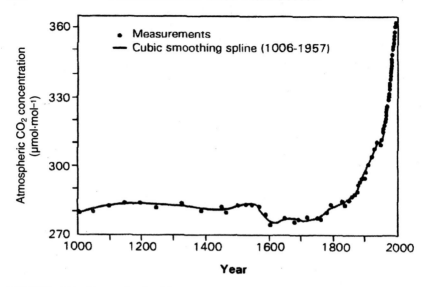

FIGURE 6.2. Atmospheric CO_2 concentrations during the past 1000 years. *Source:* Reprinted from Amthor, J. S. 1998. *Field Crops Research,* 58, J. S. Amthor, "Perspective on the relative insignificance of increasing atmospheric CO_2 concentration to crop yield," pp. 109-127, copyright 1998, with permission from Elsevier. Used by permission of Elsevier, UK.

Although it is impossible to accurately predict future CO_2 increases, CO_2 could reach alarming levels during the next 100 years if proper mitigation practices are not adopted. This is of concern because increasing CO_2 may warm the earth's surface and could alter temporal and spatial patterns of precipitation and evaporation because of enhanced global greenhouse effects (Karl, Nicholls, and Gregory, 1997).

The two major greenhouse substances of water vapor and water/ice clouds collectively are responsible for over 90 percent of the greenhouse effects that keep the earth about 33°C warmer than it would otherwise be in the absence of these substances (Prinn, 1995). These substances have short lifetimes, and quantities of these adjust fairly quickly to long-term climate represented by changes in so called long-lived greenhouse gases. These GHGs include CO_2, CO, CH_4, O_3, N_2O, NO, and NO_2. Carbon dioxide increases at 0.3 to 0.4 percent per year to air through combustion of fossil fuels and deforestation, and is removed through net uptake by land, ecosystems, and oceans. Methane increased over the past 20 years by an average of 0.8 percent per year, and is produced at poorly understood proportions by such processes as gas and coal industrial combustion, rice fields, cattle yards, and in naturally occurring wetlands. Methane is destroyed by chemical reactions driven by ultraviolet sunlight, which involve very reactive hydroxyl radicals. Methane increased at 1.2 percent per year in the late 1970s but is now increasing at less than 0.3 percent per year. Global methane emissions are decreasing, methane destruction rates are increasing, or both. Nitrous oxide has increased at 0.2 to 0.3 percent per year and, like methane, has wide ranges of poorly understood natural and anthropogenic regional sources. It is removed in the ozone layer, as are the industrial CFCs. The CFCs were increasing at 5 percent per year in the 1980s, but they are now increasing at less than 3 percent per year as a result of the Montreal Protocol. It is important to note that the potency of both N_2O and CFCs as greenhouse gases is offset significantly by the stratospheric ozone that they destroy. Although responsible for only a small percentage of the direct greenhouse warming, these gases have indirect influences. For example, increases in temperatures resulting from their presence leads to increases in water vapor and clouds, which accelerates warming. These water-related feedback processes are very important but not quantitatively well understood, and lead to much uncertainty in current climate models.

Any climatic changes associated with increasing CO_2 have the potential to affect crop physiology, growth, and yield. For example, warming might alter rates of photosynthesis and photorespiration, alter rates of plant biosynthesis, and accelerate phenological development. Increased temperature and/or altered patterns of precipitation could also modify timing for planting and/or geographic distribution of some crops (Amthor, 1998). The main objective of this chapter is to present information about influences of CO_2 on plant growth. Hopefully, the information presented will generate ideas concerning the CO_2 impacts on global climate change, and this information will have applications in many fields from agriculture to energy sources.

TOTAL CARBON IN SOILS OF THE WORLD

Interest in determining soil organic C stocks on global, continental, and regional scales has increased recently with progress in knowledge about anthropogenic climate changes (Hontoria, Rodriguez-Murilla, and Saa, 1999). Future changes in climate, atmospheric CO_2 concentrations, and land use will decisively influence the fate of C stored in soils, thereby affecting atmospheric CO_2 reservoirs and climatic conditions of earth. Climate and soil properties are key factors determining production and decomposition processes of plant litter. In general, soil organic C increases with increasing precipitation and decreases with increasing temperature (Burk et al., 1989; Hontoria, Rodriguez-Murillo, and Saa, 1999).

Soil C is a major source of CO_2 and a reservoir that responds to changes in climate and atmospheric CO_2 concentrations. Therefore, knowledge of C estimates in different groups of world soils is important for managing or manipulating atmospheric CO_2 concentrations at desirable levels. Sizes and dynamics of C pools in soils of the world are not well-known. Eswaran, Berg, and Reich (1993) cited problems for making accurate global estimates that result from

1. very high spatial variability in C contents of soils;
2. unreliable estimates of area occupied by different kinds of soils;
3. unavailability of reliable data, particularly bulk density, to compute volumetric composition; and
4. confounding effects of climate, vegetation, and land use.

Recent concerns about GHG effects and damage to the ozone layer have resulted in more concerted studies on quantities, kinds, distributions, and behavior of C in different systems (Bohn, 1976, 1982; Buringh, 1984; Johnson and Kerns, 1991; Eswaran, Berg, and Reich, 1993; Batjes, 1996).

Equilibrium of C on earth is a function of three reservoirs: Oceans have about $39,000 \times 10^{15}$ g (or Pg), the atmosphere has about 750 Pg, and terrestrial systems have about 2200 Pg (Batjes, 1996). A fourth reservoir (geological reservoirs) is estimated to have about 65.5×10^6 Pg C, which is a permanent sink (Kempe, 1979). A small fraction (about 4000 Pg) of the geological sink is present as fossil fuels from which C release takes place as a result of human mining activities (Eswaran, Berg, and Reich, 1993). Total masses of organic C stored in soils of the world is 1576 Pg (1576 × 10^{15} g), of which about 32 percent (or 506 Pg) is found in the tropics (Table 6.1). This mass is nearly three times that of C in vegetative biomasses, and about twice that of the atmosphere. Most soils contain only small concentrations of organic matter (1 to 5 percent), but these greatly modify agricultural properties of soils (Mahieu, Powlson, and Randall, 1999). Although soil-vegetation C pools are small compared with that of oceans, it is potentially more labile in the short term. The C balance of terrestrial ecosystems can be changed markedly by direct impacts of human activities, including deforestation, biomass burning, land use change, and environmental pollution, which release trace gases that enhance greenhouse effects (Batjes, 1996).

Decay of soil organic matter is one of the largest CO_2 inputs to the atmosphere. Changing climate and CO_2 concentrations may affect decay rates, and thereby affect return rates to the atmosphere for CO_2 removed by photosynthesis. This interaction with atmospheric CO_2,

TABLE 6.1. Organic carbon mass in soils of the world.

Soil order	Area (10^3 ha)		Organic C (Pg)		
	Global	Tropical	Global	Tropical	Tropical % of global
Histosols	174,500	28,600	357	100	28
Andisols	255,200	168,300	78	47	60
Spodosols	487,800	4,000	71	2	3
Oxisols	1,177,200	1,151,200	119	119	100
Vertisols	328,700	218,900	19	11	58
Aridisols	3,174,300	911,700	110	29	26
Ultisols	1,133,000	901,800	105	85	81
Mollisols	548,000	23,400	72	2	3
Alfisols	1,828,300	641,100	127	30	24
Inceptisols	2,158,000	456,500	352	60	17
Entisols	1,492,100	325,600	148	19	13
Misc. land	764,400	135,800	18	2	11
Total	13,521,500	4,966,900	1576	506	32

Source: Eswaren, Berg, and Reich, 1993. Reproduced by permission of Soil Science Society of America, Madison, WI.

and soil C accumulation, which removes C from the C cycle, should be dependent on mass of organic C in soils (Bohn, 1976).

CARBON DIOXIDE AND PLANT GROWTH

Carbon dioxide has always occupied a prominent place for consideration by plant physiologists and agronomists, since CO_2 is the substrate on which all plant growth depends (Baker, Spaans, and Reece, 1996). The current global rise in atmospheric CO_2 concentration has stimulated interest and extensive research about responses of agricultural crops to CO_2 and the effects of elevated CO_2 on plant growth (Idso and Idso, 1994; Rogers, Runion, and Krupa, 1994; Frank and Bauer, 1996; Grotenhuis and Bugbee, 1997; Heagle, Miller, and Booker, 1998; Heagle, Miller, and Pursley, 1998; Miller, Heagle, and Pursley, 1998). Plant growth stimulation responses to elevated CO_2 also vary with photosynthetic pathway. For example, growth stage (Newbery and Wolfenden, 1996), water (Patterson, 1986), nutrient availability (Duchein, Bonicel, and Betsche, 1993), and plant species (Bazzaz, 1990) affect how plants respond to enhanced levels of CO_2. In particular, C_3 plants are more responsive to CO_2 than are C_4 plants (Kimball, 1983; Kimball, Kobayashi, and Bindi, 2002). Increases in plant growth due to increased CO_2 concentration are generally, but not always (Wyse, 1980), due to higher assimilation rates per unit leaf area (Hogan, Smith, and Ziska, 1991).

Availability of N in particular has been noted to be critical in determining the magnitude of plant responses to elevated CO_2 (Bowler and Press, 1996; Lutze and Gifford, 1995). This is especially significant for native pasture species, which often grow on low-nutrient soils, and studies on these species are less common (Newbury and Wolfenden, 1996; Grayston et al., 1998). Grotenhuis and Bugbee (1997) studied the effects of near-optimal (≈ 1200 $\mu mol \cdot mol^{-1}$) and super-optimal (2400 $\mu mol \cdot mol^{-1}$) CO_2 levels on yield of two cultivars of hydroponically grown wheat. Vegetative growth increased by 25 percent and seed yield by 15 percent for both cultivars at the lower concentration of CO_2. Yield increases were primarily the result of increased numbers of heads/m^2. However, elevation of CO_2 to 2500 $\mu mol \cdot mol^{-1}$ reduced seed yield by 22 percent in one cultivar and 15 percent in the other. Toxic effects of CO_2 were similar over light level ranges from half to full sunlight. Cure and Acock (1986) reported that increased CO_2 affected plant growth mainly through changes in leaf conductance, water vapor, and CO_2 exchange. Cure and Acock (1986) concluded that leaf conductance of C_3 annual crop plants was reduced by 34 percent when CO_2 was twice current atmospheric levels.

Rice grain yields increased with enhanced CO_2 (Amthor, 1998). Increased grain yield is often, but not always, associated with increased tillering and numbers of panicles (Allen et al., 1995). Yield of rice (cultivar IR30) declined by 10 percent for each 1°C rise in day/night temperatures above 28/21°C, and elevated CO_2 had little effect in ameliorating these temperature responses (Allen et al., 1995). Sharp decreases in numbers of filled grains per panicle accompanied these yield decreases (Allen et al., 1995).

Plant responses to CO_2 depend on stage of plant development. Total dry matter in spring wheat increased only when young plants were exposed to relatively high CO_2 concentration (Sionit et al., 1981), and grain yields increased only when plants were exposed to elevated CO_2 prior to floral initiation (Krenzer and Moss, 1975). Neales and Nicholls (1978) reported that 10-day-old wheat plants responded to increased CO_2 with greater growth and higher assimilation rates, but 24-day-old plants had reverse responses. Growth and development of barley also depended on plant growth stage, with only slight responses observed in young plants. During the life cycle of barley, CO_2 at 400 and 675 $\mu mol \cdot mol^{-1}$ above ambient increased photosynthetic capacity and total dry matter production (Ingvardsen and Veierskov, 1994).

Photorespiration in C_3 plants decreases with increasing CO_2 levels because ratios of ribulose-1,5-bisphosphate (RuBP) oxygenation to carboxylation decreased (Grotenhuis and Bugbee, 1997). At 20°C, increases in CO_2 from 350 to 1200 $\mu mol \cdot mol^{-1}$ decreased rates of photorespiration in wheat from 24 to 6 percent of gross photosynthesis, and rates decreased further to 3 percent of gross photosynthesis at 2500 $\mu mol \cdot mol^{-1}$ (Brooks and Farquhar, 1985; Woodrow and Berry, 1988). Because decreases in photorespiration are asymptotic and decreases in photorespiration beyond 1200 $\mu mol \cdot mol^{-1}$ are small, 1200 $\mu mol \cdot mol^{-1}$ (1200 ppm; 0.12 percent or 100 Pa at sea level) may be considered near-optimal CO_2 levels for most crop plants (Grotenhuis and Bugbee, 1997).

Plant responses to CO_2 enrichment are generally opposite responses to elevated O_3, and include increased photosynthesis and decreased stomatal conductance (Jones et al., 1984; Jones, Jones, and Allen, 1985; Sionit et al., 1984), development of larger, thicker, and heavier leaves (Thomas and Harvey, 1983), increased branching, increased numbers of nodes (Allen et al., 1988; Rogers, Cure et al., 1984), changed root/shoot ratios (Idso, Kimball, and Mauney, 1988), and increased growth and yield (Allen et al., 1988; Rogers, Cure et al., 1984; Rogers, Cure, and Smith, 1986). Idso and Idso (1994) reported that relative growth-enhancing effects of CO_2 are greatest when resources limit growth, or when plants are grown in suboptimum environments, including those contaminated by air pollutants such as O_3.

Drake, Gonzalez-Meler, and Long (1997) noted that the primary effects of plant response to rising atmospheric CO_2 concentrations are to increase resource use efficiency. Elevated CO_2 reduces stomatal conductance and transpiration and improves water use efficiency, stimulates higher rates of photosynthesis, and increases light-use efficiency. Acclimation of photosynthesis during long-term exposure to elevated CO_2 may reduce key enzymes of the photosynthetic C reduction cycle, which could increase nutrient use efficiency. These effects have major consequences for agriculture and native ecosystems in environments with rising atmospheric CO_2 and climatic changes (Drake, Gonzalez-Meler, and Long, 1997).

CARBON DIOXIDE AND PHOTOSYNTHESIS

It is generally understood that elevated CO_2 concentrations stimulate photosynthesis in crop plants. In a survey of 60 experiments, Drake, Gonzalez-Meler, and Long (1997) reported that plant growth with elevated CO_2 concentrations increased photosynthesis by 58 percent compared to rates for plants grown with normal ambient CO_2 concentrations. In addition, CO_2 has the potential to regulate many reactions within photosynthetic systems. For example, binding Mn on the donor side of photosystem II (Klimov et al., 1995), disrupting quinone binding sites on the acceptor side of photosystem II, and activating ribulose-1,5-bisphosphate carboxylase/oxygenase (Rubisco) have been reported. (Portis, 1995). While these processes exhibit high affinity for CO_2, they become saturated at current CO_2 concentrations. Rubisco has low affinity for CO_2 in carboxylation reactions, and this reaction is not saturated at the current CO_2 (Drake, Gonzalez-Meler, and Long, 1997). Baker et al. (1990) reported that photosynthetic rates increased in rice with increasing CO_2 concentrations from 160 to 500 $\mu mol \cdot mol^{-1}$, followed by leveling off at super-ambient CO_2 (600 and 900 $\mu mol \cdot mol^{-1}$) concentrations. Yield of C_3 plants could potentially lead to average increases of >30 percent for plants grown under optimum conditions (Kimball, 1983). Increases in yield by C_4 plants would likely be less than for C_3 plants since photosynthesis is not stimulated as much in C_4 as in C_3 plants (Poorter, 1993).

Stimulated yield and/or biochemical reactions by elevated CO_2 may also be valuable for belowground ecosystem processes because 20 to 50 percent of all assimilated photosynthetic assimilates are transported to belowground roots (Merckx et al., 1986; Swinnen, Veen, and Merckx, 1995). Photosynthetic assimilates are translocated from shoots to roots and subsequently released into soil as low-molecular-weight exudates such as organic acids, sugars, phenolics, and amino acids or as high-molecular-weight solutes such as mucilage and ecto-enzymes. Elevated CO_2 stimulates root growth at similar amounts as shoot growth (Norby, Pastor, and Melillo, 1986; Curtis et al., 1990; Newton et al., 1994, 1995), and the increased C input into soil through production and turnover of root exudates and solutes can have strong effects on nonplant biological growth belowground.

In predicting responses of photosynthesis by C_3 plants to increasing concentrations of CO_2 in the atmosphere, it has been recognized that photosynthetic stimulation from increased CO_2 usually increases extensively with increasing temperature (Bunce, 1993). This temperature dependence has a firm theoretical basis related to temperature dependencies of aqueous O_2 and CO_2 solubilities and to kinetic characteristics of Rubisco. Bunce (1998) reported that short-term stimulation of photosynthesis existed for wheat and barley with doubling of CO_2 concentrations from 350 to 700 $\mu mol \cdot mol^{-1}$. However, photosynthesis decreased when temperature was lowered from 30 to 10°C at high proton fluxes.

CARBON DIOXIDE AND WATER USE EFFICIENCY

Water use efficiency (WUE) means the ratio of photosynthetic CO_2 assimilation to transpiration per unit leaf area. Increased CO_2 generally reduces stomatal conductance (Bunce, 1993; Jackson et al., 1994), which reduces plant transpiration and increases WUE (Chaudhuri,

Kirkham, and Kanemasu, 1990b; Polley, Johnson, Marino, and Mayeux, 1993; Polley, Johnson, and Mayeux, 1994; Kimball, Kobayashi, and Bindi, 2002). Decreases in leaf conductance and water requirement and increases in photosynthetic rates at increased CO_2 result in higher WUE values for most C_3 plants. Frank and Bauer (1996) reported that reductions in water used by wheat (13 percent at 650, and 29 percent at 950 compared to 350 $\mu mol \cdot mol^{-1}$ CO_2) as CO_2 increased. Nie et al. (1992) reported that native grasslands exposed to 720 $\mu mol \cdot mol^{-1}$ CO_2 had 8 and 18 percent lower evapotranspiration rates at low and high soil water levels, respectively. Polly, Johnson, Mayeux, and Malone (1993) reported that wheat grown along gradients simulating CO_2 during the past two centuries produced greater total dry matter and had higher WUE values as CO_2 increased. Baker et al. (1990) reported that CO_2 treatments in the range of 160 to 900 $\mu mol \cdot mol^{-1}$ decreased evapotranspiration and decreased WUE in rice. In studies of subambient CO_2 effects on oat, mustard, and two cultivars of wheat, WUE values increased 40 to 100 percent as ambient CO_2 was increased from about 15 to 35 Pa (Pascal) (Polley, Johnson, Marino, and Mayeux, 1993). In free air C enrichment studies with wheat, CO_2 concentrations elevated to 55 Pa increased WUE by 76 and 86 percent in cotton over two full growing seasons (Pinter et al., 1996). Increased CO_2 concentrations also increased WUE in both C_3 and C_4 wetland species (Drake, Gonzalez-Meler, and Long, 1997).

It was clear from several studies on temperate plant species that increased CO_2 can substantially reduce water loss, primarily through stomatal closure (Rogers et al., 1983; Rogers, Sionit et al., 1984). Data obtained for tropical plants (Ziska et al., 1991) indicated that WUE increased substantially under elevated CO_2 conditions. Increases in WUE were attributed to the combined effects of reduced stomatal conductance along with increased photosynthetic capacity. Observed increases were noted for both C_3 and C_4 tropical species (Ziska et al., 1991).

CARBON DIOXIDE AND RADIATION USE EFFICIENCY

Radiation use efficiency (RUE), defined as dry matter production per unit of intercepted light ($g \cdot J^{-1}$) or phytomass per unit of energy received ($J \cdot J^{-1}$), improved with increasing CO_2 concentrations (Gallo, Daughtry, and Wiegand, 1993). RUE is often crucial in crop growth models relating dry matter production to energy received. Pinter et al. (1996) noted that cotton crops grown under free air C enrichment at 55 Pa had highly significant increases in RUE values of 20 to 22 percent in consecutive years, regardless of whether crops were grown with full irrigation or with only 50 percent of optimal water supply.

MANAGEMENT STRATEGIES IN SEQUESTRATION OF CO_2

Increasing CO_2 concentrations in the atmosphere are indisputable (Keeling et al., 1995). Concentrations of CO_2 in the atmosphere are increasing at rates of about 1.5 $\mu mol \cdot mol^{-1}$ per year. The Intergovernmental Panel on Climate Change (IPCC, 1990) predicts that atmospheric CO_2 will continue to increase to about 700 $\mu mol \cdot mol^{-1}$ CO_2 by the end of the twenty-first century (Ginkel and Gorissen, 1998), mainly due to combustion of fossil fuels and deforestation. Results from climate models indicate that increased absorption of longwave radiation from increased concentrations of CO_2 and other GHGs such as CH_4 and N_2O could alter climate globally (IPCC, 1990). Managing terrestrial ecosystems, especially forests, to capture and store atmospheric C has been proposed for reducing rates of increases in atmospheric CO_2 concentrations (Schroeder and Ladd, 1991). For example, planting forests on marginal agricultural lands can provide many opportunities for C sequestration (Turner et al., 1993). Parks (1992) identified 47×10^6 ha of marginal agricultural land in the United States suited to convert to forests. Lee and Dodson (1996) suggested reducing net emissions of CO_2 by converting marginal grass pastures to pine plantations in the south-central United States. Potential C sequestration for United States croplands is presented in Figure 6.3.

Soil organic matter is associated with productivity of agroecosystems (Bauer and Black, 1994). Recent global concerns over increased atmospheric CO_2, which can potentially alter the earth's climate systems, have increased interest in studying soil organic matter dynamics and C sequestration capacity in various ecosystems (Qian and Follett, 2002). Cultivation usually causes decreases of soil organic C contents (Alvarez et al., 1998). No-tillage systems have been proposed as alternatives to conventional cropping systems for reducing

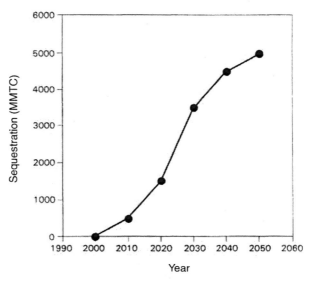

FIGURE 6.3. Projected carbon sequestration potential of U.S. croplands. *Source:* Lal et al., 1998; Reproduced by permission of Ann Arbor Press (CRC Press), Boca Raton, Florida.

soil degradation. No-tillage generally leads to increases of C in the top 5 to 10 cm of soil profiles relative to plowed soils (Kern and Johnson, 1993). Other agricultural practices such as altered management of agricultural soils and rangelands, improved efficiency of fertilizer use, and restoration of degraded agricultural lands and rangelands could reduce emissions of CO_2.

Increases in atmospheric CO_2 may be responsible for half of the so-called greenhouse effects (Houghton, Jenkins, and Ephraums, 1990). Fluxes of CO_2 between soil and the atmosphere are important in the global C cycle. It has been reported that soils, containing about three times as much C as the atmosphere, have the potential to store additional C (Campbell and Zentner, 1993). Curtin et al. (1998) reported that agricultural lands, if properly managed, could sequester 50 to 75 percent of CO_2 emissions from agricultural activities during the next 30 years in Canada. Management options to enhance C storage in Canadian prairie soils include decreasing summer fallow frequently, reducing tillage, including legumes in crop rotations, proper fertilization of crops, and incorporating grass on marginal lands

(Campbell and Zentner, 1993; Curtin et al., 1998). Table 6.2 summarizes several options for mitigation or sequestration of GHSs from cropland.

Carbon sequestration is very important in considering effects of farming practices on greenhouse gas (i.e., CO_2) emissions (Lal et al., 1998). Cropping systems and practices that enhance C sequestration of atmospheric CO_2 are beneficial. Halvorson, Reule, and Follett (1999) reported that increases in soil organic C with N fertilization contributes to improved soil quality and productivity and increased efficiency of C sequestration into soils. Carbon sequestration can be enhanced by increasing crop residue production through adequate N fertility. Sequestering more C into soils by increasing cropping intensity and adequate fertilization could contribute positively to mitigating agricultural effects on atmospheric CO_2 levels and their effects on global climate change (Lal et al., 1998).

CONCLUSION

Concentrations of GHGs in the atmosphere will probably double at a time scale of a few human generations. These increases may induce climatic changes that will exceed adaptive capacities of nature and human societies. Intensification of otherwise beneficial natural greenhouse effects causes concern because it may lead to temperature increases of the order of a few degrees centigrade (Goudriaan, 1995). Watson et al. (1990) suggested that anticipated increases in atmospheric CO_2 will increase mean global temperatures by about 4°C by the end of the twenty-first century.

Carbon contents of soils are important sources of atmospheric CO_2, and C stored in soils of the world is nearly three times that in aboveground vegetative biomass and approximately double that in the atmosphere. About 1576 Pg (Pg = petagram = 10^{15} g) of C is globally stored in soils, with about 506 Pg (32 percent) of this in soils of the tropics. It has also been estimated that about 40 percent of C in soils of the tropics is in forest soils. Studies have reported that deforestation can result in losses of 20 to 50 percent of this stored C, largely through erosion.

The primary effect of plant responses to rising atmospheric CO_2 is to increase in resource use efficiency. Elevated CO_2 reduces stomatal

TABLE 6.2. Strategies for carbon sequestration in cropland soils.

Mitigation option	Relative potential
1. Soil erosion management	High
2. Land conversion and restoration	
a. Convert marginal agricultural land to grassland, forest, or wetland	High
b. Restore degraded lands	High
c. Restrict use of organic soils	High
d. Restore wetlands	High
e. Reclaim mineland and toxic soils	Low
f. Restore salt-affected soils	Low
3. Intensification of prime agricultural land	
a. Erosion control (conservation tillage, buffers, CRP)	High
b. Supplemental irrigation	Medium
c. Soil fertility management	Medium
d. Improved cropping systems and winter cover	High
e. Elimination of summer fallowing	High
4. Biofuels	
a. Energy crops for fossil fuel	High
b. Biogas from liquid manures	High
c. Efficient grain drying	Low
d. Soil C sequestration in lands for biofuel	Low
5. Improving fertilizer use efficiency	
a. Crop rotations and tillage method	High
b. Nitrification inhibitors	Medium
c. Mixed farming	Low
d. Recycling organic material	Medium
e. Biological nitrogen fixation	Low
6. Management of rice paddies	
a. Residue management and tillage methods	Medium
b. Water management	Medium
c. Fertility management	Medium

Source: Lal et al., 1998. Reproduced by permission of Ann Arbor Press (CRC Press), Boca Raton, Florida.

conductance and transpiration, improves WUE, stimulates higher rates of photosynthesis, and increases light use efficiency. Increases in WUE in C_4 plants are due mainly to decreased stomatal conductance, while increases in WUE in C_3 plants appears to be attributed to both decreased stomatal conductance and increased C assimilation (Hogan, Smith, and Ziska, 1991). Acclimation of photosynthesis during long-term exposure to elevated CO_2 concentrations reduces key enzymes of the photosynthetic C reduction cycle, which increases nutrient use efficiency (Drake, Gonzalez-Meler, and Long, 1997).

Rising atmospheric CO_2 concentrations indicate positive implications for agriculture because most experiments have noted that plant growth will increase in response to higher CO_2 concentrations.

Average yields of most crops in many countries increased significantly in the past four decades. However, Amthor (1998) reported that although atmospheric CO_2 concentrations also increased during that period and crop growth and yields responded positively to CO_2 increases, yield increases were due mainly to factors other than increasing CO_2 concentrations. It can be concluded that future CO_2 concentrations will continue to increase, but these increases in CO_2 will have only small impacts on yields compared to increases from crop production technologies now available or will become available in the future. It is possible that continuous CO_2 increases may bring climatic changes that could have negative or positive impacts on future crop yields (Amthor, 1998). In addition, most research has been conducted under controlled conditions, and field experimental data are not sufficient to properly conclude that responses of crop plants to CO_2 are due solely to increases in the atmosphere.

Substantial amounts of C (approximately 60 to 90 Gt C) can be sequestered/conserved in forests alone over the next 50 years if deforestation is slowed and tree plantations are established.

Chapter 7

Physiology of Drought in Crop Plants

INTRODUCTION

Drought stress is a major constraint to crop production and yield stability in many regions of the world. Soil water deficits are estimated to depress agricultural crop yields in the United States by about 70 percent, compared with maximum achievable yields (Boyer, 1982). Similar problems are encountered worldwide. For example, a major constraint to bean production in many developing countries is drought, which affects 73 percent of land areas planted to beans in Latin America (Ramirez-Vallejo and Kelly, 1998). Approximately 32 percent of wheat-growing regions in developing countries experience some type of drought stress during the growing season (Ginkel et al., 1998).

Drought, whether intermittent or terminal, can be confounded with high temperatures in certain locations and aggravated by shallow soils and/or root-rotting pathogens. Droughts are inevitable and recur frequently throughout the world, despite improved abilities to predict their onset and modify their impact. Drought remains the single most important factor affecting world security and stability of land resources from which food is derived (McWilliam, 1986).

Drought is a meteorological term that means lack of precipitation over prolonged periods of time (Hale and Orcutt, 1987). It is a meteorological and hydrological event involving precipitation, evaporation, and soil water storage (McWilliam, 1986). Kobata (1995) defined droughts as environmental situations where decreases in soil moisture occur in rooting zones of crops. Droughts denote periods of time without appreciable precipitation during which water contents of soil are reduced to such extents that plants suffer from lack of water (Larcher, 1995). Frequently, but not invariably, dryness of soil is coupled with strong evaporation caused by dryness of air and high levels of radiation. However, shortages of precipitation alone are not sufficient to cause aridity. Dryness results from combinations of low precipitation and high evaporation. In dry regions, drought is of such regular and prolonged occurrence that annual evaporation may exceed total annual precipitation. Such climates are called arid as opposed to humid climates in regions with surplus precipitation (Thornthwaite, 1948). About one-third of the earth's continental area has rain deficits, and half of this (about 12 percent of land area) is so dry that annual precipitation is less than 250 mm, which is not even a quarter of potential evaporation (Figure 7.1). Reduction in soil water potential decreases water potential of whole plants or parts of plant organs. Drought influences biological and economic yields by inhibiting various physiological functions because of water deficiency. Drought is the most prevalent environmental stress and limits crop production on about 28 percent of the world land area (Dent, 1980). Most higher plants are exposed to varying degrees of water stress at some stage of their ontogeny. The type of water stress may vary from small fluctuations in atmospheric humidity and net radiation in more mesic habitats to extreme soil water deficits and low humidity in arid environments (Morgan, 1984). The impact of water stress is a function of duration, crop growth stage, type of crop species or cultivar, soil type, and management practices. Drought is sporadic in nature, resulting in drastic crop losses even in humid climates (Dunphy, 1985).

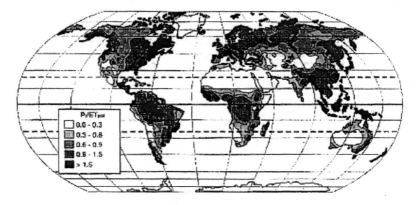

FIGURE 7.1. Dry regions of the earth. Classification is based on ratio of annual precipitation to regional potential evapotranspiration (Pr/ET$_{pot}$). Arid climate below 0.3; semiarid 0.3 - 0.6; subhumid 0.6 - 0.9; humid 0.9 - 1.5; very humid above 1.5. *Source:* Reproduced from UNESCO, 1979.

The term water deficit will be used here to indicate that water pressures in the rhizosphere are sufficiently negative to reduce water availability for transpiration and plant expansion growth to suboptimal levels (Neumann, 1995). Water deficits have major effects on plant photosynthetic capacity. Reductions in photosynthetic activity and leaf senescence are well documented and adversely affect crop yield (Gerik et al., 1996). Other adverse effects of water deficit on plant photosynthetic capacity include reduction in cell growth and enlargement, leaf expansion, assimilate translocation, and transpiration (Hsio, 1973). Protein metabolism and synthesis of amino acids are both soon impaired. One enzyme most strongly inhibited from water deficiencies is nitrate reductase. Even brief periods of negative water balance cause nitrate reductase activities to decrease by 20 percent or even by 50 percent after longer periods of water deficit (Hsio, 1973). This is why drought leads to increases in nitrate contents plants that have been supplied with N fertilizers (Larcher, 1995). Although consequences of water stress are well known, our ability to genetically manipulate these processes and improve drought tolerance is limited.

The importance of water to plant growth has been recognized for over 300 years. Despite this long time, relatively little is known about specific water requirements for most annual field and horticultural crops grown under different agroecological conditions. Another difficulty with plant water requirements is that much of the research deals with short-term responses to changing water status that may be transient and are eventually overridden by other biological changes that develop more slowly but have greater influences on yield. A sudden decrease in plant water status might induce stomatal closure, but adjustment by leaves to stresses may eventually result in stomata opening again (Passioura, 1994).

Water requirement is defined as the minimum amount of water required to provide optimal yield. The amount of water required is determined by yield levels, critical limits of deficiencies relative to yield, limits of tolerable yield reduction, size and permeability of plant evaporative surfaces, plant growth stage, and environmental factors affecting growth and transpiration (Spomer, 1985). Crop water demand and supply are influenced by meteorological, soil, and plant physiological parameters. It is now understood that shortages of water at any stage in the crop life cycle will have consequences on yield, but more serious effects are likely when shortages occur after head/ear initiation (Fischer, 1973; Day et al., 1978).

Plant water status and balances are important parameters for understanding drought physiology of crop plants. Plant water status is the quantification of plant water conditions relative to requirements. One way to quantify plant water status is to measure water potential. Water potential is the physiochemical availability of water to participate in plant functions and determines tendencies for net water movement within plant systems. Plant water balances are the differences between plant or tissue water absorption and loss. When other conditions are favorable and water balances are positive, expansion growth occurs. However, positive water balances can also cause stresses resulting from water excesses. Whenever balances are negative (less water is absorbed than used), water deficits occur and the potential for expansion growth immediately is reduced (Hsio, 1973). The nature and extent of water balance effects depend on plant species or cultivar, plant part, plant growth status, and yield degree, duration, and timing of deficits and interactions with other environmental factors (Spomer, 1985).

Progress in developing quantitative response functions to soil water deficits has been slow. Part of the problem may be that many studies have attempted to characterize water deficits with thermodynamic variables (Sinclair and Ludlow, 1985). Thermodynamic variables have not been related directly to leaf gas exchange or leaf expansion (Bennett et al., 1987; Joly and Hahn, 1989), and Ritchie (1981) proposed that responses of physiological processes to water deficits could be evaluated as functions of available soil water. This concept was refined by Sinclair and Ludlow (1986) so as to express physiological processes as functions of transpirable soil water remaining in soil. These authors defined total transpirable soil water as differences between soil water contents at field capacity and soil water contents when transpiration of drought-stressed plants decreased to 10 percent or less that of well-watered plants.

Since moisture stress is important to crop yield in many environments, this chapter discusses recent advances in drought physiology and suggestions made about how stress physiology can contribute to plant breeding programs for improving crop yields under water-limiting environments.

WATER USE EFFICIENCY

Plant water use efficiency (WUE) is generally defined as the amount of biomass accumulated per unit of water used. This definition of WUE was first applied to irrigation research, where costs of applying water to plants are major economic considerations. In physiological terms, WUE has been defined as the ratio of C assimilated to water transpired. In more recent years, WUE can be estimated by C isotope discrimination techniques (Farquhar, Ehleringer, and Hubic, 1989; Dingkuhn et al., 1991). The role of WUE in determining yield per unit land area is shown by the following equation (Passioura, 1983):

$$\text{Yield (g DM/m}^2) = \text{WUE (g DM/g H}_2\text{O)} \times \text{transpiration} \\ \text{(g H}_2\text{O/m}^2) \times \text{HI} \qquad (7.1)$$

Where DM is dry matter and HI is harvest index (HI = grain yield/ grain plus straw yield).

Water use efficiency is a physiological trait associated with drought tolerance of plants (Mian, Ashley, and Boerma, 1998). Wright, Rao, and Farquhar (1994) stated that WUE can contribute to crop productivity for plants grown in droughts. These authors also reported positive associations between WUE and total biomass yields in drought environments and suggested that WUE improvement of crop plants should result in superior yield performance if high HIs (harvest indexes) can be maintained. Water requirement, which is one index of WUE, varies among plant species and cultivars as well as among sites and years examined, even for the same plant species or cultivar. This means that WUE is determined by interactions between environmental and genetic factors. Hattendorf et al. (1988) reported that mean daily use rates for sunflower were 22 percent greater than mean rates for maize, grain sorghum, pearl millet, pinto bean, and soybean. Water use efficiency values of C_3 crops (pinto bean, soybean, and sunflower) were 90 percent less than those of C_4 crops (maize, sorghum, and pearl millet). Bauder and Ennen (1979) reported that established alfalfa and soybean used significantly greater amounts of water than spring wheat, pinto bean, or sunflower in North Dakota. In addition, Badaruddin and Meyer (1989) reported that legume crops used 10 to 25 percent more seasonal water than wheat across environments, but WUE (kg DM·ha^{-1}·mm^{-1} water) of legumes was 0 to 25 percent greater than that of wheat. Green manure and forage legumes generally had greater water use and WUE values than grain legumes, which was associated with their longer growing season and higher dry matter production.

Total water use (TWU) includes both transpired and soil evaporated water (Ehdaie, 1995). Two models for WUE have been suggested: one with two components and another with three components, so that contributions from each component relative to WUE could be evaluated (Ehdaie, 1995). The two primary components of WUE were defined as evapotranspiration efficiency (ETE: ratio of total DM to TWU) and HI (ratio of grain yield to total DM). Thus, the two-component model for WUE can be expressed as (Ehdaie, 1995)

$$WUE = ETE \times HI. \qquad (7.2)$$

This model can be extended to a three model equation when evapotranspiration can be partitioned into transpired water and soil water evaporated. This extended model contains the components uptake efficiency (UE, ratio of total water transpired to TWU), transpiration efficiency (TE, ratio of TDM to total water transpired), and HI as previously defined (Ehdaie and Waines, 1993). This extended model for WUE can be expressed as (Ehdaie, 1995)

$$WUE = UE \times TE \times HI. \qquad (7.3)$$

Many methods exist for improving efficiency of water use by plants, which have been summarized by Boyer (1996). Water use efficiency can be improved by increasing efficiency of water delivery and timing of water application. This approach is widely adopted in many developed and developing countries and in other countries where water is becoming increasingly scarce. Per capita availability of water resources declined by 40 to 60 percent in many Asian countries between 1955 and 1990 (Gleick, 1993). In 2025, per capita available water resources in these countries are expected to decline by 15 to 54 percent compared to that available in 1990 (Guerra et al., 1998). The share of water to agriculture will decline at an even faster rate because of increasing competition for available water from urban and industrial sectors (Tuong and Bhuiyan, 1994). Under these situations, the approach of adapting plants to less available water may be an important strategy for crop production. Using drought-tolerant crop species or cultivars within species depends on understanding the specific biology of those plants to be used, and whether these plants can be manipulated to achieve similar productivity with less water. This approach is complicated and will command considerably more research data for the different crops and agroclimatic regions than now available if successful crop productivity is to be achieved.

Genetic variations of WUE for many field crop plants such as soybean (Mian et al., 1996; Mian, Ashley, and Boerma, 1998), wheat (Farquhar and Richards, 1984), barley (Hubick and Farquhar, 1989), peanut (Hubick, Farquhar, and Shorter, 1986), and sunflower (Virgona et al., 1990) have been reported in recent years. Nevertheless, improvement of WUE of field crops through conventional breeding methods may not be practical because of the complexity and difficulty of measuring WUE of large numbers of breeding lines grown in field conditions (Ismail and Hall, 1992). Indirect selection for improved WUE through molecular markers linked to quantitative trait loci (QTL) conditioning WUE of crop plants may prove useful (Mian, Ashley, and Boerma, 1998). Differences among crop species for WUE are related to carboxylation pathways (e.g., C_4 species commonly have twice higher values than C_3 species) and to energy requirements for production of biomass containing different proportions of proteins, lipids, and carbohydrates (Ludlow and Muchow, 1990). Similarly, apparent differences in WUE between cultivars of the same species and among several food legumes can be related to differences in soil evaporation and chemical composition of DM.

CROP YIELD RELATIVE TO WATER STRESS

Yield performances of genotypes grown under water stresses are reflective of both plant responses to stress and of potential yield levels. Soil water deficits are common in productivity of most crops and can have substantial negative impacts on growth and development. Four main aspects of plant behavior relative to drought that can readily be linked with yield are modification of leaf area, root growth, efficiency by which leaves exchange water for CO_2, and processes involved in setting and filling of seeds (Passioura, 1994).

Since plant growth is primarily based on cell expansion, steady state equations that empirically model relative contributions of hydraulic and wall mechanical factors to rates of single cell expansion growth have been studied (Lockhart, 1965). These equations provide useful starting points for considering biophysical control of growth. The equation showing effects of hydraulic parameters on cell expansion can be expressed as (Lockhart, 1965)

$$\text{Growth rate} = L_p \, (\Psi_0 - \Psi_i) \tag{7.4}$$

where Lp is hydraulic conductivity of plasmamembranes, Ψ_0 is water pressure of external water sources, and Ψ_i is water pressure of cell solutions. According to this equation, growth rates (i.e., rates of irreversible cell volume increases) are proportional to rates of water uptake by expanding cells. Water uptake occurs when Ψ_i values are more negative than Ψ_0, and rates of water uptake are further affected by L_p.

A second equation indicates that rates of cell expansion are also limited by yielding properties of expanding cell walls (Lockhart, 1965):

$$\text{Growth rate} = m \, (P - Y) \tag{7.5}$$

where m is the extensibility coefficient of cell walls, P is cell turgor pressure, and Y is threshold pressure. In this case, growth rate is proportional to differences between P and Y. The Y values are minimum turgor pressures required to initiate irreversible expansion of cell walls. The term ($P - Y$) represents growth effective turgor pressure acting on expanding cell walls. Growth rate is also modulated by m values. These two mentioned equations can also be combined (Neumann, 1995):

$$\text{Growth rate} = (mL_p/m + L_p) \times (\Psi_0 - \Psi_\pi - Y) \tag{7.6}$$

where Ψ_π is cell osmotic pressure and the other terms as defined above. This latter equation does not directly include turgor pressure. However, cell turgor pressure is related to Ψ_π by the relationship $P = \Psi_i - \Psi_\pi$. Thus, cellular adjustment of Ψ_π will interactively affect Ψ_i and turgor. It is also important to understand that turgor pressure in protoplasts is the Newtonian counterforce of tensile forces exerted on protoplasts by expanding cell walls (wall stress). Thus, changes in cell wall characteristics (e.g., m or Y) affect growth in addition to affecting turgor pressure and Ψ_i (Cosgrove, 1993; Neumann, 1995). More detailed information growth inhibitory effects by water deficits have been discussed elsewhere (Taiz, 1984; Boyer, 1985; Cosgrove, 1993; Pritchard, 1994; Neumann, 1995).

Effects of water deficits on crop growth and yield depend upon degrees of stress and developmental stages at which stress occurs (Hsio and Acevedo, 1974; Sullivan and Eastin, 1974). Low water potentials during anthesis and early grain fill can decrease yields in grain crops. Photosynthesis is inhibited under these conditions and carbohydrate reserves may become limited for grain growth (Westgate and Boyer, 1985). Wien, Littleton, and Ayanaba (1979) reported that two-week drought stresses during vegetative or flowering growth stages had no significant effect on seed yield of cowpeas. Conversely, Shouse et al. (1981) reported that most sensitive growth stages to drought were flowering and podfilling, with yield reduction 35 to 69 percent depending on timing and length of drought treatments in cowpeas. Soil water deficits during vegetative growth had fewer effects on crop and seed yields of cowpeas than during flowering or podfilling. Turk and Hall (1980) studying moisture stress effects on field-grown cowpeas during vegetative, flowering, and pod-filling growth stages relative to plant water potential, dry matter production, seed yield, and WUE, noted that moisture stress during flowering and pod-filling growth stages had the most serious influences on reduced seed production. However, moisture stress during vegetative stages had the least effect on seed yields. Values of WUE were reduced when irrigation was withheld during flowering and pod-filling stages, and no irrigation during vegetative stages increased efficiency of water use by cowpea as seed yields were not reduced. Labanauskas, Shouse, and Stolzy (1981) reported most severe seed yield reductions in cowpeas occurred when treatment plots were subjected to water stress during both flowering and pod-filling growth stages. Seed yield reductions of 67 percent were noted when water stress occurred during flowering and pod-filling. No significant reductions in seed yield were noted when water stress occurred during vegetative stages.

In contrast to many crop plants including cowpea potato tuber quality and yield can be reduced when water stresses occur during any part of the growing season (Adams and Stevenson, 1990). Thus, continuous water supplies are recommended from tuber initiation to maturity (Miller and Martin, 1985; Porter et al., 1999). Potato plants are very sensitive to moisture stress during periods from tuber initiation to shortly before foliage senescence (Singh, 1969). Water stress during tuber initiation has been reported to reduce numbers of tubers produced per plant (Mackerron and Jefferies, 1986; Miller and Martin, 1985). Lynch and Tai (1989) reported that differential tolerances among potato cultivars to moisture stress are likely associated with differences in sensitivity during ontogeny of yield development. These authors also reported that potato cultivars were also very sensitive to moisture stress during both tuber initiation and tuber sizing growth stages, with greatest effects being during tuber sizing.

Water stress during vegetative development of maize decreased dry matter yields at harvest by reducing leaf area development, stalk dry matter accumulation, and potential grain-filling capacity of plants (Wilson and Allison, 1978; Eck, 1984; Lorens, Bennett, and Loggale, 1987). Nesmith and Ritchie (1992) reported that severe water stresses during the R3 and R5 (late grain filling) growth stages decreased total dry matter accumulation and grain yield equally. Wilson and Allison (1978) also noted that mild water stress during grain-filling reduced total dry matter accumulation but not grain yield because of remobilization of assimilates from stover to grain. These authors also reported that maize yield reductions were greatest when water potentials were low at anthesis, since embryo sacs may abort (Moss and Downey, 1971). Westgate (1994) reported that large yield reductions for maize resulted when drought occurred during flowering and early seed development. These yield reductions were primarily due to decreases in seed numbers per plant (Boyle, Boyer, and Morgan, 1991). Inhibition of flower development, failure of embryo fertilization, and abortion of zygotes all contributed to decreases in seed numbers (Westgate and Boyer, 1986). Water deficits at flowering may also decrease maize grain yields even if pollination occurs (Schussler and Westgate, 1994). Wesgate and Boyer (1985) concluded that early kernel development was highly dependent upon current supplies of assimilate in maize plants because reserves were not sufficient to maintain kernel growth during low water potentials.

Singh (1995) reported that water stress during flowering and grain-filling stages of growth reduced seed yields and weights, and accelerated maturity of dry bean. Miller and Burke (1983) noted linear relationships between water applied during flowering and grain-filling and yield of dry bean grown on sandy soil.

It is well understood that various processes of photosynthesis are affected by internal water deficits in rice. This was observed specifically by inhibition of leaf blade elongation (Turner et al., 1986), reduction of photosynthetic rates per unit leaf area (Ishihara and Saito, 1983), and acceleration of wilting (O'Toole and Moya, 1976). Relative to morphogenesis of rice panicles, which strongly affects dry matter distribution into grain, reductions in water potentials of leaf blades beyond certain limits led to delays in young panicle development (Tsuda, 1986), pollen development was inhibited by reductions of water potentials in flag leaves (Namuco and O'Toole, 1986), exertion rate of panicles during heading decreased, and numbers of sterile grains increased (Cruz and O'Toole, 1984). In plants other than rice, inhibition of various physiological processes by drought conditions was affected indirectly by dehydration of organs (Morgan, 1980; Schulze, 1986). Sensitivity of rice to water deficits was greatest during reproductive phases of growth (O'Toole, 1982). In developing practical field screening systems for reproductive phase drought resistance in rice, Garrity and O'Toole (1995) assessed canopy temperature responses among wide ranges of rice germplasm, and related their results to other plant characters related to drought resistance. Negative relationships were observed between grain yield and midday canopy temperatures on the date at which each entry flowered (Figure 7.2). Cultivars that exhibited midday temperature values >34°C at anthesis had essentially no yield. Fertility of spikelets was similarly related to midday canopy temperatures (Figure 7.3), and decreased below 30 percent as anthesis midday canopy temperatures exceeded 34°C.

Growth stages from panicle development to anthesis are the most critical growth stages for water stress effects on rice. Drought stresses during flowering caused larger decreases in yield than similar stresses during vegetative or grain ripening stages (Figure 7.4). Fukai and Cooper (1996) reported that rice grain yields decreased at rates of 2 percent per day when 15-day stress periods (when early morning leaf water potentials were less than −1.0 MPa) occurred during panicle development. Unfilled grain numbers increased sharply with stress during late panicle development. Water stresses reduced assimilate production per plant, spikelets per panicle, filled grain percentages,

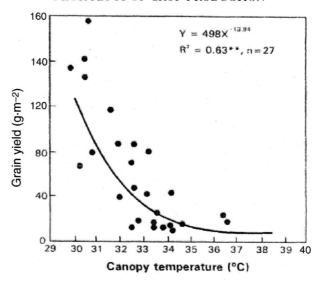

FIGURE 7.2. Relationship between rice grain yield and canopy temperature on date of 50 percent flowering. *Source:* Reproduced from Garrity and O'Toole, 1995. Used with permission from American Society of Agronomy, Madison, WI.

and individual grain weights. However, unfilled grain numbers appear most susceptible to reduced assimilate availability. Assuming reductions of 2 percent grain yield per day with delays in termination of 15-day stress periods, 20-day differences of about 40 percent were noted if plants had the same yield potential as plants grown under nonlimiting conditions.

It is likely that rice cultivars with differences in phenology will react differently to drought stresses depending on timing of stress development (Maurya and O'Toole, 1986). Phenology is important in determining grain yield responses because early-maturing cultivars often escape severe drought stresses, while late-maturing cultivars may not escape drought stresses. These results indicate that genotypes should be compared for drought resistance/susceptibility within similar phenology groups, or that genotypic variation in phenology should be corrected statistically before genotypic differences in drought tolerance. Alternatively, it may be possible to implement strategies for staggered plantings of different genotypes so that plants would flower about the same time (Fukai and Cooper, 1996).

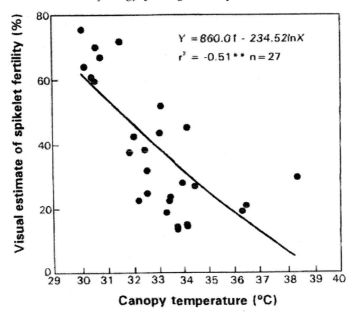

FIGURE 7.3. Relationship between spikelet fertility and canopy temperature for rice on date of 50 percent flowering. *Source:* Reproduced from Garrity and O'Toole, 1995. Used with permission from American Society of Agronomy, Madison, WI.

Yields of modern crop cultivars have generally increased with little change in above ground biomass production, and these increases are attributed to increases in HI (Gifford, 1986). Yield increases have occurred without much change in amount of water used, and this has resulted in natural improvement of WUE (Richards et al., 1993). Among factors, HI depends on relative proportions of pre- and post-anthesis biomass and mobilization of preanthesis assimilate to grain. Severe water deficits at critical growth stages at flowering can greatly decrease seed numbers and HI.

Water stress during any stage of soybean development can reduce yield, but extents of yield reductions depend on stage of development. Negative effects of stress are particularly important during flowering, seed set, and seed filling (Souza, Egli, and Bruening, 1997). Water stress during seed filling reduces yield by reducing seed size (Vieira, Tekrony, and Egli, 1992). These reductions in seed size

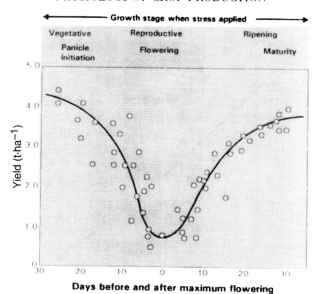

FIGURE 7.4. Effect of water stress on yield of rice. *Source:* Reproduced from O'Toole, 1982. Used with permission from International Rice Research Institute, Los Baños, Philippines.

have been associated with early maturity (Ashley and Ethridge, 1978) and shorter seed filling periods (Meckel et al., 1984; Smiciklas et al., 1989). It is generally assumed that shorter seed filling periods result from accelerated leaf senescence (Souza, Egli, and Bruening, 1997). Leaf senescence in soybean, characterized by declines in leaf N, chlorophyll, and photosynthesis, begins early during seed filling and is usually complete by physiological maturity (Egli and Crafts-Brandner, 1996). During leaf senescence, C and N are redistributed from leaves and other vegetative plant parts to developing seeds. This redistributed N can be a significant sources of N to seeds by contributing from 20 to 100 percent of mature seed N in cultivars with different maturities (Zeiher et al., 1982). Increasing source-sink ratios may reduce effects of moisture stress on seed size.

Establishment and activity of legume rhizobium symbiosis are known to be sensitive to drought stresses, and this can decrease yield (Zablotowicz, Focht, and Cannell, 1981; Kirda, Danso, and Zapata, 1989). Most research on mechanisms of drought effects on N_2 fixation has focused on nitrogenase activity rather than on nodule formation and growth. Several studies have reported that N_2 fixation by soybean is more sensitive to soil dehydration than leaf gas exchange for plants grown under controlled growth conditions (Durand, Sheehy, and Minchin, 1987; Purcell et al., 1997; Serraj and Sinclair, 1996) and field conditions (Sinclair et al., 1987; Serraj et al., 1997). Sall and Sinclair (1991) and Serraj and Sinclair (1997, 1998) reported genetic variability in N_2 fixation sensitivity to drought stress among soybean cultivars, and that plant responses of N_2 fixation rates to drought stress were related to nodule formation and growth.

Effects of drought stress on grain yield may be assessed in cereal crops in terms of yield components, some of which can be more important than others depending upon intensity of stress and growth stage at which stress develops (Johnson and Kanemasu, 1982; Giunta, Motzo, and Deidda, 1993). While intense drought mainly affects numbers of kernels per unit area through decreases in fertility, mild droughts may cause only decreases in grain weight (Giunta, Motzo, and Deidda, 1993). Reduction in numbers of spikelets per unit area and numbers of kernels per spikelet have been associated with preanthesis water stresses (Day and Intalap, 1970). Water deficits at anthesis may lead to reduced yields by decreasing spikelet numbers and fertility of surviving spikelets (Aspinall, 1984). Water stress during grain filling, especially if accompanied by high temperature, is common for plants grown in Mediterranean environments (Royo and Blanco, 1998). These stresses hasten leaf senescence, reduce duration of grain filling, and reduce grain weights (Davidson and Birch, 1978; Austin 1989), presumably by reducing assimilate supply to developing kernels. Reduced assimilate supply has been related to reduced photosynthesis (Denmead and Miller, 1976), and early senescence. When cereal crops are subjected to severe post-anthesis drought stress, kernel growth becomes increasingly dependent on contributions from vegetative reserves (Austin et al., 1980; Blum et al., 1983; Schnyder, 1993). Reductions in grain yield and yield components due to droughts can be calculated with the formula (Blum, Mayer, and Golan, 1983)

$$\text{Reduction in grain yield (percent)} = [(\ C - S\ )/C] \times 100 \quad (7.7)$$

where *C* and *S* are values of variables in control and stress treatments, respectively. Examples of variables are grain yield and yield components (such as panicles per unit area, 1000 grain weight, or panicle length in case of rice).

DROUGHT AND NUTRIENT ACQUISITION

Although nutrient and water absorption are independent processes in roots, the necessity for available water in both plant and soil for growth and nutrient transport makes them intimately related (Viets, 1972). Nutrients are absorbed by plant roots as ions and water is the transporting medium. Soil water availability in the potential tension ranges of 0.01 to 1.0 MPa is essential for nearly all processes promoting nutrient acquisition. For example, reactions affecting nutrient concentrations in soil solutions, nutrient transport by diffusion and mass flow to root surfaces, and nutrient absorption by roots are affected by water potential. Nutrient absorption by most crop plants is highest when soil water potential is near field capacity. Field capacity is defined as water that soil will hold against percolation after several days of drainage, and usually ranges from 0.01 to 0.03 MPa water potential. Soil water contents near field capacity allow for best combinations of sufficient air space for oxygen diffusion, greatest amounts of nutrients in soluble forms, greatest cross-sectional areas for diffusion of ions and mass flow of water, and most favorable conditions for root extension.

Availability of soil water affects both rates of root elongation and absorption of nutrients by roots (Baver, Gardner, and Gardner, 1972). Phosphorus compounds are commonly low in solubility and soil plow layers usually contain only small amounts of available P (e.g., few hundred g/ha of dissolved P) (Troeh and Thompson, 1993). Amounts of dissolved P and rates of dissolution are roughly proportional to the amount of water present. Potassium absorption is most rapid near field capacity (Singh, 1998). Rates of microbial activity, which are responsible for releasing soil N, S, and micronutrients for root absorption, are also dependent on availability of water (Singh, 1998).

DROUGHT RESISTANCE MECHANISMS

Ability of crop species or cultivars to remain alive, grow, and ultimately produce seed when part of their life cycle has been under water stress is known as drought resistance. Drought resistance can also be defined as relative yield of specific cultivars compared to other cultivars subjected to similar drought conditions. Drought susceptibility of cultivars is often measured as reductions in yield under drought stress (Blum, 1988), but values are confounded with differential yield potentials of cultivars. Other yield-based estimates of drought resistance are geometric means (Fernandez, 1993) or drought susceptibility indexes (Fischer and Maurer, 1978). Geometric means are often used by plant breeders interested in relative genotypic performance, since drought stress can vary in severity in field environments over years. For yield components of common bean, which exhibited the largest differential genotypic responses to water stress, were pod and seed numbers and seed size was fairly stable (Ramirez-Vallejo and Kelly, 1998). Genotypic variation was detected with partitioning indexes, chiefly HI and relative sink strength, and heritability estimates for these traits were high. Differential correlations between phenological biomass and partitioning traits and indexes for yield and drought susceptibility were effective approaches in breeding for drought resistance of common bean. Selection for high geometric yields was considered more important than selection among high-yielding individuals grown with low to moderate levels of drought susceptibility indexes (Ramirez-Vallejo and Kelly, 1998).

Resistance to drought includes both avoidance and tolerance mechanisms (Fageria, 1992). Boyer (1996) defined drought resistance as plants having improved growth when grown with limited water. Plants grown under these conditions were considered to have drought tolerance regardless of how the improvement occurred or whether WUE was affected. Drought resistance mechanisms have been divided into two groups, namely drought avoidance and resistance.

If plants have functions or adaptation abilities for reducing effects of soil and/or air dryness, plant water deficits can be alleviated. This has been called drought avoidance, dehydration postponement, or desiccation avoidance. This may be achieved by improved plant uptake of water from soil, reduced loss of water by plants, high plant water-conducting capacity, or high plant storage of water. These functional measures for avoiding desiccation are also reflected in morphology of plants (Figure 7.5). In addition, reduction in water transpiration could help conserve available water inside plants. Lower transpiration of plant water would occur with timely closure of leaf stomata. A modification in morphology may occur when leaves of plants grown under water deficiencies develop smaller and more densely distributed stomata (Tumanov, 1927). This modification makes leaves better able to reduce water transpiration from faster onset of stomatal regulation.

Ability of physiological functions inside plants to withstand internal water deficits has been called drought tolerance or dehydration tolerance (Kobata, 1995). Ability of plants to avoid drought encounters has been called drought escape. Drought escape is the ability of plants to complete their life cycle before serious water stresses

FIGURE 7.5. Modifications in plant shoot and root morphology to survive drought. a = deciduous bottle trees with water storing trunks (*Adansonia/ Chorisia* type); b = succulents storing water in the stem (*Cacti/Euphorbia* type); c = succulents with water storing leaves (*Agave/Crassulaceae* type); d = evergreen trees and shrubs with deep root systems (*Sclerophyll* type); e = deciduous often thorny shrubs (*Capparis* type); f = chlorophyllous-stemmed shrub (*Retama* type); g = tussock grasses with renewal buds protected by leaf sheaths, and with wide-ranging root systems (*Aristida* type); h = cushion plants (*Anabasis* type); i = geophytes with storage roots (*Citrullus* type); j = bulb and tuber geophytes; k = pluviotherophytes (annual plants); and l = desiccation-tolerant plants (poikilohydric type). Compiled from Larcher, 1995. Reproduced with permission from Verlag Eugen Ulmer, Stuttgart, Germany.

TABLE 7.1. Grain yield of upland rice cultivars having different growth cycles.

Growth cycle days	Flowering days	Grain yield (kg·ha⁻¹)	
		Low P (2.25 mg·kg⁻¹)	High P (4.85 mg·kg⁻¹)
110	85 (25)[a]	15910	2090
120	95 (15)	1220	1320
130	106 (30)	1070	1160
135	111 (2)	679	757

Source: Adapted from Fageria et al., 1988; Fageria, 1992.

[a]Values in parentheses represent number of cultivars tested.

damage plant tissues (Levitt, 1980). Some plant species are known to escape drought by making sure their growth and reproductive cycles occur during times when sufficient water is available. Early maturity may also provide drought escape for plants. Table 7.1 shows yield differences of upland rice cultivars planted on an Oxisol with two P levels in central Brazil. Early maturing cultivars often have higher grain yields than late maturing cultivars when droughts occur. That is usually because late maturing cultivars suffer drought stress during productive or grain filling growth periods, while early maturing cultivars are able to escape drought during these critical growth periods.

Drought resistance mechanisms in crop plants are often related to physiological traits or processes. Some of these are

1. moderated water use through reduced leaf area and shorter growth duration,
2. ability of roots to exploit deep soil moisture to provide for evapotranspirational demands,
3. capacity for somatic adjustment in leaf cells so they can retain turgor and protect meristems from extreme desiccation, and
4. control of nonstomatal water losses from leaves (Nguyen, Babu, and Blum, 1997).

In addition, genetic variation exists in and among plant species for dehydration avoidance (maintenance of relatively high leaf water potentials under conditions of soil moisture stress), osmotic adjustment, tolerance in plant or organ growth rates, plant recovery upon rehydration, tolerance of photosynthetic systems or their components, tolerance of enzyme activities, tolerance of translocation systems, stability of cellular membranes, proline accumulation, root growth attributes, plant development or morphological attributes (e.g., leaf size), leaf area per plant, leaf orientation, tiller survival, epicuticular wax content, and organ pubescence (Blum, 1983).

Roots are usually the site of highest resistance in pathways for liquid-phase movement of water through the soil-plant-atmosphere continuum (Kramer and Boyer, 1995). Efficiency of soil water uptake by roots is key for determining rates of water transpiration and plant tolerance to drought. Water uptake by roots is complex and depends on structure, anatomy, and pattern by which different root parts contribute to overall water transport (Cruz, Jordan, and Drew, 1992). Salih et al. (1999) investigated effects of soil moisture on rooting habits, leaf transpiration rates, and root and stem xylem anatomy of sorghum cultivars exhibiting different drought tolerances. Drought tolerance was associated with higher water extraction efficiencies, fewer nodal roots per plant, fewer late metaxylem vessels per nodal root, smaller leaf areas, and more well-developed sclerenchyma cells.

Discussion of some of these physiological components of drought resistance in crop plants follows.

Shortened Growth Duration and Reduced Leaf Area

Smaller leaves have lower leaf area indexes for transpiration and consequently lower water requirements compared to larger leaves. Blum, Mayer, and Golan (1989) studied growth duration effects on water use and yield of sorghum genotypes. These authors grew 26 genotypes with little or no rain during the growing season so plants were forced to use stored soil water. One physiological attribute closely associated with higher grain yield was early heading. Water was being steadily depleted from soil, and early heading favored grain growth while water was still available. The higher yielding genotypes also displayed higher HI values, higher leaf water potentials, and cooler leaves, presumably reflecting superior water acquisition compared to poorer performing genotypes (Boyer, 1996). Similar results were noted in maize and wheat studies where several genotypes of different growth durations were utilized (Jensen and Cavalieri,

1983; Sojka, Stolzy, and Fisher, 1981). It has been easy to manipulate cultivar growth duration because genetic variability among plants for phenology exists, and inheritance is known in some cases (Ludlow and Muchow, 1990).

Root Growth

Healthy and functional roots are critical to survival and productivity of crop plants, as they serve combined functions of anchorage, support, and water and mineral nutrient uptake. Root growth, development, and function are largely determined by genetic information, but respond substantially to available carbohydrate supply and local environmental conditions encountered during growth. Soil attributes that affect extent and morphology of growing roots include water content, strength or mechanical impedance, thermal regime, and structural development (Baligar, Fageria, and Elrashidi, 1998; Wraith and Wright, 1998).

Roots that extend beyond normal rooting zones for more extensive extraction of available soil water provide plants greater potential to increase yield under droughty conditions (Mambani and Lal, 1983). According to Boyer (1996), deep rooting probably accounts for major differences in drought tolerances between plant species. Maize and sorghum roots grow to soil depths of 2 to 3 m, and these plants grow and remain green when surrounding short grasses with shallow roots often become brown because of soil dehydration. Depths and extent of rooting of various plant species have been listed by Taylor and Terrell (1982). Jordan and Miller (1980) claimed that sorghum root length densities in the lower rooting zones was insufficient to meet evaporative demands for plants grown in the Great Plains region of Texas. These authors predicted that increasing root length densities to 2 cm·cm^{-3} below 50 cm soil depths would increase annual water uptake by 6 cm and reduce water stress during moisture-sensitive grain-filling stage of growth. Root length densities of 0.5 to 1.2 m were noted for drought-tolerant soybean cultivars grown under water-limiting conditions, and no root proliferation occurred below 0.8 m in drought-sensitive lines (Sponchiado et al., 1989). Greater total allocation to root biomass and greater allocation over the 60 to 120 cm depth were associated with drought tolerance among contrasting wheat cultivars (Hurd, 1974).

Genotypic differences have been reported among soybeans with more root efficiency for extracting available soil water reserves (Hudak and Patterson, 1996). It was noted that in field screening of several exotic soybean genotypes, one genotype (PI 416937) maintained leaf turgidity during drought imposed at pod-fill and yield was depressed less than that of other genotypes (Carter and Rufty, 1993). Subsequent investigations confirmed that PI 416937 root growth was tolerant to high concentrations of Al and that midday transpiration rates were higher than that of another contrasting genotype (Forrest) grown under soil water deficits (Sloane, Patterson, and Carter, 1990). Hudak and Patterson (1996) also reported that PI 416937 advantage for drought tolerance may have resided in its ability to exploit upper soil horizons (above 68 cm) for water from networks of fibrous roots. Localized measurements of soil moisture tension revealed that PI 416937 rate of soil desiccation was often slower than that of Forrest, indicating that larger total soil volumes were being explored by PI 416937 compared to Forrest. In addition, PI 416937 appeared to have greater lateral spread of roots. PI 416937 may provide soybean breeders ability to add diversity to root morphology of existing soybean cultivars. Incorporation of highly branched, fibrous roots like those of PI 416937 into other soybean cultivars might enhance ability to utilize water resources where normal root accessibility is limited. Limited water accessibility often occurs within interrow areas and/or below chemical or physical soil barriers (Hudak and Patterson, 1996).

Comparisons between monocotyledenous and dicotyledenous plant species have also pointed out discrepancies between root density and water uptake. Mason and colleagues (1983) reported that sunflower extracted as much water from irrigated clay loams as maize and sorghum plants despite having only half the total root density. Hodgson and Chan (1984) reported that safflower extracted water at greater depths than wheat when both species were grown on poorly structured vertisols. These authors inferred that safflower roots had greater capacity to penetrate and extract water from subsoils, again at lower densities. Bremmer, Preston, and Fazekas (1986) noted that sunflower extracted substantially more water from subsoils of spodosols than sorghum, although maximum depths of water extraction were similar for both species. Hamblin and Tennant (1987) compared water use of spring wheat, barley, lupin, and field pea grown on four different soil types under drought-stressed conditions, and concluded that total root lengths per unit ground area for cereals were consistently five to ten times larger than those for grain legumes.

Greater concentrations of starch in roots have been reported to be associated with drought tolerance in cotton (De Souza and Da Silva, 1987). Greater starch contents of perennial cotton appeared to be associated with greater root/shoot dry weight ratios and drought tolerance, which imparted survival mechanisms well suited to the cyclic availabilities of water in northeastern Brazil (Wells, 2002).

Osmotic Adjustment

Improving drought tolerance of crop species has long been a major objective of most breeding programs as water deficits during some part of the growing period are common to most crops (Sanchez et al., 1998). Thus, intensive studies have been conducted to identify factors involved in drought tolerance (Morgan, Hare, and Fletcher, 1986; Blum, 1989; Matin, Brown, and Ferguson, 1989). Turgor maintenance by osmotic adjustment is an important physiological adaptation to minimize detrimental effects of water deficits (Morgan, 1984). Good evidence exists that osmotic adjustment can contribute to reducing effects of drought on grain yield of wheat, sorghum, maize, chickpea, and pea (Morgan, Hare, and Fletcher, 1986; Sobrado, 1986; Santamaria, Ludlow, and Fukai, 1990; Morgan, Rodriguez Maribona, and Knights, 1991; Rodriguez, Maribona et al., 1992; Moustafa, Boersma, and Kronstad, 1996). Osmotic adjustment results from accumulation of solutes within cells, which lowers osmotic potential and helps maintain turgor of both shoots and roots as plants experience water stresses. This allows turgor-driven processes, such as stomatal openings and expansion growth, to continue, even though at reduced rates, to progressively lower water potentials (Ludlow and Muchow, 1990). In growth cells exposed to water deficits, osmotic adjustment should help restore water uptake, turgor pressure, and growth rates to pre-stress levels according to the Lockhart model (Morgan, 1991). Such predictions are consistent with reports that higher levels of osmotic adjustment are associated with increased dry matter and grain yield for crops grown in stressful environments (Boyer, 1982; Morgan, 1991). Further evidence against the concept that osmotic adjustment and of turgor pressure reduces inhibition of growth under moderate water deficits is provided by direct investigation of turgor growth relationships in developing roots, stems, and leaf tissues of plants subjected to water deficits (Frensch and Hsio, 1994; Neumann, Azaizeh, and Leon, 1994).

High concentrations of inorganic ions can be detrimental to cellular metabolism and need to be sequestered in vacuoles. To keep osmotic ions balanced, specific types of organic molecules (e.g., soluble sugars, betaines, polyols, proline) accumulate in the cytoplasm. These compounds are termed compatible solutes because they can accumulate in high concentrations without impairing normal physiological function. Typically, compatible solutes are low molecular weight compounds with highly charged polarity, highly soluble, and have larger hydration shells than denatured molecules (Bartels and Nelson, 1994). Plant species and genotypes differ in quantity and quality of compatible solutes, sugars, organic acids, amino acids, sugar alcohols, and inorganic ions like K^+, NO_3^-, and Cl^- (Morgan, 1984). Most organic solutes used for osmotic adjustment originate in current assimilation and metabolic processes. Hydrolysis of storage carbohydrates such as fructans in temperate cereals also serve as immediate sources of solutes for osmotic adjustment (Spollen and Nelson, 1994). Solutes used for cellular osmotic adjustment during plant stress conditions may be partially used for regrowth once stress has been removed (McCree, Kallsen, and Richardson, 1984). Solutes accumulated in the cytosol may also be used to maintain symplast and chloroplast volumes for plants grown with water deficits (Robinson, 1987). These specific effects together with cellular turgor maintenance are generally considered to support photosynthesis and growth processes when plants are grown with stresses (Downton, 1983; Seemann, Downton, and Berry, 1986).

Osmotic adjustment also occurs in roots and is important for root elongation and growth in dry soils (Ober and Sharp, 1994; Matyssek, Tang, and Boyer, 1991). Osmotic adjustment is also time dependent. For example, progression of water stress has to be sufficiently slow to allow solutes to accumulate and provide protection to cells (Jones and Rawson, 1979). Osmotic adjustment is widely used to describe osmoregulation relative to water stress in higher plants and is also used to describe changes in solute contents after recovery from water stresses (Morgan, 1984).

Recent osmoregulation advances relative to biochemical, physical, metabolic, and genetic aspects have been discussed in several reviews

(Kauss, 1977; Zimmermann, 1978; Hansen and Hitz, 1982; Morgan, 1984).

Increased solutes in plant parts occur with reductions in water potentials and eventually reach limits (Morgan, 1980). Typical diurnal changes in solute accumulation for plants like soybean, sunflower, and sorghum are 0.4, 0.6, and 0.7 MPa, respectively, (Takami, Rawson, and Turner, 1982; Turner and Jones, 1980; Wenkert, Lemon, and Sinclair, 1978). Solute accumulation values of 0.5 to 0.6 MPa have been measured during soil stress cycles for rice (Steponkus, Shahan, and Cutler, 1982) and from near 0.0 to approximately 2.0 MPa for wheat depending upon genotype (Morgan, 1977).

Osmotic adjustment has been recognized in plant adaptation to salinity and tolerance to low soil water potentials common to saline soils (Jones, 1985). These osmotic adjustments have been commonly recognized as plant tolerances to drought stress. Plants growing in saline soils have ready supplies of inorganic ions that can be absorbed to counteract those in soil solutions. These ions have not usually been noted for plants grown on nonsaline soils. When osmotic adjustments occur in response to salinity, halophytes (plants tolerant of saline soils) accumulate large amounts of Na^+ and Cl^-. Glycophytes (plants intolerant of salinity) often accumulate K^+ when grown on saline soils, and malate is often synthesized as a counter ion to balance changes in K^+ (Jones, 1985).

It has been reported that relationships exist for osmotic potential of plants and major ecological plant groupings when plants are studied in their natural habitats (Morgan, 1984; Ludlow and Muchow, 1990). Morgan (1984) reported levels of osmotic potential of approximately −0.5 MPa for mesophytic shade plants, −1.0 to −2.0 MPa for most crop plants, −3.0 to −4.0 MPa for xerophytes, and less than −10 MPa for some halophytes.

Osmotic adjustment is not inherited, but plant capacities to adjust to water stress is a heritable trait. Osmotic adjustment is an inducible or facultative rather than a constitutive trait. Osmotic adjustment has no effect on WUE, but contributes to grain yield for plants grown under water-limited conditions by increasing amounts of water transpired and by minimizing reductions in HI. Increases in water transpired result from stomatal adjustment, maintenance of leaf areas, and increased soil water uptake. Osmotic adjustment often reduces rates of leaf senescence because this process increases both avoidance and tolerance of dehydration. Furthermore, osmotic adjustment appears to be the major mechanism for stomatal adjustment, a process that allows stomata to remain partially open at progressively lower leaf water potentials as water stresses increase (Ludlow and Muchow, 1990).

Recent osmoregulation advances relative to biochemical, physical, metabolic, and genetic aspects of plants have been discussed in several reviews (Kauss, 1977; Zimmermann, 1978; Hansen and Hitz, 1982; Morgan, 1984).

Control of Nonstomatal Water Loss

Water vapor is lost from leaves through stomata and leaf cuticles as transpiration. When stomata are open, most water is lost through this pathway. On the other hand, the main pathway for water loss when stomata are closed is through cuticles. Some water loss may still occur through incompletely closed stomata either over whole leaf surfaces or in patches. Ludlow and Muchow (1990) called this water loss epidermal conductance rather than cuticular conductance. Cuticles are waxy layers covering outer surfaces of entire shoots. Waxes on leaves cover the epidermis and guard cells of stomata and extend into interior leaf surfaces where they become thin (Boyer, 1985). Fruits and stems often do not contain stomata, and water loss is controlled entirely by cuticles. The role of epicuticular wax load for reducing cuticular transpiration and epicuticular wax implication in improving crop drought resistance has been discussed (Blum, 1988). It was concluded that epicuticular wax varies with sorghum and wheat genotype as well as environment, and maximum values of epicuticular wax beyond which no further reduction in cuticular transpiration is attained were noted. Most cultivated dryland sorghum genotypes appear to have epicuticular values close to maximum and possess relatively high cuticular resistance (Jordan et al., 1983). Sinclair and Ludlow (1986) attributed differences in survival of soybean, cowpea, pigeonpea, and black gram cultivars during severe dehydration to differences in epidermal conductance when stomata were closed. It has been proposed that epidermal conductance variations could be used to select genotypes with decreased conductance when stomata are closed, and that water savings could be substantial (Rawson and Clarke, 1988; Sinclair and Ludlow, 1986).

An example of genetic improvement for drought tolerance based on epicuticular wax deposition has been the selection of improved seedling establishment in native range grasses grown in the western United States (Boyer, 1996). Wright and Jordan (1970) reported rapid improvement in establishment of boer lovegrass selected for seedling growth in dehydrated soil. In these experiments, stomatal conductance was relatively unimportant, while the characteristic that was most improved was thickness of cuticle coverings of shoot tissues of young seedlings (Hull, Wright, and Bleckmann, 1978). Improved cuticle coverings allowed establishment of grasses to be more reliable when rooting was shallow, rainfall was sporadic, and germination occurred with limited water (Boyer, 1996).

MANAGEMENT STRATEGIES
FOR REDUCING DROUGHT

Crop management strategies for reducing water deficits can be grouped into two major categories: (1) relating cultural practices to improved WUE of plants to reduce impacts of drought on crop production and (2) planting drought-tolerant crop species or cultivars to produce satisfactorily under drought-prone regions and soils. Under strategy number one, developing and adopting practices that will use water efficiently by plants grown with irrigation would be of concern, as the traditional solution to agricultural water shortages is irrigation. Wherever irrigation water is available, crop yields can be increased significantly by using higher technologies in crop production with less risk to farmers. Agriculture is the largest consumer of water and accounts for 72 percent of total water use in the world and 87 percent in developing countries (Barker et al., 1999). In recent years, the growing scarcity and competition for water has been witnessed worldwide, and opportunities for developing new water resources for irrigation are limited. Irrigated agriculture has traditionally consumed more than two-thirds of available water supply. As demand for industrial, municipal, and other uses rise, less water will be available for irrigation. Thus, ways must be found to increase productivity with limited amounts of water if food security is to be maintained (Barker et al., 1999). With increasing demands on water resources, greater efficiency is needed in irrigated agriculture. Overall efficiency (E_p) of irrigation systems can be defined as the ratio of water used by crops to water released at headworks. As such, E_p can be subdivided into three components: Conveyance efficiency (E_c), field channel efficiency (E_b), and field application efficiency (E_a). Conveyance efficiency (E_c) is the ratio of water received at the inlet to a block of fields to water released at the headworks; E_b is the ratio of water received at the field inlet to water received at the inlet of a block of fields; and E_a is the ratio of water used by the crop to water received at a field inlet (Doorenbos and Pruitt, 1992). The E_c and E_b values are sometimes combined and defined distribution efficiency (E_d), where $E_d = E_c \times E_b$. Guerra et al. (1998) reported that only 30 to 65 percent of water released at the headworks reaches intended field inlets in various countries of Asia. This means much improvement in irrigation efficiency on this continent could be made. Asia is where the largest irrigated land in the world is located and considerably more food could be produced with less water.

Use of sprinkler irrigation systems can reduce water loss substantially at the point of delivery to field plots and consequently improve WUE. For example, the shift from graded furrow to sprinkler irrigation (an important regional transition) of predominantly center-pivot sprinkler (Musick, Pringle, and Walker, 1988) has reduced water applications and has contributed to sustained irrigated crop productivity on the Texas High Plains in the United States (Musick et al., 1990; Howell et al., 1998). Center-pivot sprinkler irrigation is well suited to this region, where water is a far more restricted resource for irrigated agriculture than land. Widespread growth in use of center-pivot sprinkler systems in this area has made knowledge about crop-water use for management and system design even more critical, since the area is dependent on declining groundwater resources and on low, highly variable precipitation (Howell et al., 1998).

Internal drainage from root zones losses that can be reduced or managed to improve irrigation efficiency. Keys to reducing internal drainage through irrigation water management are improved uniformity of applied water and decreased average depths of water applied (Hanson, 1987). Tanji and Hanson (1990) reported that proper irrigation management can reduce overirrigation, and proper maintenance can control uniformity of application. Subsurface drip irrigation systems can be adopted to reduce root zone drainage and improved WUE. Darusman and colleagues (1997) reported that adopting subsurface drip irrigation can attain nearly maximum grain yields of maize in western Kansas with significant decreases in amount of irrigation water used (75 percent evapotranspiration) compared to full irrigation (100 percent evapotranspiration). Decreases in seasonal irrigations are accompanied by significant reductions in internal drainage losses from root zones.

Furrow irrigation is commonly used in arid, semiarid, and subhumid regions to apply supplemental water to row crops (Benjamin et al., 1997). Water is usually applied to each furrow in the field, but some researchers have proposed irrigating alternate furrows instead of every furrow in a field to increase WUE (Musick and Dusek, 1974; Crabtree et al., 1985). Small yield losses were recorded for sugarbeet, sorghum, potato, and soybean (Musick and Dusek, 1974; Crabtree et al., 1985) for alternate-furrow irrigation systems compared with every-furrow irrigation, and irrigation water use decreased by 30 to 50 percent.

Changes in water management practices in Japan from traditional continuous flooding to intermittent irrigation in flooded rice have been an important strategy for improving WUE. Some forms of intermittent irrigation similar to the Japanese systems are now practiced in many Asian rice production schemes where water supply is limited. These systems maintain yields close to maximum if weeds are satisfactorily controlled. As much as 30 to 50 percent of the water used when standing water is maintained can be saved (Greenland, 1997).

The foremost concern in arid and semiarid areas is water availability and its efficient use. In dryland areas of North America, a dominant constraint to wheat production is limited water, especially where high evaporative demand coincides with low rainfall (Campbell et al., 1993; Musick et al., 1994; Weinhold, Trooien, and Reichman, 1995). Though crop yields under dryland conditions are related to seasonal rainfall, WUE can be substantially improved by crop management practices (Cooper and Gregory, 1987; Keatinge, Dennett, and Rodgers, 1986; Harris, Cooper, and Pala, 1991). The introduction of supplemental irrigation to winter-grown cereals can potentially stabilize and increase yields, as well as increase WUE received from both rainfall and irrigation (Oweis, Zeidan, and Taimeh, 1992). Supplemental irrigation is practiced in 40 percent of rainfed wheat areas in Syria, and has contributed substantially to cereal productivity (Oweis, Pala, and Ryan, 1998).

Another significant factor to avoid water stress is planting date. Research in the Mediterranean climate of Australia indicated that delaying sowing after the optimum time (coinciding with onset of seasonal rains) consistently reduced crop yields (French and Schultz, 1984). Using a simulation model, estimated potential wheat yields would decline by 4.2 percent per week in Syria with sowing delayed after November in normal years (Stapper and Harris, 1989). Early maturing soybean cropping systems in the southeastern United States have generally called for early planting as a management practice for successful productivity. The primary rationale for producer adoption of early planting systems in this area has been to avoid late-summer droughts (Kane and Grabau, 1992; Savoy, Cothren, and Shumway, 1992; Mayhew and Caviness, 1994). In water balance studies, pearl millet WUE (defined as yield/ET) was noted to be very low compared with WUE of other cereals (Klaij and Vachaud, 1992). Low WUE values can be addressed by at least three management options: (1) moderate fertilizer addition (20 kg N/ha and 9 kg P/ha); (2) moderate increases in plant population (10,000 hills/ha); and (3) use of high water efficient genotypes (Payne, 1997).

Adopting appropriate crop rotations are other water-efficient crop management practices. Crops should be managed in rotation sequence so that root systems fully exploit available water and nutrients (Karlen et al., 1994). Sadler and Turner (1994) suggested opportunistic cropping as a means for increasing agricultural sustainability through water conservation or by increasing productivity from applied water. Norwood (1995) also reported that irrigation WUEs were highest for rotated crops, and lowest for continuous crops. Peterson et al. (1996) suggested that more efficient and profitable ways for using summer precipitation were to use spring-planted crops in rotation with winter wheat instead of summer fallow periods. Crop rotations employing diversity in plant water use, rooting patterns, and crop types (broadleaf versus grass plants) generally had increased crop yields compared with monoculture systems. Rotation effects may arise from beneficial effects on soil moisture, microbes, nutrients, and structure, and from decreases in diseases, insects, weeds, and phytotoxic compounds (Bezdicek and Granatstein, 1989; Crookston et al., 1991). In drought-prone areas of some regions, conventional approaches have been to employ fallow rotation systems, especially for wheat (Amir and Sinclair, 1995). Major benefits of fallow years are to increase water availability through storage without crop growth (Unger, 1994; Thomas et al., 1995). Wheat-fallow rotation systems increased water storage and WUE compared to continuous wheat systems (Bolton, 1981; Bonfil et al., 1999). Sunflower (deep-rooted and intermediate water user) can extract soil water from root zones below that normal for small grain crops (Alessi, Power, and Zimmerman, 1977; Unger 1984). Therefore, sunflower has the potential to improve WUE in rotations with small-grain crops. Sunflower is also a desirable warm-season crop for inclusion in more intensive dryland crop rotations because it provides diversity to rotations and is considered to be drought tolerant (Unger, 1984; Halvorson et al., 1999).

Adopting conservation tillage or minimum tillage is also an important strategy for improving WUE and reduction of drought in crop plants. No-till (NT) and minimum-till (MT) systems are more efficient than conventional-till (CT) systems for conserving precipitation in crop production (Aase and Schaefer, 1996; McGee, Peterson, and Westfall, 1997; Peterson et al., 1996; Tanaka and Anderson, 1997; Halvorson et al., 1999). Maintenance of previous crop residues on soil surfaces is associated with reduced soil erosion, soil temperature, and increased soil water content (Wilhelm, Doran, and Power, 1986; Sims et al., 1998).

More efficient water use has been reported with more intensive cropping systems (Black et al., 1981). When precipitation storage efficiency improves, as is possible with MT and NT, producers can increase successes of crop yields more intensively than with crop-fallows (Halvorson and Reule, 1994; McGee, Peterson, and Westfall, 1997; Peterson et al., 1996). Aase and Schaefer (1996) reported that NT annual-cropped spring wheat was more profitable and productive than spring wheat-fallow with 356 mm precipitation in northeast Montana. Introducing optimum combinations of improved technologies or management practices such as pest control and nutrient management can enhance crop yields and outputs per unit water transpired (Barker et al., 1999).

Matching phenology to water supply is another important strategy to reduce drought stress in crop plants. Genotypic variation in growth duration is an obvious means for matching seasonal transpiration with water supply, and thus maximizing water transpired (Ludlow and Muchow, 1990). Early flowering plants tend to provide higher yields and greater yield stability than later flowering plants if rains do not occur during the latter half of the growing season. Moreover, if cultivars can escape drought during the critical reproductive growth stages, HI is generally improved. Development of short duration cultivars provides benefits where rainfall is reasonably predictable. In unpredictable environments, potentially transpirable water may be left in the soil at plant maturity when sufficient moisture comes, and yield is sacrificed (Ludlow and Muchow, 1990).

Foliar applications of aqueous methanol have been reported to increase yield, accelerate maturity, and reduce drought stress and irrigation requirements in crops grown in arid environments under elevated temperatures and in direct sunlight (Nonomura and Benson, 1992). These authors reported that spraying C_3 plants with 100 to 500 ml·L^{-1} methanol solutions doubled plant growth and crop yields in several species.

It has been estimated that 30 to 60 percent of total water supplied to soils in semiarid regions is lost to evaporation (Cooper, Keating, and Hughes, 1983; French and Schultz, 1984). Water uptake efficiency is a measure of overall ability of plants to absorb water from soil and to reduce evaporation from soil. These traits are associated with root characteristics, early growth habits, and canopy closure. These characteristics are considered to possess adaptive values for droughty environments (Ludlow and Muchow, 1990). As water evaporation decreases, larger fractions of water will be available for plant transpiration and total dry matter production is increased.

Genetic variation is a key determinant in successful adaptation to environmental stresses (Stanca, Terzi, and Cattivelli, 1992). Sall and Sinclair (1991) reported the presence of genetic variability in N_2 fixation sensitivity to drought stress among soybean cultivars. Indications that N_2 fixation tolerance may be associated with ureide levels in plants has been noted (Serraj and Sinclair, 1997). Sinclair and Serraj (1995) reported major differences in N_2 fixation response to drought among grain legume species. Legume species producing ureides were more sensitive to drought stress than amide-producing plants, which may mean that ureides are involved in N_2 fixation sensitivity to water deficits. Selection in stress environments may not necessarily target specific individual genes governing single components of the stress response mechanism, but may act on multiple loci (Allard et al., 1993; Duncan and Carrow, 1999). Genetic improvement of plant drought resistance by selection for yield under water stress is possible, but prolonged and problematic procedures often occur. Recent developments in understanding physiological responses of plants to water stress and their associations with plant productivity allow scientists to embark on experimental selection programs that employ physiological selection criteria for drought resistance. Although appreciable numbers of plant physiological attributes have been identified as components of drought resistance, their interrelationships with plant productivity have not been clarified and should be investigated further. Even if expected progress is not attained by such selection experiments, resulting information should be extremely valuable as guidance for future research and development (Blum, 1983). Ceccarelli and Grando (1996) also reported that progress in cereal breeding for dry areas has been less successful than breeding for favorable environments. Breeding for drought resistance is complex because stress environments are intrinsically erratic in nature (Blum, Mayer, and Golan, 1983; Blum et al., 1983), and successes of cultivars are not predictable (Ceccarelli and Grando, 1996).

Planting deep-rooted crops is also an important strategy for improving crop yields of plants grown under water deficits (Sharpley et al., 1998; Halvorson, Wienhold, and Black, 2001). Sunflower can extract water at soil depths of 3 m, which indicates that this plant can be included in crop sequences where water has amassed below normal rooting depths of most crops (Jones and Johnson, 1983). Stone et al. (2002) reported that water depletion fronts advanced downward at greater rates and to deeper depths with sunflower (3.1 m) than with sorghum (2.5 m). As noted before, sunflower ability to produce under low-rainfall conditions is aided by its relatively deep root system (Connor and Hall, 1997) and relatively low water use requirement for initial seed yield (Nielsen, 1998).

Early vigorous growth has been a characteristic for increasing yields of wheat grown in rainfed environments (Asseng et al., 2003). Early vigorous growth resulting in enhanced leaf areas has been suggested as ways to increase growth when temperatures and vapor pressure deficits are low, thereby increasing transpiration efficiency of crops (Fischer, 1979). It has been reported that WUE (includes both water evaporation from soil and crop transpiration) for grain yields increased by as much as 25 percent because of early vigor (Siddique et al., 1990; López-Castañeda and Richards, 1994).

CONCLUSION

The impact of drought on agricultural production is hard to quantify because of water stress, high temperature, poor irrigation practices, and overexploitation of land, especially in overpopulated areas of developing countries. All of these interact to reduce food production. The most serious long-term consequence of continuation of events at present is desertification, which affects over 100 countries to varying degrees, and is perceived as a major threat to food security in the developing world (McWilliam, 1986).

Crop water use and WUE vary with different crop species. Plant processes that depend on increases in cell volume are particularly sensitive to water deficits. Two important examples of these sensitive processes are leaf gas exchange, which depends on guard cell volume, and leaf area increase, which depends on cell expansion. Inhibition of these processes under drought conditions can result in substantial losses in yield (Lecoeur and Sinclair, 1996). Water productivity is important, which can be defined as the amount of food produced per unit volume of water used. Water productivity can be increased by increasing yield per unit land area, either by using better crop cultivars, improved agronomic practices, or growing crops during most suitable periods. Progress for breeding drought resistance crop cultivars has been slow. Unfortunately, this ideal does not appear to have been widely achieved with commercial crop cultivars (Blum, 1988; Munns, 1988; Simane, Peacock, and Struik, 1993). A major reason for such slow progress is the complexity of drought, which often results in lack of clear identification of target environments. Another contributory factor may be that mechanisms underlying resistance to various drought regimes at the cellular and whole plant levels have not yet been clearly established (Neumann, 1995). Thus, criteria for selecting or developing plants with increased resistance to water deficits remain somewhat emperical (Blum, 1988).

Crop physiology can provide many contributions to plant breeding programs designed for improving drought resistance. Several physiological criteria for selecting resistant genotypes have been proposed. Among the physiological traits that have been examined are canopy temperature, stomatal resistance, transpiring leaf areas, and rates of water loss by excised leaves (Araghi and Assad, 1998). At present, improvements in drought resistance are largely fortuitous outcomes of empirical multienvironmental testing strategies used in plant breeding programs. Although this traditional approach has been somewhat effective, its efficiency is widely considered to be low. Nguyen, Babu, and Blum (1997) reported that because of low heritability of yield under water stress and inherent variation in the field, such selection programs are expensive and slow in attaining progress. Strong genotype by water stress interactions have been observed for putative drought resistance traits in many field studies of different crop plants. This makes the situation still more complicated for successful breeding of drought-resistant crop cultivars. Environmental variability has generally been observed from year to year and from site to site variation. Consequences of genotype by environment interactions are that particular lines often do not perform well under all conditions encountered in a target population of environments. Biological reasons for these interactions are often unknown. Therefore, more research is needed to understand physiology of drought resistance in crop plants. However, several plant traits are assumed to confer drought resistance to crop plants. Correct phenology that matches crop growth and development to water environments are important. Deep root systems and some other drought avoidance mechanisms are useful in upland conditions. Osmotic adjustment is an important drought-resistance mechanism in plants. Osmotic adjustment involves net accumulation of solutes in cells in response to decreases in water potentials of cellular environments. The consequence of net accumulation of solutes is that osmotic potentials of cells are lowered, which in turn attracts water into the cell and tends to maintain turgor pressure. Generally, osmotic adjustment contributes to turgor maintenance of both shoots and roots as plants experience water deficits. Osmotic adjustment in cells may be a promising physiological trait as it can counteract the effects of rapid declines in leaf water potential, and large genotypic variation for this trait exists (Fukai and Cooper, 1996). More agronomic research is required to define the various moisture environments, especially in terms of amount, frequency, and probability of rainfall and of expected soil moisture regime in average seasons. This is necessary so that the most appropriate phenology can be obtained (Ludlow and Muchow, 1990).

Most important strategies to reduce effects of drought are investment in agricultural infrastructures, including irrigation and improved technology, management, and more tolerant crop cultivars.

These will be required to help improve and stabilize agricultural production. To optimize crop yields in irrigated environments, irrigation should be timed in a way that nonproductive soil water evaporation and drainage losses are minimized, inevitable water deficits coincide with least sensitive growth periods. In rainfed environments, management practices should be tailored in ways that water availability matches crop water needs.

Chapter 8

Physiology of Mineral Nutrition

INTRODUCTION

Physiology of mineral nutrition refers to nutrient uptake, translocation, and assimilation or utilization by crop plants (Fageria, Baligar, and Jones, 1997). Biochemically, plant nutrition deals with complex biosynthetic events by which organic plant substances are produced from inorganic materials in the environment (Clarkson and Hanson, 1980). Factors for crop production such as water, cultivar, control of insects, diseases, and weeds, and mineral nutrition play key roles in increasing crop yields.

The term *mineral nutrition* is often used to refer to inorganic nutrients essential to plant growth. This term is a slight misnomer, because plant nutrients are not minerals. The term comes from the fact that most essential elements were combined with other elements in the form of minerals, which eventually broke down into individual component elements (Fageria, Baligar, and Jones, 1997). Plants contain small amounts of 90 or more elements, but only 16 elements are known to be essential for normal plant growth (Epstein, 1972; Fageria, 1984). These nutrients are C, H, O, N, P, K, Ca, Mg, S, Mn, Fe, Zn, Cu, B, Mo, and Cl. The discovery of essential nutrients, their chemical symbols, and their principal forms for uptake are presented in Table 8.1. Even though some of these elements have been known since ancient times, their essentiality has been established only within the past century (Tamhane et al., 1966; Marschner, 1995). Carbon, H, and O are responsible for about 95 percent of the weight of plants, and these nutrients are supplied through air and water to plants. Chlorine may also be supplied through atmospheric air. Humans normally furnish the other 12 nutrients, at least if good crop yields are to be obtained. Hence, nature is responsible for producing nearly all of the weight to plants as far as nutrients are concerned. The remaining

TABLE 8.1. Essential mineral nutrients for plant growth, principal forms for nutrient uptake, and date and person(s) discovering each nutrient.

Nutrient differences	Chemical symbol	Principal forms of uptake	Year nutrient discovered/by whom	
			Year	Person discovering
Carbon	C	CO_2	1882	J. Sachs
Hydrogen	H	H_2O	1882	J. Sachs
Oxygen	O	H_2O, O_2	1804	T. De Saussure
Nitrogen	N	NH^+_4, NO_3^-	1872	G. K. Rutherford
Phosphorus	P	$H_2PO_4^-$, HPO_4^{2-}	1903	Posternak
Potassium	K	K^+	1890	A. F. Z. Schimper
Calcium	Ca	Ca^{2+}	1856	F. Salm-Horstmar
Magnesium	Mg	Mg^{2+}	1906	Willstatter
Sulfur	S	SO_4^{2-}, SO_2	1911	Peterson
Iron	Fe	Fe^{2+}, Fe^{3+}	1860	J. Sachs
Manganese	Mn	Mn^{2+}	1922	J. S. McHargue
Boron	B	H_3BO_3	1923	K. Warington
Zinc	Zn	Zn^{2+}	1926	A. L. Sommer & C. B. Lipman
Copper	Cu	Cu^{2+}	1931	C. B. Lipman & G. MacKinney
Molybdenum	Mo	MoO_4^{2-}	1938	D. I. Arnon & P. R. Stout
Chlorine	Cl	Cl^-	1954	T. C. Broyer et al.

Source: Fageria, 1984, 1992; Fageria, Baligar, and Jones, 1997.

nutrients have to be supplied to plants for adequate growth. Nitrogen, P, K, Ca, Mg, and S are known as "macronutrients," while, Mn, Fe, Zn, Cu, B, Mo, and Cl are known as "micronutrients." All nutrients are equally important for plant growth, even though some are required in only small amounts. Differences for dividing nutrients into micronutrients and macronutrients are related to their quantities required by plants. The low requirement of micronutrients for plants can be accounted for by their participation in enzymatic reactions and as constituents of growth hormones rather than as components of major plant structural and protoplasmic products, as is the case for macronutrients (Stevenson, 1986; see Table 8.2). Essential plant nutrients can also be classified as metals or nonmetals. The nutrients considered metals include K, Ca, Mg, Fe, Zn, Mn, Cu, and Mo. The nonmetals include N, P, S, B, and Cl (Bennett, 1993).

Classification of plant nutrients according to their biochemical behavior and physiological functions seems to be an appropriate way to consider nutrients required for plant growth (Mengel et al., 2001). For physiological considerations, plant nutrients may be divided into four groups:

1. C, H, O, N, and S: Major constituents of organic materials and involved in enzymatic processes and reduction-oxidation (redox) reactions.
2. P and B: Involved in energy transfer reactions and esterification with native alcohol groups.
3. K, Ca, Mg, Mn, and Cl: Involved in osmotic and ionic balances and have specific functions in enzyme conformation and catalysis.
4. Fe, Cu, Zn, and Mo: Involved as structural chelates and complexes or metalloproteins and in electron transport by changing valency.

Three important criteria are used to confirm essentiality of nutrients for growth of higher plants: (1) Plants cannot complete their life cycle (germination to seed production) if the nutrient is lacking in the growth medium; (2) the nutrient cannot be replaced by another, and it has direct effects on growth or metabolism rather than indirect effects on growth (e.g., antagonism of another element present at toxic levels); and (3) the nutrient forms part of molecules or constituents of plants (e.g., the nutrient itself is essential in plants). Examples for essentiality of N and Mg are that N is a constituent of proteins and Mg is a constituent of chlorophyll. Some nutrients such as Co, Na, and Si have been cited as being essential to some plants. These elements are usually not essential but may be beneficial for growth of some plants. Examples of this are Co as a component of vitamin B_{12}, which is essential for microorganisms associated with N_2 fixation (e.g., *Rhizobia* in root nodules of leguminous plants and by some species of

TABLE 8.2. Functions of essential nutrients in plants.

Nutrient	Function
Carbon	Basic molecular component of virtually all molecules (e.g., carbohydrates, proteins, lipids, nucleic acids).
Hydrogen	Component of most molecules. Important in ionic balance, acts as major reducing agent, and is key to energy relations of cells.
Oxygen	Component of virtually all molecules like carbon.
Nitrogen	Component of most organic compounds, only in lower quantities than C, H, and O.
Phosphorus	Central role in plants is in energy transfer and protein metabolism and component of many molecules.
Potassium	Osmotic and ionic regulation and functions as cofactor or activator for many enzymes of carbohydrate and protein metabolism.
Calcium	Involved in cell division and has major role in maintenance of membrane integrity and detoxify Al.
Magnesium	Component of chlorophyll and as cofactor for many enzymatic reactions.
Sulfur	Like phosphorus as it is involved in plant cell energetic and component of many molecules.
Iron	An essential component of many haem and nonhaem Fe enzymes and carriers, including cytochromes (respiratory electron carriers) and ferrodoxins. The latter are involved in important metabolic functions (e.g., N_2-fixation, photosynthesis, and electron transfer).
Manganese	Involved in O_2-evolving system of photosynthesis and is a component of enzymes arginase and phosphotransferase.
Boron	Specific biochemical functions of B are unknown, but it may be involved in carbohydrate metabolism and synthesis of cell wall components.
Zinc	Essential component of several dehydrogenases, proteinases, and peptidases (e.g., carbonic anhydrase, alcohol dehydrogenase, glutamic dehydrogenase, malic dehydrogenase)
Copper	Constituent of some important enzymes (e.g., cytochrome oxidase, ascorbic acid oxidase, and laccase).
Molybdenum	Required for normal assimilation of N in plants. Essential component of nitrate reductase and nitrogenase (N_2-fixation enzyme).
Chlorine	Essential for photosynthesis and as activator of enzymes involved in splitting water. It also functions in osmoregulation of plants growing on saline soils.

Source: Compiled from Oertli, 1979; Ting, 1982; Stevenson, 1986.

N_2-fixing blue green algae) (Liu, Reid, and Smith, 1998). In the case of Na, some plant species, particularly *Chenopodiaceae* and other species adapted to saline conditions, absorb this element in relatively high amounts (Epstein, 1972). Silicon is beneficial for several crops and is especially associated with plant ability to tolerate abiotic and mineral toxicity stresses and provide resistance to many biotic stresses (Clark, 2001). Even though plants receive some beneficial effects from some elements, this does not prove essentiality of the element(s) to higher plants. For example, Al, like Si, has provided some beneficial effects to plant growth (Foy, 1992), but this does not mean that Al is essential for plants. In the case of rice, Al up to 3.8 $cmol_c \cdot kg^{-1}$ concentration improved growth (Fageria and Santos, 1998). Aluminum is considered toxic for most plants grown in acidic soils (pH below ~5) when present in sufficiently high concentrations, and plant species and cultivars vary widely in tolerance to Al toxicity (Foy, 1992).

Deficiencies of any given nutrient often occur in plants if the nutrient in question has low concentrations in soil, and plants are unable to absorb the nutrient in sufficient quantities because of dry or other adverse conditions inhibiting absorption. For plants to achieve well-balanced metabolism, high production, and unimpeded development, nutrients must not only be absorbed in sufficient quantities but also must be absorbed in balanced proportions with other nutrients.

Visual symptoms and soil and plant analyses are the usual methods to identify nutritional deficiencies in crop plants. Chlorosis and/or browning of leaves are major visible symptoms for detection of nutrient deficiencies. Degree of visual symptoms and growth stage at which leaf symptoms of mineral deficiencies occur vary among species and nutrients. Maize is one of the most sensitive crops to many nutrient deficiencies, and symptoms appear as early as the four- to five-leaf stage of growth. In rice, barley, wheat, maize, and sunflower, N, P, Ca, Mg, Fe, and S deficiency symptoms normally occur in early growth stages.

An appropriate criterion for diagnosing nutritional deficiencies in annual crops is evaluation of crop responses to applied nutrients. If given crops respond to an applied nutrient in a given soil, that nutrient is most likely limiting for that crop even though symptoms may not have appeared at the time of observation. Studies have been conducted in Brazil to provide evidence for which nutrient is most yield limiting to annual crops grown on an Oxisol (Table 8.3). Based on

TABLE 8.3. Relative dry matter yield (percent) of shoots for five annual crop species grown with different fertility levels on an Oxisol.

Fertility treatment	Rice	Common bean	Maize	Soybean	Wheat
AFL [a]	100	100	100	100	100
–N	86	96	70	88	88
–P	7	28	45	44	55
–K	85	96	63	96	58
–Ca	98	30	50	52	42
–Mg	83	36	85	80	82
–S	95	95	78	84	94
–Zn	67	75	67	52	91
–Fe	92	94	85	80	94
–Cu	73	92	96	68	91
–B	77	37	78	80	99
–Mo	57	99	81	88	91
–Mn	87	96	93	72	76

Source: Adapted from Fageria and Baligar, 1997.

[a] AFL = adequate fertility level, and minus (–) sign against each nutrient means without application of that nutrient.

information from these studies, research priorities can be made to correct nutrient deficiencies to improve crop yields on specific soils. The magnitude of dry weight reductions for five annual crop species not receiving application of any of the 12 essential nutrients to an Oxisol varied among crops and nutrients. Phosphorus and Ca were the most yield-limiting nutrients. Among micronutrients, B and Zn caused decreases in dry weight for common bean. Molybdenum and Zn were the most dry-weight-reducing nutrients for upland rice. Among crops tested, susceptibility to P deficiency based on aboveground plant dry weight was of the order upland rice > common bean > soybean > maize > wheat. The order of susceptibility to Ca deficiency was common bean > wheat > maize > soybean > upland rice. Among crops tested, upland rice had the most tolerance to soil acidity, and common bean had the least tolerance. Loneragan (1997) emphasized that the contribution of added nutrients during the twentieth century for increasing crop yields was as high as 50 percent worldwide. The science of plant nutrition had great achievements in the twentieth century. Plant nutrition became established as an individual discipline based on scientific achievement, clearly defined chemical elements required for plant growth, and contributed to improvements of plant production and environmental quality. Informational seeds sown by Justus von Liebig have produced beautiful flowers in the green revolution of the twentieth century, and mature plants have many buds preparing to bloom in our present century. Further, discoveries during the twentieth century about new essential nutrients and of legume symbiotic fixing of N_2, soil chemistry, soil microbiology, plant physiology, plant genetics, fertilizer technology, and ability to diagnose and correct soil problems have enhanced food production extensively (Loneragan, 1997). Worldwide nutrient deficiencies of essential macro- and micronutrients and toxicities of Al, Mn, Fe, S, B, Cu, Mo, Cr, Cl, Na, and Se have been reported to help farmers know where trouble spots may occur (Table 8.4) (Baligar and Fageria, 1997; Baligar, Fageria, and He, 2001).

The objective of this chapter is to discuss some of the latest research advances relative to absorption, translocation, function, and assimilation of essential plant nutrients in annual crop plants and to indicate appropriate management strategies for improving nutrient uptake and utilization efficiencies for maximizing crop yields. These aspects are central to a broad understanding of mineral nutrition of higher plants. Nutrient acquisition by plants is complex and dynamic since soil, climate, and plant factors and interactions are involved. Most research relative to physiology of nutrient acquisition has been conducted under controlled environments using specific nutrient cultures in the growth medium. Such research is important to understand basic principles of nutrient acquisition but may lack some important aspects when plants are grown under field conditions for economic purposes. Nutrient concentrations, temperature, pH, humidity, diseases, insects, and weeds can be adequately controlled when plants are grown in controlled environments. Controlling these factors at appropriate levels is difficult in field conditions. Therefore, the question frequently asked is whether controlled conditions or greenhouse results can be applied to field conditions. The answer to this question is that some results can be applied, but not all. Results of nutrient

TABLE 8.4. Potential element deficiencies and toxicities associated with major soil orders.

Soil order (U. S. taxonomy)	Soil group (FAO)	Element Deficiency	Toxicity
Andisols (Andepts)	Andisol	P, Ca, Mg, B, Mo	Al
Ultisols	Acrisol	N, P, Ca, and most other	A, Mn, Fe
Ultisols/Alfisols	Nitosol	P	Mn
Spodosols (Podsols)	Podsol	N, P, K, Ca, micronutrients	Al
Oxisols	Ferrasol	P, Ca, Mg, Mo	Al, Mn, Fe
Histosols	Histosol	Si, Cu	
Entisols (psamments)	Arenosol	K, Zn, Fe, Cu, Mn	
Entisols (fluvents)	Fluvisol		Al, Mn, Fe
Mollisols (aqu), Inceptisols, Entisols	Gleysol	Mn	Fe, Mo
Mollisols (borolls)	Chernozem	Zn, Mn, Fe	
Mollisols (ustolls)	Kastanozem	K, P, Mn, Cu, Zn	Na
Mollisols (aridis) (udolls)	Phaeozem		Mo
Mollisols (rendolls)	Rendzina	P, Zn, Fe, Mn	
Vertisols	Vertisol	N, P, Fe	
Aridisols	Xerosol	K, P, Mg, Fe, Zn	Na
Aridisols/Arid Entisols	Yermosol	Mg, K, P, Fe, Zn, Co, I	Na, Se
Alfisols/Ultisols (albic) (poorly drained)	Planasol	Most nutrients	Al
Alfisols/Aaridisols/Mollisols (natric) (high alkali)	Solonetz	K, N, P, Zn, Cu, Mn, Fe	Na
Aridisols (high salt)	Solonchak		B, Na, Cl

Source: Dudal, 1976; Clark, 1982; Baligar, Duncan, and Fageria, 1990; Baligar and Fageria, 1997.

deficiency symptoms created under controlled conditions usually coincide with field symptoms. Confirmation of controlled environmental and field symptoms were made for rice and common beans grown on an Oxisol in central Brazil (Fageria and Barbosa Filho,

1994; Fageria, Oliveira, and Dutra, 1996). It was also possible to evaluate whether a given crop would respond to a given nutrient when grown in a given soil. When these crops responded to applied nutrients under controlled environments, they also responded to the same nutrients in the field. However, response magnitudes varied. Similarly, it has been possible to test nutrient uptake efficiency by crop genotypes grown in controlled conditions so that reduced numbers of genotypes or more promising genotypes can be taken to the field for reduction of costs of field experimentation. However, it has not been possible to extrapolate nutrient application rates used in controlled conditions to those applied in the field and vice versa. Under controlled conditions, nutrient application rates per unit soil volume are much higher than under field conditions. An example of this is presented in Figure 8.1, where maximum grain yield of upland rice grown under greenhouse conditions on an Oxisol was about 400 mg N/kg soil. If this rate of N was transformed to the field, it would require about 800 kg N/ha. When upland rice was grown on an Oxisol in the field, adequate N rate was near 120 kg N/ha (Figure 8.2).

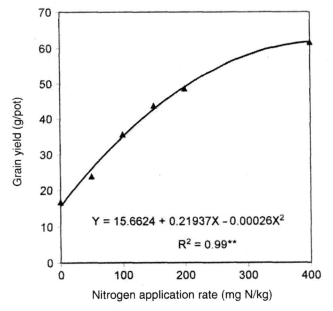

$$Y = 15.6624 + 0.21937X - 0.00026X^2$$

$$R^2 = 0.99^{**}$$

FIGURE 8.1. Response of upland rice to N application rates when grown on an Oxisol in a greenhouse. *Source:* Adapted from Fageria, 2000.

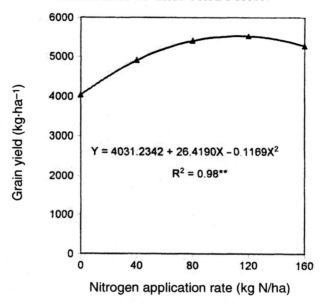

FIGURE 8.2. Response of upland rice to N application rates when grown on an Oxisol in the field. *Source:* Reproduced from Stone et al., 1999.

Nutrient acquisition by plants depends on ion concentrations at root surfaces, root absorption capacity, and plant demand. In physiological terms, nutrient absorption by plants is usually referred to as ion uptake or ion absorption, because nutrients in almost all cases are in ionic forms before absorption can occur (Hiatt and Leggett, 1974; Barber, 1995).

ROOT MORPHOLOGY

Nutrient absorption through roots is the primary mechanism to satisfy nutritional requirement of plants, although small amounts of some nutrients may be absorbed through leaves if applied as foliar sprays or are contaminants in the atmosphere (Barber, 1995). For example, C as CO_2 is absorbed by leaves almost exclusively from the atmosphere, while O as O_2 and S as SO_2 are common in the atmosphere and considerable amounts are available for leaf absorption. Small amounts of other ions such as Fe, Mn, Cu, and Zn may be absorbed from the atmosphere through leaves, but these nutrients are absorbed mainly by roots. Under special circumstances such as where deficiencies must be alleviated immediately, these micronutrients are applied as foliar sprays. Because the bulk of nutrients are absorbed by roots, some discussion of root morphology and cell structure is warranted to better understand nutrient uptake processes in plants. A root cross section of cells and tissues involved in ion absorption is presented in Figure 8.3. Important features of root morphology are root hairs, epidermis, cortex, and stele. During absorption, ions have to move through the epidermis, cortex, endodermis, and stele and empty into the xylem. From the xylem, ions are transported to shoots. Phloem tissue transports photosynthetic assimilates from leaves and shoots to roots during plant development. Two parallel pathways for solute movement across cortex cells exist to reach steles: One is passage through extracellular space or apoplast (cell walls and intercellular spaces) and the other is passage from cell to cell in the symplast through plasmodesmata to cellular particles and to vacuolo compartments within cells (Marschner, 1995).

Active nutrient uptake occurs across cells surrounded by membranes. Therefore, cell structure and membranes that facilitate nutrient uptake mechanisms are important. Figure 8.4 shows a simplified

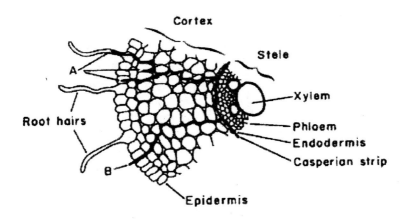

FIGURE 8.3. Transverse section of maize root showing symplastic (A) and apoplastic (B) pathways of ion transport across roots. *Source:* Reprinted from *Mineral Nutrition of Higher Plants,* Marschner, H., p. 60, copyright 1986, with permission from Elsevier.

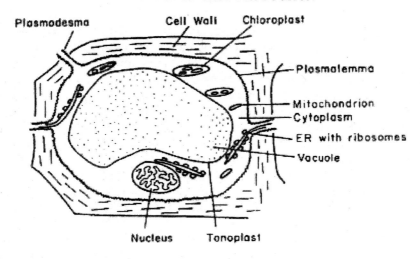

FIGURE 8.4. Simplified scheme of a mesophyll cell. *Source:* Reproduced from Mengel and Kirkby, 1978, page 97. Used with permission from International Potash Institute, Basel, Switzerland.

diagram of a mesophyll cell and its components. Vacuoles, nuclei, chloroplasts, ribosomes, and mitochondria are embedded within cells. Plasma membranes separate the cytoplasm and cell wall, while tonoplast membranes separate the cytoplasm from the vacuole. Plasmalemma membranes form boundaries between cells and the outer medium, and it is these membranes (not cell walls) that make effective barriers against uptake of ions and molecules dissolved in aqueous outer media (Mengel and Kirkby, 1978).

Each plant cell organ has specific functions to facilitate plant growth and development. Vacuoles are important in water economy of cells as well as providing sites for segregation of water and end products of metabolism. Light energy conversion and CO_2 assimilation occur in chloroplasts. Enzymes controlling various steps of metabolism such as the tricarboxylic acid (TCA) cycle, respiration, and fatty acid metabolism occur in mitochondria. Ribosomes are supermolecular assemblies composed of ribosomal nucleic acids and proteins enabling synthesis of polypeptides from free amino acids. Plasmodesma connect cells (Mengel and Kirkby, 1978).

Biological membranes consist of protein and lipid molecules in approximately equal proportions and are about 7 to 10 nm thick (Mengel and Kirkby, 1978). Danielli and Davson (1935) proposed a unit membrane model whereby each unit consists of two lipid molecular layers with their hydrophobic tails (fatty acids) oriented inward (Marschner, 1995). Both outer boundaries of lipid layers are coated with protein layers. It was argued that this kind of structure could serve well as barriers because protein layers would enhance rigidity and lipidic fractions would prevent penetration of membranes by hydrophilic particles, including hydrated inorganic ions (Mengel and Kirkby, 1978).

Based on electromicroscopic studies, Singer (1972) proposed a membrane model consisting mainly of liquid amphiphilic bilayers and proteins. Amphiphilic indicates the presence of both hydrophilic (OH, NH_2, P, and carboxylic groups) and hydrophobic (hydrocarbon chains) regions in membranes. Lipids and proteins may be bound by electrostatic, H, and hydrophobic bonds (Mengel and Kirkby, 1978). No coating protein on the outer sides of membranes exists in the Singer (1972) model, but coating and globular proteins embedded in lipid bilayers exist in the Danielli-Davson (1935) model. Some of these proteins may even extend through membranes to form protein channels from one side of a membrane to another. These protein channels can be considered hydrophilic pores through which polar solutes such as ions can be transported (Walker, 1976). Biological membranes are not completely impermeable. Membranes may allow diffusion of hydrophilic ions and molecules, but the degree of permeability depends on components making up the membranes. In addition, enzymes present in biological membranes may directly or indirectly be involved in transport of ions and molecules across membranes (Mengel and Kirkby, 1978).

ACTIVE AND PASSIVE ION TRANSPORT

Ion movement in plant cells is both active and passive. In passive processes, ions move from higher to lower concentrations or down chemical gradients of potential energy. For active uptake, ions move against concentration gradients and ion movement depends on electrochemical potential gradients. In active processes, cations are attracted to negative electropotentials, and anions are attracted to positive electropotentials. Electrochemical potentials are established across membranes because of unequal charge distributions. Difference between membrane potentials and actual potentials created by nonequilibrium distributions is a measure of quantity of energy required. A modified Nernst equation can be used to calculate electrical charge, which is described as follows (Ting, 1982):

$$\psi = (- RT/ZF) \ln (a_i/a_o) \qquad (8.1)$$

where ψ = electrochemical potential between root cells and external solutions in millivolts (mV); R = gas constant (8.3 J/mol K); T = absolute temperature (K); Z = net charge on ion (dimensionless); F = Faraday constant (96,400 J/mol); a_i = activity of ion on inside of tissue; and a_o = activity of ion outside of tissue. For quick calculations, it is convenient to remember that $RT/F = 26$ mV.

Measurement of electrochemical potentials in cells and the outer media as well as ionic concentrations in cells and the outer media can give indications whether ions move actively or passively. Values of these measurements can be put into the Nernst equation to calculate electrical potentials. Negative values indicate passive uptake and positive values indicate active uptake for cations. Negative values indicate active transport and positive values indicate passive transport for anions (Mengel and Kirkby, 1978). These measurements are valid only when equilibrium conditions are maintained in the system, which is difficult under practical conditions. Electrical potential differences across plasma membranes are in the range of –60 to –200 mV (cytoplasmic negative), and electrical potential differences across tonoplasts are relatively low at 0 to –20 mV, with cytoplasm values being negative relative to vacuole values (Hodges, 1973).

ION UPTAKE MECHANISMS

Concentrations of ions in the cytoplasm are often much higher than in soil solutions, and in extreme cases may be 10,000-fold higher. Therefore, roots must be able to absorb ions against extensively different concentration gradients. Uptake against concentration gradients or against electrochemical gradients requires metabolic energy and is commonly considered active uptake. At present, two principal theories of ion transport across membranes exist. They are known as the carrier theory and the ion pump theory (Mengel and Kirkby, 1978).

Carrier Theory

Carriers are commonly used to refer to agents responsible for transporting ions from one side of membranes to the other. Even though carrier properties are similar to that of enzymes, they have not been isolated and characterized. Isolation of carriers will not necessarily entail their removal from membranes, but at present measuring their activities has not been successful. Mengel and Kirkby (1978) presented a hypothetical scheme for carrier ion transport across membranes. In this transport process, carriers meet particular ions for which they have affinity, form carrier ion complexes, and move across membranes. The enzyme phosphatase located at inner membrane boundaries splits off phosphate from carrier complexes and ions are released. In this transport process, energy is required and involvement of adenosine triphosphate (ATP) is normally reported (Marschner, 1995). The high energy molecule ATP is generated from adenosine diphosphate (ADP) + inorganic P (Pi) from respiration (oxidative phosphorylation) reactions.

Uptake may be described by the following equations (Mengel and Kirkby, 1978):

$$\text{Carrier} + \text{ATP} \xrightarrow{\text{Kinase}} \text{Carrier P} + \text{ADP} \qquad (8.2)$$

$$\text{Carrier P} + \text{Ion} \longrightarrow \text{Carrier P} - \text{Ion} \qquad (8.3)$$

$$\text{Carrier P} - \text{Ion} \xrightarrow{\text{Phosphatase}} \text{Carrier} + \text{Pi} + \text{Ion} \qquad (8.4)$$

$$\text{Net: Ion} + \text{ATP} \xrightarrow{\text{Transport}} \text{Ion} + \text{ADP} + \text{Pi} \qquad (8.5)$$

ATPase Theory of Ion Transport

Hodges (1973) proposed an ATPase theory of ion transport in plants. ATPase is a group of enzymes having the capacity to dissociate ATP into ADP and Pi. Energy liberated from these reactions can be utilized in ion transport across membranes. In plants, this phenomenon is known as activity of ATP, which is associated with the plasmalemma and is activated by cations (Hodges et al., 1972). Detailed descriptions of this process for ion transport have been provided (Hodges, 1973; Clarkson, 1984; Mengel and Kirkby, 1978).

Ion Uptake Kinetics

Epstein and Hagen (1952) formulated the enzyme kinetic hypothesis of ion transport and carrier function. Only a brief description of this hypothesis is provided because this enzyme kinetic hypothesis of membrane transport has been extensively reviewed (Clarkson and Hanson, 1980; Epstein, 1973; Hodges, 1973; Nissen, 1974; Nissen et al., 1980; Fageria, 1984). According to Epstein and Hagen (1952), transport of ions into plant cells may be analogous to relationships between binding of substrates to enzymes and release of their products after catalysis. The overall sequence of events during enzyme catalyzed reactions,

$$S \xleftrightarrow{E} P, \tag{8.6}$$

may be depicted as shown (Segel, 1968):

$$E + S \leftrightarrow ES \leftrightarrow EX \leftrightarrow EY \leftrightarrow EZ \leftrightarrow E - P \leftrightarrow P + E. \tag{8.7}$$

The enzyme *(E)* first combines with substrate *(S)* to form an enzyme substrate complex *(ES)*. On the surface of enzymes, substrates may go through one or more transitional forms *(X, Y, Z)* and finally be converted to a final product *(P)*. The final product dissociates allowing free enzymes *(E)* to begin again (Segel, 1968). Ion uptake by plants follows hyperbolic relationships with increasing concentrations (up to ~200 mmol/m^3) in the growth medium (Ingestad, 1982), and can be explained by Michaelis-Menten kinetics. Uptake rates at given concentrations can be predicted using the following equation:

$$V = (V_{max}C_i) / (K_m + C_i) \tag{8.8}$$

where V_{max} = maximum velocity, C_i = concentration of ion in the growth medium, and K_m = Michalis constant equal to substrate ion concentrations giving half maximal rates of uptake. Small values for

K_m imply high affinity between ion and carrier. If uptake rates and concentrations are plotted as reciprocals, straight lines are normally obtained. Extrapolation of these lines provide intercepts at $1/V_{max}$, the concentration at half maximal velocity corresponding to K_m. This method of calculating K_m is known as the Line-Weaver-Burk plot method. Alternative procedures for plotting rates of uptake *(V)* against *V/C* and V_{max} exist so that V_{max}/K_m can be obtained by extrapolation of experimentally determined slopes to the ordinates and abscissas, respectively (Clarkson, 1974). This method is known as the Hofstee plot method.

Claassen and Barber (1976) described an ion absorption isotherm relative to I_{max}, K_m, and E with the proposed equation

$$I_{max} = (I_{max} C_i) / (K_m + C_i - C_{min}) \qquad (8.9)$$

where I_{max} = maximum net influx of ions into roots, $nmol \cdot m^{-2} \cdot s^{-1}$, C_{min} = value of C_{10} (C_{10} = ion concentration in solution at the root surface, $mmol \cdot L^{-1}$) where influx = eflux and In = 0, $mmol \cdot L^{-1}$, and the other terms as defined above (Barber, 1995). Since uptake rates were zero at concentrations above zero, the lines were extended to the ordinate giving negative uptake at zero ion concentrations, and was termed efflux *(E)* from roots (Barber, 1995). Nielsen and Barber (1978) made further modifications by using concentrations in solution where net influxes reached zero rather than *E* as a third point describing the lower end of the absorption curve. This values was termed C_{min} and the equation written as follows:

$$I_{max} = [I_{max} (C_i - C_{min})] / [K_m + C_i - C_{min}]. \qquad (8.10)$$

It should be kept in mind that ion uptake kinetic values vary with plant age, nutrient concentration, temperature, root morphology, plant demand for nutrients, and analytical technique used to make measurements.

Hai and Laudelout (1966) and Fageria (1973, 1976) proposed continuous flow techniques to measure nutrient uptake kinetics. The basic principle of continuous flow systems is that rates of nutrient uptake (U) are equal to products of flow rates (F) and differences between concentrations of solution entering systems (C_o) and those of outgoing solutions (C_s). The mathematical equation can be written as follows:

$$U = F (C_o - C_s) \tag{8.11}$$

The rate of ion uptake (mg/g·h or μg/g·h root weight [fresh or dry]) is calculated as follows:

$$\text{Rate of ion uptake} = (1 - C_s/C_o) \times F \times C \, / \, \text{root weight} \tag{8.12}$$

where C_s = concentration of ion in the outgoing solution, C_o = concentration of ion in the ingoing solution, C = concentration of stable ion in nutrient solution (ppm or mmol/L).

Flow rates through systems in continuous flow culture techniques is an important parameter to be considered in ion uptake. Edwards and Asher (1974) discussed the significance of solution flow rates in flowing culture experiments and concluded that actual flow rates required for a particular experiment will depend upon nature and concentration of ion, plant age, efficiency of roots in absorbing tested ions, and conditions of experimentation.

The response of I_{max} and K_m in the transport process to physical and metabolic factors can provide some insight into the general nature of processes moving ions across membranes (Clarkson, 1984). Specificity of mechanisms involved can be inferred from effects of competing ions on V_{max} values. Studies such as this led to the conclusion that two different sets of binding sites existed for K+ and Na+ (Epstein, 1966). At low concentrations of K+, Na+ had no effect on V_{max} of K+, but at higher concentrations of K+, I_{max} was increasingly depressed by increasing Na+.

Epstein, Rains, and Elzam (1963) noted relationships between ion concentration and uptake rate to be more complex when concentrations varied over wide ranges. These authors attributed kinetics to simultaneous functioning of two different carriers for the same ion and suggested that both mechanisms are located in the plasmalemma. On the other hand, Laties (1969) suggested that mechanisms function in series, one at the plasmalemma and one at the vacuolar membrane or tonoplast. Nissen (1971, 1974) suggested that uptake mechanisms remain unchanged over wide concentration ranges, but their characteristics change at certain discrete external concentrations. Nissen (1973) and Nissen et al. (1980) reexamined ion absorption data in many plant species and concluded that ion uptake in higher plants can be described by multiphasic uptake mechanisms, which account for apparent contradicting evidence for parallel and series models. In contrast to other uptake kinetic models, which may all be termed continuous, the multiphasic model predicts discontinuous relationships between ion concentration and rate of uptake. Fitting kinetic models to data for different ions and for different plant species, cultivar, or tissue showed that multiphasic models fit better than continuous models (Nissen et al., 1980).

ION ABSORPTION MEASUREMENT

Ion uptake measurements are mostly made by tracer techniques using excised roots. Rates of uptake measured in this way tend to ignore large amounts of ions transported across roots into the xylem and finally to shoots. In practice, only small fractions of nutrients absorbed are retained in roots, and major portions are exported to shoots (Asher and Ozanne, 1967; Loneragan and Snowball, 1969). Thus, measurement of ions in both roots and shoots should be performed in ion absorption studies.

One way to measure rates of nutrient absorption by roots is to estimate changes in nutrient contents in roots and shoots by chemical analyses. The values obtained by these procedures should be averaged over several days and provide useful indications about the absorption process (Pitman, 1976). Equations to measure nutrient uptake from plant analysis data have been discussed elsewhere (Williams, 1948; Loneragan, 1968; Pitman, 1976).

Rate of monovalent cation uptake is normally higher that of polyvalent cations in crop plants (Fageria, 1973). An example of this is presented in Figures 8.5 and 8.6. Uptake rates from nutrient solution of 100-days old rice plants were much higher for K^+ (Figure 8.5) compared to Mg^{2+} (Figure 8.6). These differences in uptake rates demonstrate that uptake mechanisms for K^+ are more effective and selective than for Mg^{2+}. Similar effective uptake mechanisms are supposed to exist for NO_3^- and HPO_4^-, and most likely for NH_4^+ and Cl^-. Plant cells need these active uptake mechanisms to pump sufficient inorganic ions within a short time; these are indispensable for high plant growth rates. While NO_3^-, NH_4^+, and HPO_4^- are needed

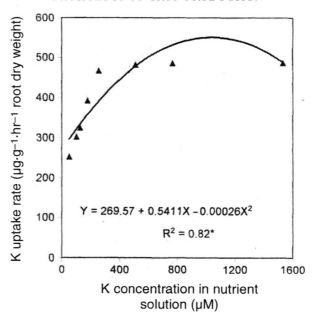

FIGURE 8.5. Uptake rate of K+ by 100-days old rice plants grown in nutrient solution (uptake determined by continuous flow method). *Source:* Adapted from Fageria, 1973.

for synthesis of various organic compounds, high quantities of K+ are prerequisite for optimal activation of numerous enzymes and balance of cellular osmotic concentrations. On a dry matter basis, younger tissues have higher concentrations of N, P, and K than older tissues (Mengel, 1974b). In addition, Hiatt and Leggett (1974) reported that certain physical properties of ions influence their rates of uptake. The most important of these properties are charge and hydrated radii. Monovalent ions are absorbed faster than divalent or polivalent ions of similar hydrated radii, and ions of small radii are absorbed faster than ions with large hydrated radii.

ION TRANSLOCATION FROM ROOTS TO SHOOTS

After absorption by roots, long-distance transport of ions to shoots occurs in the vascular system of the xylem and phloem, and

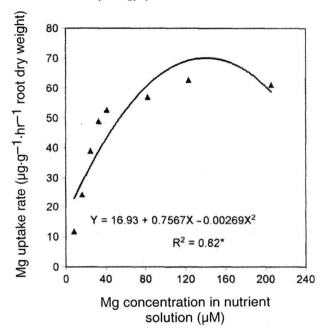

FIGURE 8.6. Uptake rate of Mg^{2+} by 100-days old rice plants in nutrient solution (uptake determined by continuous flow method). *Source:* Adapted from Fageria, 1973.

predominantly in the xylem vessels with water being the transporting agent. Xylem transport is driven by gradients in hydrostatic pressure in roots. Even though gradients in water potential and solute flow in the xylem from roots to shoots is unidirectional, long-distance transport of solutes in the phloem, which has living sieve tubes, is bidirectional. Increases in transpiration rates of plants enhance both uptake and translocation of mineral nutrients in the xylem (Marschner, 1995). Transpiration of water speeds movement of most elements in the xylem, but transpiration is not necessary for movement of ions in the xylem (Tanner and Beevers, 1990; Blevins, 1994). A common misconception is that xylem tissue is responsible for most mineral element movement in plants. Nevertheless, concentrations of many elements (e.g., K and P) were higher in phloem sap than in xylem sap (Blevins, 1994). Nitrate, Mn, and B are relatively phloem immobile.

Rate-limiting stages in the nutrient translocation chain are uptake, conduction, and release of ions by the symplast into roots, as the transpiration stream is usually capable of carrying high quantities of mineral nutrients. Even if velocity of the transpiration stream in the xylem is low, it is sufficient to move nutrients absorbed by roots to shoots. The other long-distance transport system is the phloem, which plays an equally important role in distribution of nutrients. Phloem and xylem tissues are linked at many sites along the transport system, particularly in roots and in nodes of stems (Larcher, 1995).

PHYSIOLOGICAL FUNCTIONS OF NUTRIENTS

Physiological functions of each essential nutrient is important and is discussed to provide information about the role each nutrient plays in metabolic processes and how each affects crop yield. Although some functions of each essential nutrient are provided in Table 8.2, some additional discussion about physiological production aspects are mentioned.

Nitrogen

Nitrogen is the most essential nutrient in determining yield potential of crops grown in intensive agricultural systems, and fertilizer N is a major input for crop production (Fageria and Baligar, 2005). Among the essential plant nutrients, N is quantitatively the most important nutrient required for plant growth. Nitrogen deficiency is more widespread among crop plants than deficiencies of other nutrients, and productivity of most agroecosystems is limited by availability of N. Very few soils can sustain satisfactory crop production without addition of N from some source. Except for legumes, most crops require large N fertilizer applications to obtain high crop yields. Hence, amounts and potential availability of N are key components of high-yielding crop management systems (Brye et al., 2002). Plants acquire relatively high amounts of N even though soils generally contain relatively low amounts. Most N in soils is in organic fractions, and most organic matter is located in upper soil horizons. Organic matter itself contains an average of about 50 g N/kg (w/w), but plow layers of cultivated soils (0 to 20 cm) usually contain only from 0.2 to 4 g·kg^{-1} N (w/w) (Barber, 1995). Nitrogen greatly influences leaf growth, leaf area duration, and photosynthetic rate per unit leaf area to control production of carbohydrates and other photosynthetic products (source activity), and influences numbers and sizes of vegetative and reproductive storage organs (sink capacity) (Engels and Marschner, 1995). With limited N, these physiological processes and plant organs are restricted, resulting in lower dry matter and/or grain/fruit yields. Tables 8.5 and 8.6 provide information about N influences on productivity of crops.

Effects of N availability on grain or dry matter yields can be assessed through crop physiological components such as radiation interception, radiation use efficiency, and dry matter partitioning to reproductive

TABLE 8.5. Dry matter yield of lowland rice grown with different rates of N (three years of trials).

N rate	Days after sowing[a]					
	22 (IT)	35 (AT)	71 (IP)	97 (B)	112 (F)	140 (PM)
	kg·ha^{-1}					
0	313	815	3065	5650	7694	5278
30	320	860	3709	6913	8953	6764
60	342	1230	3721	8242	11056	7294
90	374	1044	4164	8695	10758	7303
120	380	1229	4313	9570	13378	8215
150	452	1207	4893	10031	12745	8624
180	351	1294	5077	11290	13682	9060
210	351	1130	5841	10384	13490	9423
Mean	360	1101	4348	8847	11470	7745
F-test (year)	**	**	**	**	*	**
F-test (N)	**	**	**	**	**	*8
F-test(Y x N)	NS	NS	NS	NS	NS	NS
CV, %	24	22	21	22	19	17
R^2	0.5647NS	0.7555*	0.9719**	0.9676**	0.9416**	0.9647**

Source: Reproduced from Fageria and Baligar, 2001. Used with permission from Marcel Dekker.

[a] IT = initiation of tillering; AT = active tillering; IP = initiation of panicle; B = booting; F = flowering; PM = physiological maturity.

* Significant at P = 0.05 ** Significant at P = 0.01.

TABLE 8.6. Significance of F values derived from analysis of variance for lowland rice grain yield and yield components measured after three years of trials and grown with eight N rates.

Yield and yield Components	Years	N rates	Y × N	CV (percent)
Grain yield	**	**	*	10
No. of panicles	**	**	NS	9
1000-grain wt.	**	*	NS	6
Panicle length	**	**	NS	4
Spikelet sterility	**	NS	NS	17

Source: Reproduced from Fageria and Baligar, 2001. Used with permission from Marcel Dekker.

*, ** Significant at P = 0.05 and P = 0.01, respectively.

organs. Leaf area indexes, leaf area duration, crop photosynthetic rates, percent radiation interception, and radiation use efficiency are increased when adequate N is available to plants (Uhart and Andrade, 1995).

Major forms of N compounds within plants are proteins (including amino acids and chlorophyll) and nucleic acids. These compounds are major constituents of plant protoplasm, so shortages of N inhibit cell division and consequent reductions in growth. Shortages of protein compounds increase C/N ratios, which result in excess carbohydrate accumulation to increase contents of cellulose and lignin. Under such conditions, cell membranes thicken to increase lignified tissues and enhance early maturity (Ishizuka, 1978). Swank et al. (1982) reported that N plays a predominant role in photosynthesis and has close links between C and N metabolism in achieving maximum yields of maize. Murata and Matsushima (1975) attributed two major roles of N in rice to (1) establishment of yield capacity and (2) establishment and maintenance of photosynthetic capacity.

Nitrogen in agricultural soils is mainly absorbed as NO_3^- and NH_4^+ by crop plants. Plants can absorb both forms equally, and the N form absorbed is mainly determined by what form is abundant and/or accessible at the time. For instance, NO_3^- dominates in well-drained soils, while NH_4^+ dominants in anaerobic conditions or in cold climates. In addition, absorption of NH_4^+ occurs faster than absorption of NO_3^- (Gaudin and Dupuy, 1999). Under agricultural conditions, soil NO_3^- concentrations generally range between 0.5 and 10 mM (7 to 140 mg·kg^{-1}), while NH_4^+ concentrations are often 10 to 1000 times lower, reaching millimolar ranges unless exceptional conditions exist such as just after fertilization (Wiren, Gazzarrini, and Frommer, 1997). Nevertheless, autotrophic nitrification of NH_4^+ can be retarded by low soil pH and low soil temperatures, which favors NH_4^+ uptake by plants grown on acidic and cold soils. Accordingly, calcifuge plant species which are adapted to acidic soils often absorb NH_4^+ in preference to NO_3^- (Haynes and Goh, 1978). Hydroponic and greenhouse studies have indicated that wheat, maize, and several other crops grow best when given mixtures of NH_4^+ and NO_3^-, a concept termed mixed N nutrition, enhanced NH_4^+ supply (EAS), or enhanced NH_4^+ nutrition (EAN) (Goos et al., 1999). Bock et al. (1991) concluded that wheat had the physiological potential to respond favorably to EAN as opposed to exclusively NO_3^-. Wheat N uptake increased by 35 percent when one-fourth of N was supplied as NH_4^+, compared to 100 percent NO_3^- (Wang and Below, 1992). High-yielding maize genotypes were unable to absorb NO_3^- during ear development, which limited yields that would otherwise have increased if some N had been applied as NH_4^+ (Pan et al., 1984). Assimilation of NO_3^- required energy equivalents up to 20 ATP mol^{-1}·NO_3^-, whereas NH_4^+ assimilation required only 5 ATPmol^{-1}·NH_4^+ (Salsac et al., 1987). Potential energy savings for yield could be obtained if plants were supplied only NH_4^+ (Huffman, 1989). This concept has not been consistently observed, nor is it easy to conduct experiments on this given the nature of N-cycle dynamics (Raun and Johnson, 1999).

Uptake capacity for NO_3^- in barley cultivars or accessions from regions with low soil temperatures has been reported to be lower than for those originating from warm regions (Bloom and Chapin, 1981). In studies comparing N uptake of two barley cultivars of different origins, the cultivar adapted to cold soils not only had higher N uptake capacity at low temperatures but also preferred NH_4^+ compared to NO_3^- uptake (Bloom and Chapin, 1981). The cultivar adapted to warm soils had similar V_{max} values for both NO_3^- and NH_4^+ uptake. Besides differences among genotypes for preference of NH_4^+ or NO_3^- uptake (Smart and Bloom, 1988), increases in proportion of

NH_4^+ uptake were noted in genotypes of gramineous species supplied equimolar concentrations of NH_4^+ and NO_3^- as root-zone temperatures decreased (Macduff and Hopper, 1985; Clarkson, Hopper, and Jones, 1986). However, stronger inhibitory effects of low root-zone temperatures on NO_3^- compared to NH_4^+ uptake were not observed in solutions containing either NO_3^- or NH_4^+ alone (Macduff et al., 1987; Clarkson, Jones, and Purves, 1992). This indicated that uptake systems for NO_3^- were not temperature dependent but that enhanced supply of NH_4^+ decreased NO_3^- absorption more at low than at high root-zone temperatures (Clarkson, Jones, and Purves, 1992). Bufogle et al. (1998) reported that $(NH_4)_2SO_4$ and urea $[CO(NH_2)_2]$ were equally effective sources of N for flooded rice production. Gaudin and Dupuy (1999) reported that utilization of $CO(NH_2)_2$ instead of $(NH_4)_2SO_4$ made little difference because $CO(NH_2)_2$ was quickly hydrolyzed by the following reaction:

$$CO(NH_2)_2 + 2H_2O \leftrightarrow HCO_3^- + NH_4^+ + NH_3 \qquad (8.13)$$

Nitrogen uptake depends upon adequate supplies of carbohydrate translocated to roots from vegetative tissues. Nitrogen absorbed during reproductive growth and remobilization of N accumulated during vegetative growth both contribute to grain N (Moll, Jackson, and Mikkelsen, 1994). Nitrogen uptake in many plant species has been primarily governed by demand for N created by growth, provided sufficient N is available in the growth medium, and uptake by roots is not inhibited by unfavorable environmental factors.

Models of N uptake demand by shoots that regulate N uptake by roots have been proposed (Ben Zioni, Vaadia, and Lips, 1971; Touraine, Muller, and Grignon, 1992). Cooper and Clarkson (1989) proposed a model that integrates N uptake by roots with N utilization by shoots independent of location of N assimilation (roots or shoots) and form of N supplied (NO_3^- or NH_4^+). In this model, N uptake by roots is regulated by amount of shoot demand. With low demand, large fractions of N translocated to shoots via the xylem were retranslocated to roots via the phloem, and this served as the signal to reduce uptake. With high demand, large fractions of N translocated via the xylem were utilized for shoot growth and small fractions of retranslocation served as the signal to increase uptake. Nitrogen translocated to shoots in excess of requirements might be retranslocated to roots and serve as signals to decrease N uptake (Engels and Marschner, 1995).

Yield components affecting cereal yields are head yield per unit area, kernels per head, and kernel weight. Increasing N generally increases tiller numbers and kernels per head but reduces kernel weight (Pearman, Thomas, and Thorne, 1978; Bruckner and Morey, 1988). However, changes in one factor will often be reciprocated by changes in the other two factors, giving cereals the ability to compensate for stress without losing yield (Rasmussen, Rickman, and Klepper, 1997).

Knowledge of N uptake requirements at each stage of plant development is fundamental for effectual management of N inputs for achieving appropriate yield levels. Figures 8.7 and 8.8 show accumulation of N in flooded rice grown on an Inceptisol and maize grown on an Oxisol in central Brazil. In both crops, relationships between growth cycle and N accumulation were quadratic. For lowland rice, about 110 kg N/ha produced 5580 kg grain and 7740 kg shoot dry matter at harvest. The distribution of N within these plant parts was 46 kg N in shoots and 58 kg N in grain. The highest demand for N by lowland rice was between 20 to 100 days after sowing and was demonstrated by linear uptake. For maize, 210 kg N/ha produced 8150 kg grain and 13,700 kg shoot dry matter at harvest, with 82 kg N in

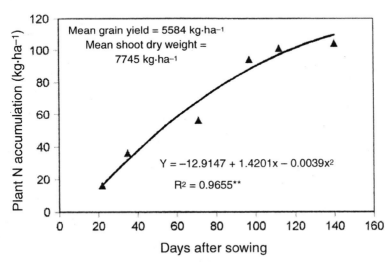

FIGURE 8.7. Relationship between flooded rice age and N accumulation in shoot and grain. *Source:* Reproduced from Fageria, 1999.

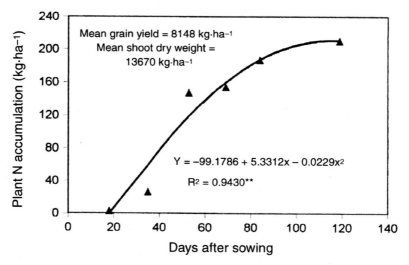

FIGURE 8.8. Relationship between maize age and N accumulation in shoot and grain. *Source:* Adapted from Fageria and Baligar, 1999.

shoots and 128 kg N in grain. Maize had highest N uptake demand from 20 to 90 days after planting. Considerable, and in many cases most, N is absorbed by small grain crops before anthesis (Austin et al., 1977). For instance, remobilization and translocation of N from vegetative to reproductive tissue accounted for more than 75 percent of grain N in oat (Cataldo et al., 1975) and wheat (Dalling, Boland, and Wilson, 1976). Wheat and barley accumulated 80 to 90 percent of total plant N before anthesis, but dry matter produced before anthesis contributed little N to grain (Austin et al., 1980; Ma, Dwyer, and Smith, 1994). Grain protein for many cereal crops depends on plant ability to transfer N from vegetative tissues to grain.

Nitrogen harvest index (NHI = N accumulation in grain/N accumulation in grain and straw) measures N partitioning to provide indications on how effectively plants utilize acquired N for grain production. Genetic variability for NHI exists within small grain species (Halloran, 1981). The NHI values ranged from 57 to 74 percent in spring wheat (Loffler and Busch, 1982), 57 to 86 percent in durum wheat (Desai and Bhatia, 1978), 25 to 51 percent in wild oat, and 42 to 67 percent in common oat (Fawcett and Frey, 1982). High NHI values are associated with efficient utilization of N (Fawcett and Frey,

1983) and grain protein yield (Welch and Yong, 1980). Selecting for high NHI values may give simultaneous improvement for grain yield and grain protein percentages or increases in grain yield with constant grain protein percentages. Grain yield and NHI values were positively associated in three wheat populations (Loffler and Busch, 1982). Selection for high NHI values increased grain yields 4.7 to 11.3 percent in hard red spring wheat (Loffler and Busch, 1982). Rattunde and Frey (1986) reported that NHI values were positively associated with mean oat grain yield and response of oat grain yields to environmental productivity, but inversely related to mean straw yield. Efficiency of N translocation was depressed with high levels of N (Fawcett and Frey, 1982; Halloran, 1981). Table 8.7 shows NHI values for lowland rice grown with different timing of N fertilization. The NHI values were significantly higher when N was applied at the

TABLE 8.7. Nitrogen harvest index (NHI) in flooded rice under different N timing treatments.

Treatment	NHI (percent)
T_1	47a
T_2	44a
T_3	29ab
T_4	12c
T_5	36ab
T_6	15bc
F-test	***
CV (%)	15

Source: Adapted from Fageria and Baligar, 1999b.

Note: Means followed by same letter within column are not significantly different at P = 0.05 by Tukey's test.

T_1 = total N application at sowing; T_2 = 1/3 N application at sowing + 1/3 N application at active tillering + 1/3 N application at panicle initiation; T_3 = 1/3 N application at sowing + 1/3 N application at panicle initiation + 1/3 N application at booting; T_4 = 1/3 N application at sowing + 1/3 N application at panicle initiation + 1/3 N at flowering; T_5 = zero N application at sowing + ½ N application at the initiation of tillering + ½ N application at panicle initiation, and T_6 = zero N application at sowing + 1/3 N application at initiation of tillering + 1/3 N application at booting + 1/3 N application at flowering.

***Significant at P = 0.01.

T_1, T_2, and T_5 treatments (highest grain-producing treatments) compared to the T_4 and T_6 treatments (lowest grain-producing treatments). These differences indicated that partitioning efficiency of N depends upon when N is acquired. Physiological processes that have important influences on N partitioning efficiency were established before the booting and flowering growth stages for rice. Westcott et al. (1986) reported NHI values to range from 68 to 74 percent for lowland rice, and Dingkuhn et al. (1991) reported NHI ranges from 60 to 72 percent for three semi-dwarf rice cultivars differing in growth duration.

Energy and molecular structure required for incorporation of N into plant compounds depend on C metabolism, especially photosynthesis (Larcher, 1995). Hence, increases in plant dry matter are chiefly limited by N supply. Overfertilization with N results in very rapid growth of shoots, inadequate supporting tissues, poorly developed root systems, delayed reproductive development, insufficient resistance to climatic stress, and greater susceptibility to parasitic fungi and insects (Larcher, 1995).

Efficiency in utilization of absorbed nutrients within plants is as important as efficiency of nutrient absorption when evaluating plant biomass production relative to nutrient supply (Siddiqi and Glass, 1981). Nutrient uptake efficiency and nutrient utilization efficiency are governed by different physiological mechanisms. Physiological factors associated with N use efficiency include

1. Root proliferation (absorption efficiency, selective ion absorption, and tolerance to NH_4^+);
2. NO_3^- uptake efficiency (uptake induction, stimulation/inhibition, and growth stage partitioning);
3. N translocation efficiency (nitrate reductase regulation, remobilization efficiency, and protein concentration);
4. nitrate reductase enzyme efficiency (induction, rate of activity, accumulation, and distribution); and
5. NH_4^+ tolerance (NH_4^+ root assimilation, enzyme metabolism, and K accumulation) (Duncan and Carrow, 1999).

Each stage of N assimilation is under genetic control (Duncan and Carrow, 1999). Nitrate is reduced to NH_3 by two soluble enzymes, i.e., nitrate reductase (NR) and nitrite reductase (NiR). These metalloenzymes require the cofactors Mo-pterin for NR and Fe-hydrochlorine (siroheme) for NiR. The NR enzyme is localized in cytosols, whereas NiR is located in leaf and root tissue plastids (Duncan and Carrow, 1999).

Rice hybrids have mean yield advantages of 10 to 15 percent over inbred genotypes (Yang and Sun, 1988). This yield advantage by hybrid rice over inbred rice is mainly related to uptake and N use efficiency. Total N uptake by hybrid rice shoots is greater than that of conventional cultivars, especially from transplanting to tillering and from panicle emergence to grain-filling growth stages (Yang, 1987). Hybrid rice absorbs about 15 to 20 percent of its total accumulated N after heading, and responds well to N applications at flowering. Nitrogen absorption in conventional genotypes after heading is about 6 to 7 percent of total uptake (Yang, 1987). Hybrid rice had greater N efficiencies (defined as grain yield per unit N fertilizer applied) than conventional lines (Lin and Yuan, 1980; Yang, 1987). This increased N efficiency was not due to greater internal N uptake in dry matter production (defined as unit dry matter produced per unit N accumulated in plant) (Yang, Zhang, and Ni, 1999). Higher N recovery efficiency in hybrid rice has been related to greater root N absorption potential, greater shoot N use capacity (N demand by shoots [i.e., how much and how fast shoots can use N]), and greater N remobilization efficiency (N translocation to grain [i.e., NHI]). These processes are major factors causing higher N efficiency in hybrid rice. Nitrogen use efficiency in maize has been strongly associated with numbers of ears per plant (Anderson, Kamprath, and Moll, 1985). Evidence indicates that prolificacy in maize is associated with both greater N uptake and more efficient utilization of accumulated N (Anderson, Kamprath, and Moll, 1984; Moll, Kamprath, and Jackson, 1987).

Tracer techniques using ^{15}N have been used extensively to identify and measure different absorption and loss pathways to assess use efficiency of fertilizer N. Nitrogen use efficiency by crop plants has been defined in five different ways (Fageria, 1992). Each of the five N use efficiency definitions were calculated in lowland rice using the following formulas (Fageria, Baligar, and Jones, 1997):

1. Agronomic efficiency:

$$(AE) = (G_f - G_u / N_a) = kg \cdot kg^{-1} \tag{8.14}$$

where G_f = grain yield of fertilized plots (kg), G_u = grain yield of unfertilized plots (kg), and N_a = quantity of N applied (kg).

2. Physiological efficiency:

$$(PE) = (Y_f - Y_u / N_f - N_u) = kg \cdot kg^{-1} \qquad (8.15)$$

where Y_f = total biological yield (grain plus straw) of fertilized plots (kg), Y_u = total biological yield of unfertilized plots (kg), N_f = nutrient accumulation of fertilized plots (kg), and N_u = nutrient accumulation of unfertilized plots (kg).

3. Agrophysiological efficiency:

$$(APE) = (G_f - G_u / N_{tf} - N_{tu}) = kg \cdot kg^{-1} \qquad (8.16)$$

where G_f = grain yield of fertilized plots (kg), G_u = grain yield of unfertilized plots (kg), N_{tf} = N accumulation by straw and grains in fertilized plots (kg), and N_{tu} = N accumulation by straw and grains in unfertilized plots (kg).

4. Apparent recovery efficiency:

$$(ARE) = (N_f - N_u / N_a) \times 100 = \% \qquad (8.17)$$

where N_f = N accumulation by total biological yield (straw plus grain) in fertilized plots (kg), N_u = N accumulation by total biological yield (straw plus grain) in unfertilized plots (kg), and N_a = quantity of N applied (kg).

5. Utilization efficiency:

$$(UE) = PE \times ARE = kg\ kg^{-1} \qquad (8.18)$$

Each N use efficiency definition for lowland rice decreased with increasing N rates, except PE (Table 8.8). Across different N rates, AE was 23 kg grain produced per kg N applied; PE was 146 kg biological yield (straw plus grain) per unit of N accumulated; APE was 63 kg grain produced per kg of N accumulated in grain and straw across N rates; ARE was 39 percent; and UE was 58 kg grain produced per kg of N utilized. Decreasing N use efficiency with increasing N rates seemed to indicate that plants grown at lower N rates actually produced lower yields due to N deficiency. The apparent high efficiency of N use at lower N rates resulted from dilution of N by

TABLE 8.8. Nitrogen use efficiencies of flooded rice grown on an Incepitsol with different N rates (three years of trials).

N rate (kg·ha⁻¹)	Agronomic efficiency (kg·kg⁻¹)	Physio-logical efficiency (kg·kg⁻¹)	Agrophy-siological efficiency (kg·kg⁻¹)	Apparent recovery efficiency (percent)	Utilization efficiency (kg·kg⁻¹)
30	35	156	72	49	76
60	32	166	73	50	83
90	22	182	75	37	67
120	22	132	66	38	50
150	18	146	57	34	50
180	16	126	51	33	42
210	13	113	46	32	36
R^2	0.9292*	0.6162NS	0.8723*	0.8226*	0.9025**

Source: Reproduced from Fageria and Baligar, 2001. Used with permission from Marcel Dekker.

* Significant at $P = 0.05$ **Significant at $P = 0.01$.

limited amounts of grain or dry matter yields. In experiments with nutritional variables, plants grown with the lowest nutrient concentrations will inevitably have the highest utilization quotients because of dilution effects (Jarrell and Beverly, 1981).

The AE in lowland rice grown in the tropics has been reported to be in the range of 15 to 25 kg grain produced per kg applied N (Yoshida, 1981). Results reported in Table 8.8 are within these ranges. Higher PE (146 kg·kg⁻¹) compared to APE values (63 kg·kg⁻¹) across N rates may have been due to inclusion of dry matter in calculating efficiency values. Singh et al. (1998) reported APE values of 64 kg grain per kg of N uptake and AE values of 37 kg grain per kg of N in 20 lowland rice genotypes. The ARE values of 39 percent across N rates was considered to be quite low.

Percentages of N recovery vary with soil properties, methods, amounts, and timing of fertilizer applications, and management practices. Percentage values usually range from 30 to 50 percent for crops grown in the tropics (Prasad and De Datta, 1979). Studies conducted in the southern United States relative to influence of different application timings and N management strategies had N recovery efficiency of 17 to 61 percent of applied N at maturity for rice (Norman, Wells, and Moldenhauer, 1989; Westcott et al., 1986). Using 20 lowland rice cultivars, Singh et al. (1998) reported N recovery efficiency of 37 percent. The low N recovery efficiency for lowland rice may be related to N losses from soil via nitrification-denitrification processes, NH_3 volatilization, or leaching (Craswell and Vlek, 1979). Efficiency of N utilization for grain production in the tropics is about 50 kg grain per kg N absorbed, and this efficiency appears to be almost constant regardless of rice yields achieved (Yoshida, 1981). Efficiencies of 58 kg grain per kg N absorbed across N rates were within this limit (Fageria and Baligar, 2001). The amount of accumulated N needed to produce one ton of rough rice grain was 15 to 17 kg N (Wada, Shoji, and Mae, 1986). This value took into consideration that an average yield of 5 to 6 t·ha^{-1} and 19 kg N was needed for higher yielding rice genotypes. However, absolute amounts of N absorbed are not the only factors to be considered when growing high yielding rice genotypes. Rather, N supply patterns and uptake processes of N throughout the whole plant life cycle are equally important. This is because formation (number, size, or extent) of each yield component strongly depends on amount of N supply at each crucial growth stage for each yield component (Mae, 1997).

Nitrogen use efficiency and N translocation to grain of upland rice, common bean, maize, and soybean grown on an Oxisol in central Brazil were determined (Table 8.9). Nitrogen use efficiency was higher in cereals compared to legumes. Dry matter production efficiency per unit N absorbed was reported to be lower in legumes than in cereals (Nakamura et al., 1997). This was ascribed to low growth efficiency of shoots, but not to nodulation and low growth efficiency of grain (Shinano, Osaki, and Tadano 1995). Shinano, Osaki, and Tadano (1994) suggested that when $^{14}CO_2$ was fed to leaves of rice and soybean, the amounts of $^{14}CO_2$ released from soybean leaves were higher than for rice grown under similar light conditions because of higher photorespiration in soybean. In addition, photosynthesized C in soybean leaves was more distributed into organic compounds and amino acids than for rice (Shinano, Osaki, and Tadano, 1994). Accordingly, it has been hypothesized that photosynthesized C distribution mechanisms appear to be different between cereal andlegume crops. For example, photosynthesized C is actively

TABLE 8.9. Nitrogen use efficiency and N translocation in grains of four crop species.

Crop species	N use efficiency (kg grain/kg N accumulated[a])	N translocation to grain (percent of total uptake[b])
Upland rice	30	50
Common bean	19	78
Maize	39	61
Soybean	9	61

Source: Adapted from Fageria, 2001b.

[a] Total N accumulated in the straw and grain.

[b] Total N uptake by straw and grain.

distributed into tricarboxylic and amino acid pools for legumes grown under light conditions resulting in high respiratory rates, whereas high proportions of photosynthesized C is distributed into carbohydrate pools for cereal crops. Photorespiratory activity is also higher for legumes than for cereals (Nakamura et al., 1997). On the other hand, N translocation into grain was higher for legumes compared to cereals. Duncan (1994) summarized physiological and morphological factors associated with N use efficiency, and these factors included root proliferation, uptake rate per unit root mass or area, ion concentration (balance among influx, efflux, and growth), translocation, and utilization.

Management Strategies to Improve N Nutrition

Soil-crop management affects soil N balances, which have direct bearings on soil productivity. Conventional N fertilizers are applied as NH_3, NH_4^+, NO_3^-, $CO(NH_2)_2$ (urea), or combinations of two or more of these forms. However, plants absorb N from soil as NH_4^+ and NO_3^-, which are the available forms of N. Uptake of fertilizer N by crops is generally relatively low, commonly ranging from 25 to 50 percent of that applied (Peterson and Frye, 1989). Worldwide, N recovery efficiency for cereal crops (wheat, rice, barley, maize, rye, oat, and millet) is approximately 33 percent (Raun and Johnson, 1999).

With increasing costs of N fertilizer (e.g., increased costs of natural gas from which N fertilizers are usually manufactured), the unaccounted 67 percent losses have been estimated to be worth more than $20 billion annually (Raun and Johnson, 1999; Raun et al., 2002). Improved fertilizer N usage in the world is needed, and some methods used to enhance this have been to use improved N, efficient crop cultivars, and improved management practices (Raun and Johnson, 1999). Cooperative studies are ongoing between the Consultative Group on International Agricultural Research (CGIAR) linked with advanced research programs at universities and research institutes worldwide to refine agricultural practices in both developed and developing countries. Leaching of N, denitrification, NH_3 volatilization, surface runoff, immobilization of N, N transformation factors and sources, placement, adequate levels, and timing of management practices influence N fertilizer efficiency. Some of the best N management practices that might be adopted to improve N use efficiency and consequently improve crop production are

1. optimum rates of application;
2. appropriate sources;
3. appropriate methods of application;
4. appropriate times of application;
5. adequate plant densities;
6. balanced nutrient supplies;
7. use of green manures;
8. use of legumes in crop rotations;
9. use of crop and animal wastes;
10. use of slow N release fertilizers;
11. use of nitrification inhibitors;
12. adequate moisture supplies;
13. control of insects, diseases, and weeds;
14. control of erosion; and
15. use of efficient cultivars.

Phosphorus

Principal forms of P in soils are Ca-phosphates, adsorbed phosphates, occluded phosphates, and organic phosphates (Mengel, 1985). Proportions of these fractions on total soil phosphate differ greatly for various soil types. According to Walker and Syers (1976),

Ca-phosphates decrease and occluded phosphates increase during pedogenesis. Thus, highly weathered acidic soils are rich in occluded phosphates while Ca-phosphate contents are almost nil. Organic phosphates may differ considerably according to organic matter content of soils. In highly weathered podsolic soils, proportions of organic phosphate may be as low as 5 percent of total soil phosphate, while 90 percent of total soil phosphate may be organic in humus-rich alpine soils (Dalal, 1977). Forms of phosphate in soil solutions also vary according to pH. In dilute solutions, phosphoric acid dissociates as follows (Stevenson, 1986):

$$H_3PO_4 \xleftrightarrow[+H^+]{-H^+} H_2PO_4^- \xleftrightarrow[+H^+]{-H^+} HPO_4^{2-} \xleftrightarrow[+H^+]{-H^+} PO_4^{3-} \quad (8.19)$$

In pH ranges of most soils (5 to 8), the amounts of undissociated H_3PO_4 and trivalent PO_4^{3-} are negligible. Thus, essentially all phosphate consumed by plants is in $H_2PO_4^-$ and HPO_4^{2-} forms. At pH 6, about 94 percent of phosphate occurs as $H_2PO_4^-$, but the percentage drops to 60 percent at pH 7 (Stevenson, 1986).

Phosphorus is required for synthesis of phospholipids, nucleotides, ATP, glycophosphates, and other phosphate esters. Phosphorus deficiency decreases unit leaf area photosynthetic activity for several plant species (Morison and Batten, 1986; Israel and Rufty, 1988). When P supply is inadequate, cell division slows and whole plants become dwarfed, as in N deficiency. Unlike N, P is not a constituent of chlorophyll so that concentrations of chlorophyll in leaves remain comparatively high and leaves often become darker green when P is deficient. For legumes, P is important for effective symbiotic N_2-fixation. Other visual symptoms of P deficiency are that leaves may become reddish, purplish, or orangish from accumulation of anthocyanins (Hewitt, 1963) and leaves may turn straw color and die (Clark and Baligar, 2000). Adequate P increases numbers of panicles or heads in cereals and numbers of pods in legumes grown on Oxisols (Fageria, Baligar, and Jones, 1997). Currently, considerable cultivated lands throughout the world are deficient in available phosphate, and this in a major reason that food production in many countries is limited (Wasaki et al., 1997).

Plants differ widely in ability to grow in soils with low P, and this has been attributed to several factors including differences in root system morphology and root hair density (Randall, 1995) and root associations with mycorrhiza (Clark and Zeto, 2000). One major benefit of mycorrhizal associations with roots is ability of these fungi to enhance acquisition of P (Clark and Zeto, 2000). Some mycorrhizal isolates are also very effective in enhancing P acquisition by plants grown in low pH soils (e.g., pH 4.0) where P availability is usually low (Clark, Zeto, and Zobel, 1999; Clark, Zobel, and Zeto, 1999). Mycorrhizal roots also have ability to enhance other essential nutrients and alleviate some elemental toxicities (Clark, Zobel, and Zeto, 1999).

Another important factor associated with plant ability to grow with low available soil P is the formation of root exudates that enable plants to make insoluble or low solubility P compounds such as Fe- or Al-phosphates available for plant acquisition and probably at the same time protecting against Al toxicity by chelation. White lupin and pigeon pea are well adapted to acidic P-deficient soils (Hocking et al., 1997). White lupin forms proteoid roots (bottlebrush-like clusters of rootlets covered with dense mats of root hairs), which are considered to be responsive to low P availability (Marschner, 1995). Recent research showed that proteoid roots occur on white lupin at concentrations of inorganic P normally found in agricultural soils (Hocking et al., 1997). Proteoid roots of white lupin secrete large quantities of citric acid, which may solubilize insoluble P compounds (Gardner, Barber, and Parbery, 1983) to increase P uptake by plants. For pigeon pea, secretion of piscidic, malonic, and oxalic acids appears to be the mechanism by which this species is able to release P from Fe- and Al-phosphates (Otani, Ae, and Tanaka, 1996) and protect against Al (or Fe) by chelation.

Phosphorus efficiency in crop plants includes absorption (uptake) efficiency, translocation (partitioning), and internal utilization (redistribution) efficiency. Phosphorus use efficiency and P translocation to grain of four crop species is presented in Table 8.10. Phosphorus use efficiency was much higher in cereals such as rice and maize than in legumes such as common bean and soybean. In addition, higher portions of accumulated P were translocated to grain in each of these crop species than soybean. However, maximum translocation of P into grain was noted for common bean followed by maize, rice, and soybean. Figures 8.9 and 8.10 show P accumulation by maize, upland rice, soybean, and common bean during their growth cycles on an Oxisol in central Brazil. The demand of P by maize was high throughout its

TABLE 8.10. Phosphorus use efficiency and P translocation in grains of four crop species.

Crop species	P use efficiency (kg grain/kg P accumulated[a])	P translocation to grain (percent of total uptake[b])
Upland rice	360	70
Common bean	251	79
Maize	361	76
Soybean	78	54

Source: Adapted from Fageria, 2001b.

a Total P accumulated in the straw and grain.

b Total P uptake by straw and grain.

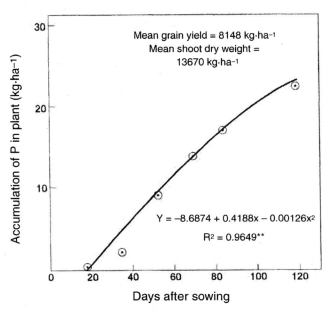

FIGURE 8.9. Accumulation of P in maize during growth cycle when grown on an Oxisol in central Brazil. *Source:* Adapted from Fageria, 1999.

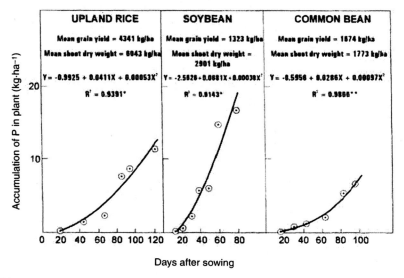

FIGURE 8.10. Accumulation of P in upland rice, soybean, and common bean during their growth cycles. *Source:* Adapted from Fageria, 1999.

growth cycle. At harvest, maize accumulated 23 kg P and produced grain yields of 8150 kg·ha⁻¹ and straw yields of 13,670 kg·ha⁻¹. Of the 26 kg P accumulated, 24 percent was retained in straw and the remaining 76 percent translocated to grain. In upland rice, P uptake was slow the first 40 days of growth and was essentially linear during that time. About 11 kg P accumulated in grain and straw of upland rice at maturity, and plants produced 4340 kg·ha⁻¹ grain and 6040 kg·ha⁻¹ straw. Of the 11 kg P accumulated in plants, 16 percent was retained in straw and the remaining 84 percent was translocated to grain. Accumulation of P in soybeans and common beans was slow the first 40 days of growth but increased as plants aged (Figure 8.10). About 17 kg P accumulated in grain and straw of soybean at harvest. Of this amount, 46 percent was retained in straw and the remaining 54 percent translocated to grain. Soybeans produced 1320 kg·ha⁻¹ grain and 2900 kg·ha⁻¹ straw. Common beans accumulated about 7 kg P in grain and straw and produced 1670 kg·ha⁻¹ grain and 1770 kg·ha⁻¹ straw. Of 7 kg P, 21 percent was retained in straw and the remaining 79 percent translocated to grain. From these results, it was concluded that major portions of accumulated P were translocated to grain in both cereals and legumes. Various soils and plant mechanisms and processes that contribute to genetics of P use efficiency are provided in Exhibit 8.1.

Management Strategies to Improve P Nutrition

Since P acquisition by plants rarely exceeds 20 percent of total fertilizer P applied, improving plant potential for greater P efficiency is important (Friesen et al., 1997). Low recovery of P is related to P fixation by active Al and Fe in neutral to acidic soils. The amount of active Al and Fe is much greater than that of P applied to soils. Hence, P concentrations in soil solutions are much lower than those predicted from dissolution equilibrium of pure precipitates of phosphate minerals. Adoption of appropriate management practices is important for improving P efficiency in crop production. Most of these practices are related to modification of soil environment to improve P utilization efficiency by crops. Important management practices are

1. optimum rates of application;
2. appropriate sources;
3. appropriate methods of application;
4. optimum soil pH;
5. adequate moisture supplies;
6. consideration to residual value of P fertilizers;
7. addition of plant organic matter;
8. placement of KCl with P fertilizers;
9. addition of synthetic iron and Al chelating agents;
10. adequate soil aeration;
11. use of P-efficient and Al-tolerant cultivars;
12. addition of animal manures; and
13. inoculation with mycorrhiza.

Potassium

Physiological functions of K in plants are enzyme regulation, osmotic regulation and movement, formation of carbohydrates, nucleic acids, and proteins, photosynthesis, enhancement of rooting, early establishment, heat/cold/drought tolerance, wear resistance, and

EXHIBIT 8.1. Soil and plant mechanisms, processes, and other factors that influence genotypic differences in P use efficiency.

A. Phosphorus acquisition
1. Diffusion and mass flow (buffer capacity, ionic concentration, ionic properties, tortuosity, soil moisture, bulk density, temperature)
2. Root morphological factors (numbers, length, root hair density, root extension, root density)
3. Physiological [root:shoot ratios, root microorganisms (e.g., mycorrhizal fungi and bacteria), nutrient status, water uptake, nutrient influx and efflux, rate of nutrient transport in roots and shoots, affinity for uptake (Km), threshold concentration (Cmin)]
4. Biochemical (enzyme secretion as phosphate, complexing compounds, phytosiderophores), proton exudates, organic acid production (e.g., citric, trans-aconitic, malic acid exudates)

B. Phosphorus movement in roots
1. Transfer across endodermis and transport within roots
2. Compartmentation/binding within roots
3. Rate of nutrient release to xylem

C. Phosphorus accumulation and remobilization in shoots
1. Demand at celular level and storage in vacuoles
2. Retransport from older to younger leaves and from vegetative to reproductive parts
3. Rate of complexing agents in xylem transport

D. Phosphorus utilization and growth
1. Metabolism with reduced nutrient concentration in tissue
2. Lower element concentration to support structure, particularly stems
3. Elemental substitution (e.g., Na for K function)
4. Biochemical nitrate reductase for N use efficiency, glutamate dehydrogenase for N metabolism, peroxidase for Fe efficiency, pyruvate kinase for K deficiency, metallothionein for metal toxicities

E. Other factors
1. Soil factors
   a. Soil solution (ionic equilibria, solubility, precipitation, competing ions, organic ions, pH, phytotoxic ions)

   b. Physico-chemical properties of soil (organic matter, pH, aeration, structure, texture, compaction, soil moisture)
2. Environmental effects
   a. Intensity and quality of light (solar radiation)
   b. Temperature
   c. Moisture supply
3. Plant diseases, insects, and allelopathy

Source: Compiled from various sources by Baligar, Fageria, and He, 2001.

maintenance of crop quality (Duncan and Carrow, 1999). An important function of K is to keep plant cells turgid by controlling osmotic processes in cell sap. This abundant, nontoxic, monovalent cation is important in maintaining ionic balances or electrical neutrality in plants. Its role in ionic balances may be one reason that K is found in higher concentration in both xylem and phloem saps. Evidence is mounting that K is a key component in mass flow of materials in the phloem (Lang, 1983; Vreugdenhil, 1985).

Potassium regulates normal functioning of metabolic processes within cells. One principal role of K is as an activator of numerous enzymes. Potassium is essential for protein synthesis (Peoples and Koch, 1979). Potassium deficiency decreases rates of net photosynthesis and translocation, and increases rates of dark respiration. Potassium also affects plant growth mainly by affecting cell extension. Adequate K supplies result in greater synthesis of cell wall materials and creating cell walls that are thicker to help crops resist lodging and attacks by diseases and insects. Morphogenic effects of K include increased numbers of storage cells to affect sink capacity of grains (Shuman, 1994).

Table 8.11 points out some of the effects that K has on grain yield and yield components of flooded rice grown on an Inceptisol in central Brazil. Yield and yield components were significantly affected by K fertilization. Correlations between grain yield of upland rice and five K use efficiency traits is presented in Figure 8.11. Methods of calculating these efficiencies have been listed in the section on N. Each K use efficiency trait was correlated with grain yield. However,

TABLE 8.11. Grain yield and yield components of flooded rice (cultivar IR 36) as influenced by K fertilization.

K level (g/pot)	Grain yield (g/pot)	Panicles (number/pot)	Panicle grain (number/panicle)	1000-grain wt. (g)
0.0	11	50	36	15
0.37	18	51	43	17
0.74	18	57	47	18
1.12	22	58	48	18
1.49	21	56	47	18
Regression	Q*	L*	Q*	Q*

Source: Reproduced from Fageria, 1989.

*Significant at P = 0.05.

Q = quadratic equation L = linear equation.

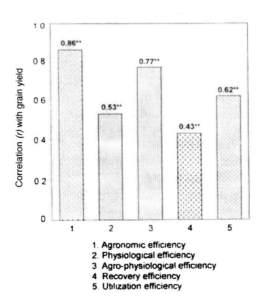

1. Agronomic efficiency
2. Physiological efficiency
3. Agro-physiological efficiency
4. Recovery efficiency
5. Utilization efficiency

FIGURE 8.11. Correlation values *(r)* for grain yield of upland rice relative to K use efficiency. *Note:* **Significant at P = 0.01. *Source:* Reproduced from Fageria, 1998.

AE values had the highest correlation, followed by APE and UE, while ARE values had lowest correlations with grain yield.

Deficiency of K may lead to lodging, increased water stress, reduced photosynthetic rates, and decline in plant and grain quality (Richardson and Croughan, 1989). Potassium is important in many plant functions and can influence susceptibility of plants to diseases. K can not only increase resistance of plant tissues but may also reduce fungal populations in soil, reduce fungal pathogenicity, and promote more rapid healing of injuries (Huber and Arny, 1985). Potassium is absorbed by crop plants in large quantities and makes up about 80 percent of total cationic ion species in the phloem (Mengel and Kirkby, 1978). Potassium is highly mobile in the phloem. Potassium utilization is efficient in the sense that it is readily redistributed from older leaves to younger leaves and growing organs. As a result, symptoms of K deficiency usually appear first in older leaves.

Potassium deficiency is not as widespread as N and P deficiencies. It has been postulated that the widespread K deficiency in the U.S. Cotton Belt has been related to early maturing, high yielding, fast fruiting cotton cultivars creating greater K demands than what the plant root system is capable of supplying (Oosterhuis, 1995). Cotton appears to be more sensitive to low K availability than most other major field crops, and often symptoms of K deficiency occur on plants grown in soils not considered to be K deficient (Cope, 1981; Oosterhuis and Bednarz, 1997). Kim and Park (1973) reported that split applications of K increased rice yields for plants grown on heavy sulfate soils in South Korea. From experiments conducted in Japan and Taiwan, Su (1976) reported beneficial effects of split applications of K on rice. Research covering 25 years from Japan indicated remarkable grain yield responses to K topdressings on podzolized and poorly drained rice fields (Nogurchi and Sugawara, 1966). Fageria (1989) reported significant yield increases in upland and lowland rice cultivars grown on Oxisols and Inceptisols in Brazil. Similar benefits of K topdressings for rice have been reported in India (Singh and Singh, 1979).

Potassium efficiency in crop plants is linked to root growth and morphology, uptake (influx) efficiency, efflux, translocation, and utilization efficiency. Gene-controlled K transport carrier synthesis has been responsible for variation in maximum K uptake efficiency among crop species (Duncan and Carrow, 1999). Information about

K use efficiency and translocation to grain of four crop species is presented in Table 8.12. Potassium use efficiency was higher in cereals than in legumes. This trait was highest in maize and lowest in soybean. However, K translocation to grain in cereals was low compared to legumes. In upland rice, only 8 percent of accumulated K in plants was translocated to grain, whereas translocation of K was 15 percent in maize. In common bean, 46 percent of absorbed K was translocated to grain and 36 percent translocated to grain in soybean. This means that major portions of accumulated K have been retained in straw or stems of cereal and legume crops. This means that recycling of crop straw can supply substantial amounts of K to growing crops. Potassium uptake at different growth stages is presented for soybean in Figure 8.12. Accumulation of K was almost linear from 27 to 120 days of plant growth before leveling off. At harvest, soybean accumulated 60 kg K in straw and grain and produced 1320 kg·ha^{-1} grain and 2900 kg·ha^{-1} straw. Potassium accumulation in upland rice and common bean is presented in Figure 8.13. In upland rice, accumulation of K was highest from 20 days after sowing to maturity. In common bean, uptake was linear up to about 60 days of growth before leveling off. Upland rice accumulated 137 kg K at harvest and produced 4340 kg·ha^{-1} grain and 6040 kg·ha^{-1} straw. In common bean, K accumulation at harvest was 68 kg, producing 1670 kg·ha^{-1} grain and 1770 kg·ha^{-1} straw.

TABLE 8.12. Potassium use efficiency and K translocation in grains of four crop species.

Crop species	K use efficiency kg grain/kg K accumulated[a]	K translocation to grain percent of total uptake[b]
Upland rice	25	8
Common bean	24	46
Maize	64	15
Soybean	19	36

Source: Adapted from Fageria, 2001b.

a Total K accumulated in straw and grain.

b Total K uptake by straw and grain.

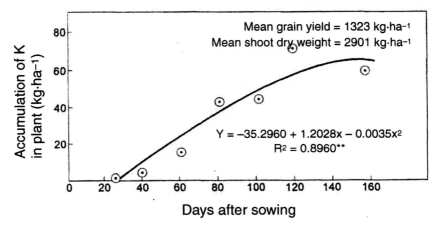

FIGURE 8.12. Accumulation of K in soybeans crop during its growth cycle. *Source:* Adapted from Fageria, 1999.

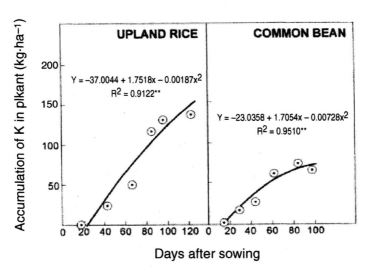

FIGURE 8.13. Accumulation of K in upland rice and common bean during their growth cycles. *Source:* Adapted from Fageria, 1999.

Management Strategies to Improve K Nutrition

As for N and P, appropriate management practices can enhance K use by crop plants and consequently help in efficient crop production. The following important practices could be adopted:

1. adequate rates of application;
2. appropriate time of application;
3. appropriate methods of application;
4. use of efficient cultivars;
5. incorporation of crop residues;
6. adequate moisture supplies; and
7. liming acidic soils,

Calcium and Magnesium

Calcium is essential for healthy plant growth and has two main functions. One is similar to that of K relative to regulation of osmotic pressure of cell sap to maintain turgidity. The other function is as a constituent of plant fabrics. Calcium is essential for growth of plant meristems, and particularly for proper functioning of root tips (Ishizuka, 1978). Calcium is a constituent of calcium pectate, which is found in the middle lamella of cell walls. Once Ca enters into middle lamella of cell walls, it has finished its metabolic function. Movement of Ca from cell walls is irreversible, so Ca is essentially immobile. Consequently, if plants suffer from shortages of Ca at any stage of growth, newly growing plant parts cannot receive Ca from older tissues. Thus, Ca deficiency symptoms normally appear in new tissues that are growing most vigorously (Ishizuka, 1978). High amounts of Ca in plants are a result of high levels of Ca in soil solution rather than efficiency of uptake. The Ca contents of monocots are generally lower than that of dicots (Table 8.13). Calcium activates some enzymes, particularly those associated with membranes (Mengel and Kirkby, 1978). However, Ca influences activities of only a few enzymes compared to Mg (Marschner, 1995).

One of the most important and best-known functions of Mg is as a constituent of the chlorophyll molecule. Along with K, Mg is also a component of ribosomes and chromosomes (Shuman, 1994). Magnesium also activates many enzymes associated with phosphorylation (Marschner, 1995). Considerable amounts of Mg in plants act to

TABLE 8.13. Calcium and Mg concentrations in shoots of upland rice and common bean at different growth stages.

Crop age (days)	Upland rice		Crop age (days)	Common bean	
	Ca (g·kg⁻¹)	Mg (g·kg⁻¹)		Ca (g·kg⁻¹)	Mg (g·kg⁻¹)
19	3.2	2.7	15	15	4.1
43	4.0	2.8	29	21	4.3
68	3.4	2.6	43	20	4.1
90	2.8	2.1	62	19	4.1
102	2.9	2.2	84	13	3.8

Source: Reproduced from Fageria, 1999.

regulate pH in cells and to balance cations and anions. If normal metabolic processes are to be maintained, it is necessary that Ca and Mg levels be balanced in plants. Magnesium uptake by plants is greatly influenced by other competing cations such as K^+, NH_4^+, Ca^{2+}, Mn^{2+}, and H^+ (Kurvits and Kirkby, 1980; Heenan and Campbell, 1981). Unlike Ca, Mg is readily translocated from mature to young actively growing regions of plants. Hence, Mg deficiency first appears in older leaves of crop plants (Fageria and Barbosa Filho, 1994).

Accumulation of Ca and Mg during the growth cycles of upland rice and common bean grown on an Oxisol of central Brazil are presented in Figures 8.14 and 8.15. In upland rice, Ca and Mg accumulation was linear during the growth cycle. At harvest, about 30 kg Ca/ha accumulated, and of this quantity 27.5 kg was in straw and the remaining 2.5 kg was in grain. For Mg accumulation, about 24 kg Mg/ha accumulated at harvest and of this quantity 18.6 kg was in straw and the remaining 5.4 kg accumulated in grain. This means that major portions of Ca and Mg accumulated in rice remain in straw and small portions translocated to grain. Rice produced 221 kg straw and 160 kg grain per kg Ca accumulated at harvest. Accumulation of Ca and Mg followed quadratic patterns for common bean (Figure 8.15). Maximum accumulation of Ca and Mg occurred at about 84 days after sowing in common bean and decreased at harvest. The decreases may have been related to degradation of leaves at harvest. Decreases in Ca were more drastic compared to Mg. This meant that common

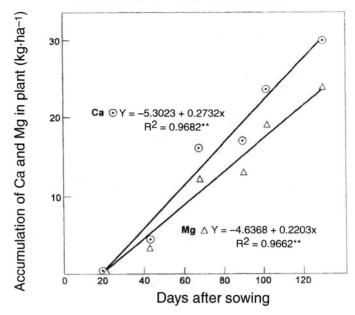

FIGURE 8.14. Accumulation of Ca and Mg in upland rice during its growth cycle. *Source:* Adapted from Fageria, 1999.

bean leaves most likely had higher concentrations of Ca compared to Mg. Common beans produced 68 kg straw and 64 kg grain per kg of Ca accumulated in plants.

Management Strategies for Improving Calcium and Magnesium Nutrition

To improve Ca and Mg uptake efficiency by crop plants, adoption of some soil and crop management practices are essential and are summarized as follows:

1. application of optimum rates;
2. appropriate sources;
3. appropriate Ca/Mg and Ca/K ratios; and
4. use of efficient cultivars.

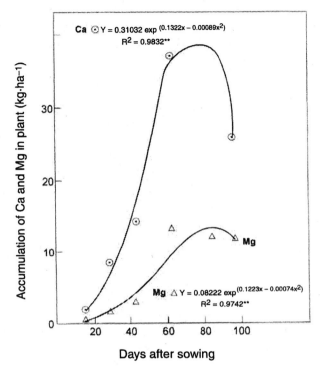

Ca ⊙ $Y = 0.31032 \exp^{(0.1322x - 0.00089x^2)}$
$R^2 = 0.9832^{**}$

Mg △ $Y = 0.08222 \exp^{(0.1223x - 0.00074x^2)}$
$R^2 = 0.9742^{**}$

FIGURE 8.15. Accumulation of Ca and Mg in common bean during its growth cycle. *Source:* Adapted from Fageria, 1999.

Sulfur

Sulfur is an essential nutrient for plants because it is a constituent of the essential amino cysteine and methionine and several other compounds such as coenzymes, thioredoxins, and sulpholipids. Many other S-containing compounds that occur in plants are involved in defense mechanisms against herbivores, pests, and pathogens, or contribute to special tastes and odors of food plants (Bennett and Wallsgrove, 1994). Sulfur has important influences on quality of final plant products because of its role in synthesis of amino acids, proteins, and some secondary metabolites.

Sulfur deficiency reduced grain yield of flooded rice by reducing tillering, numbers of panicles, grains per panicle, and grain weight

(Blair, 1987). Sulfur deficiency also decreased wheat mesophyll cell division, leading to smaller leaf sizes and decreased chlorophyll and protein contents of individual chloroplasts, but had no effect on chloroplast numbers per cell (Burk, Holloway, and Dalling, 1986). Numbers of chloroplasts per spinach mesophyll cell were also directly related to leaf S contents (Dietz, 1989). Chlorophyll contents and rubisco activity per chlorophyll unit both decreased about 70 percent, while CO_2 exchange rates declined 80 percent, as leaf S concentrations declined in sugarbeet from 2.5 to 0.5 mg S per g dry matter (Terry, 1976). In addition, Xu et al. (1996) observed that mild S deficiency caused leaf protein contents, rubisco activities, and CO_2 exchange rates to decline about 20 percent in tomato. Critical levels of S below 35 mg per kg dry matter, specific leaf S (mg·m^{-2} leaf area) declines caused reduced CO_2 exchange rates, quantum efficiency, dark respiration, leaf S, leaf N, and rubisco contents in soybean (Sexton, Batchelor, and Shibles, 1997). On average, most S-deficient plants had 70 percent lower CO_2 exchange rates, 42 percent less quantum efficiency, and 50 percent decline in dark respiration relative to plants that received adequate S. Total leaf N and rubisco contents were related to leaf S levels, and declined an average of 62 and 95 percent, respectively, as S declined from 120 to 40 mg S/m^2 leaf area. Sulfur deficiency also limited protein synthesis and amounts of photosynthetic apparatus present by limiting amounts of methionine and cysteine available for assembly of new proteins (Anderson, 1990). Sulfur deficiency also caused increased rates of protein degradation within chloroplasts (Dannehl, Herbik, and Godde, 1995). Nitrogen assimilation has also been linked to S metabolism and protein synthesis, so as S metabolism slows, N assimilation also is reduced (Anderson, 1990).

Uncontrolled amounts of S from atmospheric deposition often exceed those required by crops so that S deficiency has rarely been observed in crop plants until recent years. Due to concern about effects of acid rain on natural ecosystems, which often exist on poorly buffered soils and catchments, steps have been taken to reduce emissions of sulfur dioxide (SO_2) in various parts of the world. For example, total emissions of SO_2 in the United Kingdom decreased from 6.4 million tons in 1970 to 2.9 million tons in 1995 (Zhao et al., 1997). Some other European countries achieved even more dramatic decreases during this same period. These changes have resulted in increased incidence of S deficiency on arable crops (McGrath and Zhao, 1995;

Zhao et al., 1997). Sulfur deficiency has been reported with increasing frequency in various parts of the world because of increases in use of high purity, low S fertilizers and use of high-yielding cultivars that have higher S requirements. A particular example of this has been S deficiency reported in many countries for oilseed rape (McGrath and Zhao, 1995).

Management Strategies for Improving Sulfur Nutrition

Efficient fertilizers and fertilizer management techniques are essential to maximize S benefits for farmers. Sulfate fertilizers such as gypsum or single superphosphate are not effective in areas of high rainfall, and S inefficiency has been attributed to high leaching losses of SO_4^{2-}. Lysimeter studies have generally shown that effectiveness of gypsum for supplying S to plants depends on agroclimatic factors, particularly rainfall intensity, timing of application, SO_4^{2-} adsorption capacity, and soil liming. Jones, Martin, and Williams (1968) estimated that 80 percent of applied gypsum S was lost by leaching during one crop season from soils with low SO_4^{2-} adsorption capacity. Leaching losses of sulfate fertilizers may be reduced by increasing gypsum particle size (Korentajer, Byrnes, and Hellums, 1984) or by using sparingly soluble sources (e.g., anhydrite) of sulfate. Dehydration of gypsum at 400°C resulted in formation of anhydrite, which is less susceptible to leaching than gypsum. Topdressings of $(NH_4)_2SO_4$ can furnish N as well as S requirements to crops and avoid much N and S leaching. Ammonium sulfate, single superphosphate, and triple superphosphate contain approximately 24 percent, 11.9 percent, and 1.4 percent S, respectively. In addition, more effective and efficient use of S fertilizer requires better information about factors controlling crop responses to S. Improved plant and soil tests for S are also needed.

Micronutrients

Micronutrient deficiencies in field crops have increased markedly in recent years because of intense cropping systems, loss of top soil by erosion, liming of acidic soils, and use of marginal lands for crop production (Fageria, Baligar, and Clark, 2002). Micronutrient deficiency problems are also aggravated by high demands of modern crop cultivars for these nutrients. Field research about micronutrient nutrition of crop plants is relatively limited compared to that for macronutrients.

Zinc

Zinc deficiency is widespread for crops grown in soils of arid and semi-arid regions of the world, and severe reductions in plant growth and yields have been reported (Graham, 1984; Torun et al., 2003). Plants grown in soils exhibiting Zn deficiency are generally low in organic matter (<2 percent) and high in $CaCO_3$ (Sillanpaa and Vlek, 1985). Low organic matter in soils has been proposed as a major reason for widespread occurrence of Zn deficiency in plants (Cakmak et al., 1996). Several reports have noted beneficial effects of soil applications of organic materials on Zn nutrition of plants (Arnesen and Singh, 1998; Torun et al., 2003). Additions of organic materials to soil can increase concentrations of soluble organic ligends that form readily soluble Zn complexes to enhance Zn uptake by plants (Torun et al., 2003). Other mechanisms responsible for increasing Zn uptake by plants have been decreases of its adsorption onto soil clay minerals (Shuman, 1995), reduced soil pH during organic matter decomposition (Arnesen and Singh, 1998), increased Zn diffusion in soil (Alvarez, Rico, and Obrador, 1997), and released Zn during organic matter degradation (Qiao and Ho, 1997).

Zinc deficiency decreased yields of upland rice grown on highly weathered Oxisols and Ultisols (Fageria, 1992; Fageria, Baligar, and Clark, 2002). Root weights and lengths increased for common bean, soybean, and wheat with increases in soil Zn from 0 to 80 mg·kg^{-1}, but not for maize or rice (Baligar, Fageria, and Elrashidi, 1998). Wheat responses to Zn was pH dependent since plants did not have enhanced concentrations of Zn when grown at pH 4.0 to 4.6, but did when grown at pH 5.2 (Chairidchai and Ritchie, 1993). Figure 8.16 notes Zn uptake relative to soil pH for common bean grown on an Inceptisol in central Brazil. Quadratic decreases in Zn uptake were noted as soil pH increased from 4.9 to 7.0. Common beans contained 71 mg Zn/kg shoot dry weight when grown at pH 4.9, but only had 13 mg·kg^{-1} shoot dry weight when grown at pH 7.0. That is, Zn decreased by 446 percent when grown on higher pH soils compared to lower pH soils.

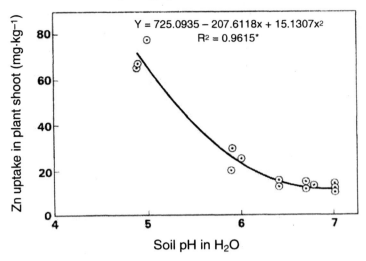

FIGURE 8.16. Relationship between soil pH and Zn uptake by common bean. *Source:* Adapted from Fageria and Baligar, 1999a.

Barber (1995) noted that soil pH had strong effects on Zn adsorption. This occurred because concentrations of Zn in soil solutions decreased 30-fold for every unit of pH increase between pH ranges of 5 to 7. Reductions of Zn concentrations in soil solution were believed to be due to Zn adsorption on hydrous oxide surfaces. Levels of Zn in soil solutions at specific pH values depended on nature of soil surfaces and level of Zn in soil. Where hydrous oxide surfaces were present, Zn levels in solution were usually lower. When $CaCO_3$ was present in soil at high pH, solution Zn levels were also lower. At soil pH values above 7.5, levels of complexed Zn in solution depends on solubility of organic matter and presence of Ca and other cations that have ability to suppress Zn solubility (Barber, 1995). Table 8.14 provides information on decreases in Zn uptake of four crop species grown on a Brazilian Oxisol with increased levels of $CaCO_3$. Increase in Zn uptake has also been related to applied P for each of the species.

Zinc is a constituent of some nonenzyme proteins and also an activator of some enzymes. Several enzymes involved in carbohydrate metabolism and protein synthesis in crop plants are activated by Zn, including most dehydrogenases (Fageria, Slaton, and Baligar, 2003). Zinc is involved either directly or indirectly in starch formation, since

TABLE 8.14. Zinc uptake by four crop species grown with different levels of lime and P.

Lime level g·kg soil	P level mg·kg^{-1} soil	Upland rice g/pot	Wheat g/pot	Common bean g/pot	Maize g/pot
0	0	201	69	458	408
0	50	3324	758	982	346
0	175	3910	1090	683	1280
2	0	142	28	158	270
2	50	2740	410	373	796
2	175	1871	496	393	1522
4	0	39	12	130	161
4	50	902	232	324	345
4	175	1226	292	379	739
F-test					
Lime		**	**	**	**
Phosphorus		**	**	**	**
Lime × P		**	**	**	NS

Source: Reproduced from Fageria, Zimmermann, and Baligar, 1995. Used with permission from Marcel Dekker.

** Significant at P = 0.01; NS Not significant.

Zn-deficient plants often have reduced starch concentrations. Zinc is involved in N metabolism and acts as a catalyst to enhance plant regeneration frequency of indica rice (Fageria, Slaton and Baligar, 2003). Zinc is involved in phytochrome activity, membrane integrity, and synthesis of tryptophan (precursor to the hormone indoleacetic acid [IAA]) (Sabrawat and Chand, 1999). Zinc is also vital for oxidation processes in plant cells, is involved in transformation of carbohydrates, and regulates sugars in plants. Its deficiency retards photosynthesis and N metabolism (Gupta, 1995).

Zinc deficiency symptoms occur in both young and old leaves, because it can be remobilized from lower to upper leaves. Zinc deficiency symptoms are generally characterized by so called rosette-like stands of thick small leaves. In addition, Zn deficiency often has fairly pronounced chlorosis between leaf veins similar to that in Fe deficiency.

Copper

Copper deficiency has been reported in annual crops grown on highly weathered acidic soils (Fageria, 2001a). This deficiency has been related to intensive cropping systems and to use of liming to raise soil pH. About 4.5 million ha of agricultural land in southwest Australia are neutral to alkaline, and when these lands are newly cleared for agricultural use, wheat commonly becomes Cu deficient (Brennan, 1994). Faba bean, chickpea, lentil, and wheat have also responded to Cu applications when grown on these soils (Brennan and Bolland, 2003).

Copper contents of crops normally range from 5 to 20 mg·kg^{-1}, which is the second lowest concentration of micronutrients next to Mo (Obata, 1995). Properties whereby Cu is essential to plants are somewhat similar to those for Fe. Except for certain amine oxidases and galactose oxidase, Cu participates directly in redox enzymes (Clark, 1991). Copper may also be involved in synthesis or stability of chlorophyll, and is involved in both protein and carbohydrate metabolism. Protein synthesis is reduced, causing soluble amino nitrogen compounds to build up in Cu-deficient plants. Specific Cu requirements have also been noted for N_2 fixation (Hallsworth, Wilsen, and Greenwood, 1960; Snowball, Robson, and Loneragan, 1980; Mengel and Kirkby, 1978), and absence of Cu markedly depressed nodule development (Hallsworth, Wilsen, and Greenwood, 1960). However, information is lacking about how Cu is involved in N_2 fixation processes (Snowball, Robson, and Loneragan, 1980). Copper deficiency affects grain formation more than vegetative growth in cereal crops.

Reduced seed set has been reported in Cu-deficient plants, which might result from inhibition of pollen release from stamina (Dell, 1981). Lignification of anther cell walls is required for rupture of stamina and subsequent release of pollen, and lignification of anther cell walls is reduced or absent in Cu-deficient plants (Dell, 1981). Impaired lignification of cell walls is the most typical anatomical change induced by Cu deficiency in higher plants. Lack of lignification exhibited in the characteristic distortion of young leaves (bending and twisting of stems and twigs) increases lodging susceptibility of cereals, particularly when combined with high N (Marschner, 1986). In plants suffering from severe Cu deficiency, xylem vessels are not usually sufficiently lignified. Copper has relatively high stability constants with chelating compounds (compared to Zn), so that deficiency symptoms occur mainly on young leaves.

Availability of Cu depends largely on soil pH. Figure 8.17 shows relationships between soil pH and Cu uptake by common bean. With increasing soil pH, quadratic decreases in Cu uptake were noted. Decreases were 125 percent at pH 7.0 compared to pH 4.9. Decreases in Cu uptake with increasing soil pH may be related to adsorption by soil rather than precipitation (Barber, 1995). Critical deficiency levels of Cu in plants range from 3 to 5 mg·kg^{-1} dry weight. Critical toxic levels of Cu are above 20 to 30 mg·kg^{-1} dry weight and may include Fe deficiency symptoms (Shuman, 1994).

Boron

Boron deficiency in crop plants has been reported throughout the world (Gupta, 1979; Blevins and Lukaszewski, 1998). Deficiency of B is generally common for plants grown on light textured soils in

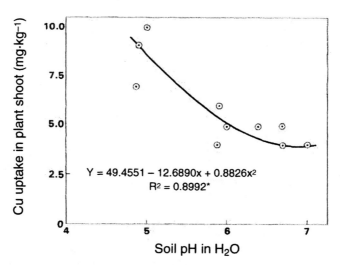

FIGURE 8.17. Relationship between soil pH and Cu uptake by common bean. *Source:* Adapted from Fageria and Baligar, 1999a.

humid climates where B is readily leached from soil. Boron in soil solutions usually occurs as the undissociated boric acid (H_3BO_3). Adequate B levels are essential for higher crop yields and quality. Boron improved root development in common bean, soybean, and wheat grown on an Oxisol of central Brazil (Baligar, Fageria, and Elrashidi, 1998). Boron has been reported to counteract toxic effects of Al on root growth of dicotyledonous plants (Blevins and Lukaszewski, 1998). Boron requirements for reproductive growth are much higher than for vegetative growth of most crop species (Loomis and Durst, 1992). Boron has also been reported to be essential for N_2 fixation (Mateo et al., 1986). Warington (1923) is credited with the first definitive proof that B is required by higher plants. Later, Sommer and Lipman (1926) established B requirements for six nonleguminous dicots and for one graminaceous plant (barley). However, B requirements are highly variable, and optimum quantities for one plant species could be either toxic or insufficient for other plant species. Based on B requirements, plants can be divided into three general groups: (1) graminaceous plants that have lowest B demands; (2) remaining monocots and most dicots with intermediate B requirements; and (3) latex-forming plants with highest B requirements (Mengel and Kirkby, 1978). Boron deficiency affects development of meristems or actively growing tissues so that deficiency symptoms are death of growing points of shoots and roots, failure of flower buds to develop, and ultimately blackening and death of these tissues. Boron uptake normally decreases with increasing soil pH (Figure 8.18). In common bean grown on an Oxisol, decreases were 120 percent when soil pH was raised from 4.9 to 7.0. This decrease in B uptake may have been related to adsorption processes as soil pH increased. Boron adsorption increased as pH increased above 4 and reached a maximum at pH 8 to 9 before decreasing at higher pH values (Barber, 1995). Soil texture also affects B adsorption; it is higher in heavy textured soils compared to light textured soils.

Manganese

Manganese is essential as a metal activator for several enzymes, including oxidase, peroxidase, dehydrogenase, decarboxylase, and kinase (Fageria, Slaton, and Baligar, 2003). Manganese is also important in photosynthesis and in nitrate reduction. Because of the role

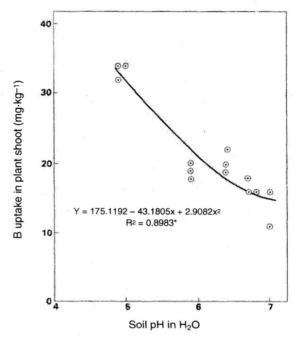

FIGURE 8.18. Relationship between soil pH and B uptake by common bean. *Source:* Adapted from Fageria and Baligar, 1999a.

Mn has in photosynthesis, its deficiency lowers levels of soluble carbohydrates, particularly in roots. Manganese is necessary for photosynthetic O_2 evolution where it participates in the Hill reaction to split water so that O_2 is formed during photosynthesis, In Mn-deficient plants, photosynthesis capability decreases as chlorophyll contents decrease (Shuman, 1994). Manganese is immobile in plant tissues, hence its deficiency first appears in young tissues. Visual Mn deficiency usually is noted as graying of young leaves, especially between veins. This was followed by reduced growth and dry matter yields. Oat is very sensitive to Mn deficiency and is an excellent indicator plant for this disorder.

Manganese uptake is highly pH dependent. Figure 8.19 illustrates Mn uptake by common bean grown on an Inceptisol under different soil pH values. As soil pH increased, quadratic decreases in Mn uptake occurred. For plants grown in the lowest pH soil (4.9), Mn

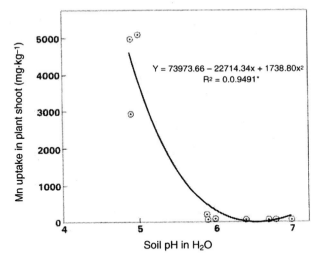

FIGURE 8.19. Relationship between soil pH and Mn uptake by common bean. *Source:* Adapted from Fageria and Baligar, 1999a.

uptake was 4420 mg·kg^{-1} shoot dry weight and decreased to 174 mg·kg^{-1} when grown in the highest pH soil (7.0). Plants decreased 2440 percent in Mn when grown in pH 7 soil compared to pH 4.9 soil. Manganese uptake can also be reduced by higher concentrations of Fe, Cu, and Zn (Fageria, Baligar, and Clark, 2002). Iron-Mn interactions occur both at uptake and inside plants at metabolic sites, whereby Mn can inactivate Fe metabolic activities by decreasing Fe concentrations in plants (Shuman, 1994). In flooded rice, Fe toxicity is very common (Fageria, Slaton, and Baligar, 2003). Iron toxicity can be alleviated by application of appropriate levels of Mn fertilizer.

Iron

Iron stress (deficiency or toxicity) in crop plants often represents serious constraints for stabilizing and/or increasing crop yields (Fageria, Baligar, and Wright, 1990). Factors that decrease availability of Fe in soil or compete with Fe in plant absorption contribute to Fe deficiency. Iron deficiency has been common for many crops such as maize, sorghum, peanut, soybean, common bean, oat, barley, and upland rice (Fageria, Baligar, and Clark, 2002). Iron deficiency (commonly called Fe chlorosis) is a common symptom for many soybean- and sorghum-producing areas throughout the United States (Heitholt et al., 2003). Iron toxicity is restricted mainly to flooded or lowland rice (Fageria, Baligar, and Edwards, 1990). Monocotyledonous plant species are generally less "Fe efficient" than dicotyledonous species. Plants have been classified as Fe efficient if they respond to Fe deficiency stress by inducing biochemical reactions that make Fe available to plants and "Fe inefficient" if they do not (Marschner, 1995). These induced reactions or compounds are release of H^+ and reducing compounds from roots, reduction of Fe^{3+} to Fe^{2+} by roots, and increased organic acids produced by roots (Brown, 1978).

Iron is required in higher plants for chlorophyll synthesis and is a component of many enzymes that catalyze metabolism in plants. Enzymes containing metals are generally divided into two groups: (1) metal activated and (2) metalloenzymes. In the first group, Fe acts as a temporary link between enzyme and substrate during biochemical reactions. Examples of this are activation of several oxidases. The majority of Fe enzymes belong to the metalloenzyme group in which Fe is firmly bound to proteins. Frequently, Fe is complexed by or attached to small molecules called prosthetic groups. Examples of these are peroxidases, catalase, and cytochrome oxidase. Each of these enzymes contains Fe bound as haem. Thus, Fe has close relationships with oxidase activities (Okajima, Uritani, and Huang, 1975). Among Fe enzymes, the ferredoxins are of special interest because of their importance in photosynthesis. Ferredoxins are small protein molecules containing Fe and labile sulfides that catalyze phosphorylation of ADP in the presence of light to ATP.

Iron is immobile in plant tissues and its deficiency appears first in young leaves as interveinal chlorosis. Leaf veins lose chlorophyll last so symptoms are commonly streaked providing a "Christmas tree–like pattern." At advanced stages of deficiency, entire leaf blades may become yellow or white. Approximately 80 percent of Fe in green leaves is located in chloroplasts (Terry, 1980), so Fe deficiency strongly affects processes located in chloroplasts. Since an obvious effect of Fe deficiency in higher plants is development of chlorotic leaves, this chlorosis is associated with loss of not only chlorophyll but also of thylakoid constituents. Reduction in thylakoid membranes during Fe deficiency is accompanied by decreases in photosynthetic pigments (Terry and Abadia, 1986; Monge, Vale, and Abadia, 1987).

Some environmental factors that can induce Fe deficiency in crop plants are

1. low available Fe contents of soil;
2. high levels of lime application or "lime-induced chlorosis";
3. poor root growth;
4. high P concentration;
5. high levels of Mn, Zn, and Cu;
6. high light intensity;
7. high levels of nitrate;
8. high or low temperatures;
9. unbalanced cation ratios;
10. addition of organic matter to soil; and
11. virus infections (Hale and Orcutt, 1987).

Strong interactions between Fe and P exist in plant metabolism, and this sometimes results in P deficiency when relatively high concentrations of Fe exist in the growth medium. Tissue analysis commonly performed does not usually distinguish between functional Fe and that inactivated by precipitation or complexation. Active Fe has been termed as that fraction that can be extracted by 1 N HCl from ground dry plant tissue (Hale and Orcutt, 1987). Iron uptake is significantly reduced with increasing soil pH (Figure 8.20). Iron uptake by common beans grown in pH 4.9 soil was 328 mg·kg^{-1} shoot dry weight, while it decreased to 96 mg Fe/kg when grown in pH 7.0 soil. This meant that shoot Fe decreased 242 percent when grown in pH 7.0 soil compared to pH 4.9 soil.

Using plants with resistance to Fe deficiency or genetically breeding for resistance to Fe deficiency have generally been accepted as the most important economic approach to overcome Fe deficiency (Fageria, Baligar, and Clark, 2002; Heitholt et al., 2003). Where genetic resistant materials are not available, foliar sprays can be used to overcome Fe deficiency in crop plants (Fageria, Baligar, and Clark, 2002). Plants have different strategies for solubilization and uptake of Fe. For example, graminaceous plants exude phytosiderophores or Fe bearers (e.g., mugeneic or avenic acids) that transport Fe into roots (Römheld, 1987). Phytosiderophores have been defined as a group of root exudates exhibiting strong complexing properties for Fe^{3+}, whichh ave been identified as nonproteinogenic amino acids such as

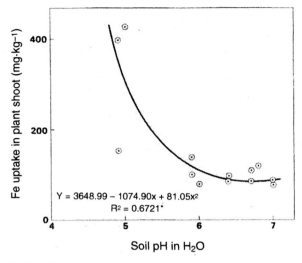

FIGURE 8.20. Relationship between soil pH and Fe uptake by common bean. *Source:* Adapted from Fageria and Baligar, 1999a.

mugineic acid and its derivatives (Takagi, Nomoto, and Takemoto, 1984). In this respect, phytosiderophores are analogues of microbial siderophores that are literally Fe bearers (Hinsinger, 1998). The synthesis and release of phytosiderophores in the rhizosphere are stimulated by Fe deficiency (Römheld, 1991), and described as Strategy II systems for Fe acquisition and developed exclusively by graminaceous species (Marschner, 1986). Graminaceae species differ widely in their ability to produce phytosiderophores, both quantitatively and qualitatively. Among the ranges of graminaceous species studied (Marschner, Treeby, and Römheld, 1989), enhancement in release of phytosiderophores by Fe-deficient species has been reported to increase according to resistance of species to lime-induced chlorosis (Figure 8.21). These compounds, their complexes, and transport of Fe in plants have been reviewed extensively (Römheld and Marschner, 1986; Marschner, Treeby, and Römheld, 1989; Römheld, 1991).

Molybdenum

Arnon and Stout (1939) established that Mo was an essential nutrient for higher plants using tomato as test plant. Molybdenum is

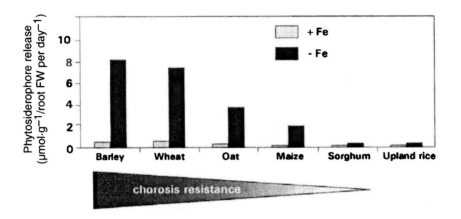

FIGURE 8.21. Amount of phytosiderophore released by roots of Fe-sufficient and Fe-deficient seedlings of graminaceae species differing in tolerance to lime-induced chlorosis. The most chlorosis-resistant species such as barley and wheat exhibited highest rates of phytosiderophore release and greatest re- sponse to Fe deficiency. *Source:* Reprinted from *Advances in Agronomy,* 64; Hinsinger, P., "How do plant roots acquire mineral nutrients? Chemical process involved in the rhizosphere," pp. 225-265, copyright (1998), with permission from Elsevier.

indispensable in N_2 fixation processes for aerobic and anaerobic N fixers (*Azotobactor* and *Clostridium*) and for symbiotic *Rhizobium.* Molybdenum is an essential component of the important enzymes nitrogenase and nitrate reductase (Beevers and Hagenman, 1983). The basic mechanism of N_2 fixation by nitrogenase and Mo function is similar for free living N_2 bacteria as for N_2 fixing microorganisms living in symbiosis with higher plants (Mengel and Kirkby, 1978). Molybdenum has been reported to be essential for absorption and translocation of Fe in plants (Rao and Adinarayana, 1995).

The most important soil factor affecting availability of Mo to plants is soil pH. As an acid former, Mo increases in solubility with increases in soil pH (Barber, 1995). Soil solution Mo levels increase tenfold for each unit increase in soil pH. Where Mo has been added to soil, amounts adsorbed will decrease as pH increases (Barber, 1995). Liming of acidic soils is generally sufficient to correct Mo deficiency. Soils where Mo deficiencies have been reported have been acidic, and excess amounts of Mo normally occur in alkaline soils (Okajima, Uritani, and Huang, 1975).

Plant requirements for Mo are lower than for any other mineral element essential to plants. Critical deficiency levels range from 0.1 to 1 mg·kg^{-1} plant dry weight. The deficiency and toxicity range for Mo in crop plants is narrow. It has been reported that Mo toxicity in many crop plants may occur when Mo exceeds 2 mg·kg^{-1} of dry weight (Rao and Adinarayana, 1995). Molybdenum requirements are much greater for legumes than for cereals and nonlegumes because Mo is involved in nodule development and N_2 fixation. Thus, concentrations needed for nonlegumes are much lower than those needed by legumes. Antagonistic relationships between Mo and Cu and sulfate have been reported, whereas phosphate promotes Mo absorption. Nitrogenase and nitrate reductase both depend on specific Mo valencies for appropriate function (Shuman, 1994). It has been generalized that the major function of Mo is closely associated with N metabolism in most crop plants (Blevins, 1994) by reducing NO_3^- to NO_2^-. Thus, Mo functions much differently from the other elements in plants. If plants absorb N as NO_3^-, Mo is indispensable for normal growth. On the other hand, if the N has been supplied as NH_4^+, Mo is not necessary (Ishizuka, 1978). Molybdenum is immobile in plants and Mo deficiency first appears in younger leaves. Molybdenum deficiency symptoms are similar to those of Fe deficiency.

Chlorine

The essentiality of chlorine for plant growth was confirmed in 1954 (Broyer et al., 1954). It was generally assumed for many years that field-grown crops would not benefit from application of Cl fertilizers because of Cl being so ubiquitous in the environment (Fixen, 1993). The first evidence that plants could benefit from Cl fertilization (field-grown wheat) has been reported only in recent years (Taylor et al., 1981; Christensen et al., 1981). Engel, Bruckner, and Eckhoff (1998) reported wheat grain yield increased by an average of 417 kg·ha^{-1} (9.7 percent) in 86 cases where responses to Cl were assessed. Largest grain yield responses (>800 kg·ha^{-1}) occurred at sites with lowest plant Cl concentrations (<0.50 g·kg^{-1}), and kernel size was the most important yield component affected by applied Cl.

Approximately 73 percent of yield responses to applied Cl could be accounted for by larger kernel size. Biological functions of Cl in plants are presumed to require concentrations of no more than 0.10 g·kg^{-1} (Fixen, 1993). The beneficial effects of Cl are likely due to its osmolregulatory role in plants (Flowers, 1988). Chlorine is also involved in several plant processes, including photosynthesis, sugar translocation, and maintaining or increasing water potential (Voss, 1993). The importance of Cl in plant growth and grain yield is commonly highly dependent on growing environments (e.g., water and temperature conditions).

Chlorine is absorbed in greater quantities by crop plants than other micronutrients. Chlorine functions in capture and storage of light energy through its involvement in photophosphorylation reactions in photosynthesis. It is not present in plants as a true metabolite but as a mobile anion. It is involved with K in regulation of osmotic pressure, acting as an anion in counterbalances to cations. General plant deficiency symptoms are chlorosis in younger leaves and wilting.

Management Strategies for Improving
Micronutrient Uptake

Recovery of applied micronutrients by crop plants is in the range of 5 to 10 percent (Mortvedt, 1994). Some reasons for low recovery of applied micronutrient fertilizers are poor distribution of the low rates applied to soils, fertilizer reactions with soil to form unavailable reaction products, and low mobility in soil, especially those of Cu, Fe, Mn, and Zn. The following management practices should improve uptake and use of micronutrients in crop plants:

1. use of adequate levels, forms, and methods of application;
2. use of efficient species or cultivars within species;
3. avoidance of overliming acidic soils;
4. control of insects, diseases, and weeds;
5. maintenance of adequate balances of macro- and micronutrients;
6. proper diagnosis of deficiencies or toxicities through soil and plant analysis and visual symptoms.

BENEFICIAL AND TOXIC ELEMENTS

In addition to the essential nutrients discussed above, some elements such as Si can be beneficial to plant growth but are not considered essential. Aluminum is one of the most yield-limiting elements for plants grown in acidic soils, but under some circumstances it can benefit plant growth.

Silicon

Silicon has not been proven to be an essential element for most higher plants, but has been considered essential to some terrestrial plants such as species within the *Equisitacae* family and some algaelike diatoms and members of the yellow-green and golden group (Clark, 2001). However, Si has been reported to provide beneficial growth effects for many crops, including rice, wheat, barley, cucumber, sugarcane, and tomato (Savant, Snyder, and Datnoff, 1997). Silicon fertilizers are applied to crops in several countries for increased plant productivity and sustainable production. Plants absorb Si in the form of silicic acid $[Si(OH)_4^0]$. This compound, like B, is absorbed as an undissociated molecule. After $Si(OH)_4^0$ is transported to shoots, it concentrates through loss of water and is polymerized as silica gel on surfaces of leaves and stems. Silicon can also interact with absorption and reduce toxicities of high Mn and Al (Clarkson and Hanson, 1980; Galvez et al., 1987) and reduce plant phosphate requirements (Clarkson and Hanson, 1980). Evidence is lacking concerning physiological roles of Si in plant metabolism, and its beneficial effects are usually observed only in plants that accumulate Si. Silica gel deposited on plant surfaces is thought to contribute to the beneficial effects of Si. Beneficial effects of Si are small under conditions of optimal growth but become obvious when plants are stressed.

Nearly all plants absorb some Si, but certain groups of plants such as certain grasses, sedges, nettles, and horsetails accumulate 2 to 20 percent of foliage dry weight as hydrated polymers or silica gel (Clark, 2001). This silica gel primarily impregnates walls of epidermal and vascular tissues, and appears to strengthen tissues, reduce water loss, and retard fungal infection and insect attack (Clark, 2001). Silicon forms Si-enzyme complexes that act as protectors and regulators of photosynthesis and other enzyme activities, Si is also important in structural rigidity of cell walls (Takahashi, 1995; Savant, Snyder, and Datnoff, 1997).

Silicon applications to organic soils increased Si concentrations in leaves and enhanced grain yields of rice and sugarcane (Snyder, Jones, and Gascho, 1986; Anderson, 1990; Savant, Snyder, and Datnoff, 1997). Silicon also benefits rice in several other ways, including improved photosynthetic activity, improved water use efficiency, increased resistance to fungal diseases and insect damage, increased straw strength, improved leaf turgor, improved P metabolism, increased grain and milling yields, and improved plant nutrition (Yoshida, 1981). Rice diseases have been reduced when Si concentrations in tissue increase (Datnoff, Deren, and Snyder, 1997). Grain discoloration associated with several different pathogens, including *Curvularia, Helminthosporium,* and *Fusarium,* decreased as Si application rates increased from 0 to 960 kg·ha^{-1} (Korndörfer, Datnoff, and Corrêa, 1999). Grain discoloration is associated with reduced grain weight, reduced seed germination, and lower market value of grain. It has also been reported that Si applications may reduce Fe and Mn toxicity in flooded rice by increasing oxidizing power of roots to make Fe and Mn less soluble and by increasing internal tolerance of plants to excess levels of these elements (Korndörfer, Datnoff, and Corrêa, 1999). Amounts of Si in rice leaves were directly related to amounts of Si applied to soil. This may explain why plants were more resistant to fungal diseases (Adatia and Besford, 1986). Mechanical barriers are believed to be created by Si deposition in leaf epidermis cells. Silicon also can reduce amino acids and starch formation, which promote fungal growth (Takahashi, 1995).

Research has reported that Si can ameliorate toxic effects of Mn (Galvez et al., 1987) and Al (Galvez et al., 1987; Bloom and Erick, 1995; Hodson and Evans, 1995; Cocker, Evans, and Hodson, 1998). The mechanism for ameliorating these toxicities is unclear, but three suggestions have been put forward: (1) Si induces increases in solution pH; (2) reduces availability of Al due to formation of hydroxyaluminosilicate species; and (3) detoxifies Al inside plants (Cocker, Evans, and Hodson, 1998).

Although mechanisms by which Si can ameliorate diseases on plants are unknown, ability to prevent fungal germ tubes from penetrating silicated cells has been accepted as a Si benefit (Savant, Snyder, and Datnoff, 1997). Recent studies indicated that plants treated with Si also formed phenolic compounds, which could reduce fungal development and growth (Cherif, Asselin, and Belanger, 1994). Silicated plant cells also prevented or slowed insect mandible penetration and/or damaged insect mandibles to the extent that insect damage was reduced (Ou, 1985; Sawant, Patil, and Savant, 1994).

Symptoms of Si deficiency in rice are soft droopy leaves, increased lodging, and increased incidence of brown spots, which could result from infestation by *Helminthosporium oryzae* (Elwad and Green, 1979). Silicon deficiency normally occurs on low Si organic soils or on highly weathered soils that have been depleted of Si. Application of Si-containing fertilizers such as furnace, dolomitic, and calcium silicate slags will usually correct Si deficiency (Wells et al., 1993).

Aluminum

Aluminum toxicity is a common growth-limiting factor for plants grown in acidic soils (Foy, 1992). Aluminum concentrations increase considerably in plants grown in acidic soils with pH below 5.0 to 5.5 (Ahlrichs et al., 1990). Aluminum toxicity symptoms appear first on roots, as length is retarded and root tips thicken, swell, become stubby, coralloid, brittle, and dark (Clark, 1993). Shoot symptoms appear on older leaves, growth is reduced, and leaves have light, diffuse chlorosis. Tips and margins of leaves normally turn brown and die as symptoms advance from tip to base. Since Al toxicity affects roots more dramatically than shoots, uptake of nutrients and water are often restricted to enhance deficiencies of many other nutrients. Genotypes also vary in Al toxicity symptoms. Symptoms of P, Ca, and Mg deficiencies often occur on plants that are exposed to toxic amounts of Al in soils.

Excesses of Al in the growth medium influence several physiological and biochemical processes in which growth and development are affected. The most important processes affected by excess Al are summarized as follows (Fageria, Baligar, and Wright, 1988):

1. interference with cell division in roots;
2. increase in cell wall rigidity by cross-linking pectins;
3. reduced DNA replication by increasing rigidity of DNA double helixes;
4. alteration in root membrane structures and functions;

5. interruption of enzymes functions governing sugar phosphorylation and deposition of cell wall polysaccharides;
6. decreased cell permeability through protein coagulation and inhibited cell division;
7. uptake and utilization of most essential nutrients are inhibited and growth of roots and shoots are reduced;
8. Al interferes with water use by plants to reduce crop yields;
9. reduced root respiration to reduce uptake of water and nutrients;
10. precipitation of nucleic acids by forming strong complexes;
11. abnormal distribution of ribosomes on the endoplasmic reticulum of root cells so that protein synthesis is reduced;
12. trivalent Al coordination forms complexes with carboxyl and sulfhydryl groups of proteins producing cross linkages; and
13. Al binds to either proteins or lipids depending on pH and other conditions.

Liming acidic soils is still a common, effective, and the most predominant method of reducing Al toxicity. However, development of Al-tolerant genotypes is becoming an attractive and promising strategy to cope with Al toxicity problems for crops grown in acidic soils. Identification of Al-resistant genotypes for major crops of the world has been reported (Fageria, Baligar, and Wright, 1988; Clark, 1991; Baligar et al., 1993). Fageria and Santos (1998) evaluated responses of common bean and rice to Al concentrations ranging from 0 to 3.83 $cmol_c \cdot kg^{-1}$ with Al saturations varying from 0 to 30 percent of CEC grown on an Inceptisol (Figure 8.22). Rice grain yields increased with increasing KCl extractable Al concentrations and Al saturations, but common bean yield decreased when grown with two of the Al toxicity treatments. This indicated that plants differ in Al tolerance. Plant species with greater tolerance to Al should be explored in practical farming systems to reduce cost of liming and improve yields of crops grown on acidic soils. Carver and Ownby (1995) reported that some wheat genotypes are highly tolerant to soil acidity (especially to Al) and their adaption to soil acidity is genetically controlled.

The mechanisms by which certain plants tolerate high levels of Al are still being assessed. Several hypotheses have been suggested, but more research needs to be conducted to verify these. Following are some hypotheses that need to be pursued (Fageria, Baligar, and Wright, 1988):

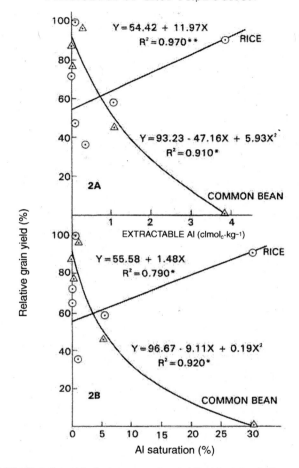

FIGURE 8.22. Relationship between extractable Al and relative grain yield of rice and common bean (A) and relationship between Al saturation and relative grain yield of rice and common bean (B). *Source:* Reproduced from Fageria and Santos, 1998. Used with permission from Marcel Dekker.

1. Al-tolerant plants either prevent excess absorption of Al by roots or detoxify Al after it has been absorbed.
2. Al-tolerant plants or cultivars have higher rates of root growth, thereby uptake of water and nutrients are greater.
3. Al-tolerant plants may have higher cellular respiration, which reverses uptake of ions.

4. Al-tolerant plant species or cultivars increase growth media pH and reduce Al solubility and toxicity. In contrast, Al-sensitive species or cultivars may lower pH of growth media and increase Al solubility and toxicity.
5. Aluminum increases viscosity of protoplasm in plant root cells and decreases overall cell permeability to salts. Tolerant cultivars have reduced viscosity compared to sensitive cultivars.
6. Al-tolerant species may prevent excess Al from getting into roots and also restrict its transport to shoots. Rice, rye, and alfalfa are in this group.
7. Al-tolerant species contain high levels of organic acids or compounds that complex Al to detoxify it within the soil and plant tissues.
8. Al tolerance in some plants is associated with ability to absorb and metabolize P.
9. Al-tolerant plant species have higher root phosphatase activities and absorb lower levels of organic or inorganic P more efficiently than Al-sensitive plants.
10. Superior Al tolerance in certain pasture species coincides with more efficient uptake and transport of P and Ca.
11. Al tolerance among certain cultivars of wheat, barley, and soybean is associated with ability to resist Al-induced Ca deficiency or Ca transport difficulties.

CONCLUSION

New and exciting developments in mineral nutrition research in the past few decades have greatly contributed to our understanding of the role of essential nutrients in improving yields of annual crops. There are 16 nutrients essential for plant growth, and deficiency of any essential element may drastically reduce growth and yield of crop plants. Mineral deficiency causes visible abnormalities in pigmentation, size, and shape of leaves, tillering in cereals, branching in legumes, and leaf photosynthetic rate.

Nutrient acquisition by plants is dynamic and complex, and absorption kinetics are usually more important than thermodynamics in describing uptake. Rates of nutrient absorption by roots depend upon nutrient supplies to root surfaces, active absorption by roots, and plant demand for nutrients. Nutrients are transported to roots by mass flow, diffusion, and root interception. After reaching root surfaces, nutrients move into the xylem through various root cells. From the xylem, ions are transported to growing organs in shoots for metabolic processing. Ion concentrations in cell sap are much higher than in outside media, and ions have to move against concentration gradients. Energy is required for this process, which is supplied through root respiration. Kinetics of ion absorption by roots is similar to kinetics of enzyme-catalyzed reactions. Kinetic information is providing insights into the nature of ion carriers. Most ion uptake studies have been of short duration in solution culture experiments using excised roots. These studies need to be conducted for longer durations using intact plants grown in soil to provide more practical application of experimental results.

An important component of ion uptake in higher plants is the presence of electrogenic proton pumps, which involve plasmalemma ATPase (Hodges, 1973). Electrochemical potentials inside root cells are generally in the range of -100 to -200 mV to create gradients for movement of cations into cells (Higinbotham, Etherton, and Foster, 1967). Movement of cations such as K^+ likely occur through membrane channels and down electrochemical gradients (Bentrup, 1990). Anion uptake occurs against electrochemical gradients and has been termed active (Fageria, Baligar, and Jones, 1997). Although not yet fully characterized, carriers for anions such as nitrate exist. Nitrate uptake is inducible in roots of higher plant (Minotti, Williams, and Jackson, 1968). Uptake of anions involves proton cotransport, which explains increases in pH around roots when anion exceeds cation uptake (Ulrich and Novacky, 1990).

Ions that have penetrated the plasmalemma by active or passive transport are translocated into the cytoplasm, probably by plasmatic streaming. These ions can be transported to plasmodesmata from cell to cell. Another problem that needs more study is ion transport through casparian strips. It has been supposed that ions in the symplasm pass this barrier by transport through plasmatic fractions of the endodermis. Thus, transport of water and hydrophylic particles into cell walls is considerably restrained by lignin and lipid materials of casparian strips (Mengel, 1974a). Vacuoles of plant cells play important roles in homeostasis of cells. Vacuoles are also involved in regulation of cytoplasmic pH, sequestration of toxic ions and xenobiotics, regulation of cell turgor, storage of amino acids, sugars, and CO_2 as malate, and possibly as sources for elevating cytoplasmic Ca. These activities are driven by two primary active transport mechanisms present in vacuolar membranes (tonoplast). These two mechanisms employ high-energy metabolites to pump protons into vacuoles to establish proton electrochemical potentials that dictate transport of many solute concentrations (Barkla and Pantoja, 1996).

Plant cells can preferentially absorb certain ions required in larger amounts. Thus, cations have preference over anions for uptake, and among cations some are accumulated in higher concentrations than others. When necessary, electrical neutrality can be maintained by ion exchange of H^+ and HCO_3^- (Larcher, 1995).

Each of the essential nutrients is important in attaining higher crop yields. Even so, N plays a major role in assuring high yields by enabling rapid establishment of large canopies for photosynthesis, high leaf area duration, and establishing large sink capacity. Isotopic ^{15}N tracers have been important tools in understanding N cycling in agricultural and natural soil-plant systems.

Increases in crop yields from application of micronutrients have been reported in many parts of the world. Soil factors such as pH, redox potentials, biological activities, organic matter, cation exchange capacities, and clay contents are important for determining availability of micronutrients in soils. Further, root-induced changes in the rhizosphere affect availability of micronutrients to plants. Major root-induced changes in the rhizosphere are pH, reducing capacities, redox potentials, and root exudates that sparingly mobilize soluble mineral nutrients and detoxify excess toxic elements such as Al. These exudates may make elements like Fe more available, but they can also produce water-soluble metal (Al) complexes, which reduce affiliation of metals with roots. Compared to macronutrients, micronutrients are required for crop growth in lower amounts and serve mainly as constituents of prosthetic groups in metalloproteins and as activators of enzyme reactions. However, their requirement is as important as macronutrients for crop production. Plant analyses have been recognized as useful tools to assess plant nutrition status of crops, and their importance has been recognized (Fageria, Baligar, and Jones, 1997). Soil testing is the most widely used diagnostic tool to assess fertilizer needs of crops, and recommendations for liming and fertilization usually are based on soil tests. To reduce costs for soil analyses, emphasis should be given to develop multinutrient extractants, but accuracy for prediction of specific nutrients must not be compromised in development of such extractants. Micronutrient application rates range from 0.2 to 100 kg·ha^{-1} depending on micronutrient, crop requirement, and methods of application. Higher rates are required for soil broadcast applications than for banded applications or foliar sprays. More field data are needed to improve fertilizer recommendations, especially for long-term availability of micronutrients. Research is also needed to evaluate effects of other fertilizers on availability of micronutrients.

Because recommended application rates are considerably lower for micronutrients, most micronutrient sources are applied to soils with NPK fertilizers, or may be incorporated or bulk blended with granular fertilizers or mixed with fluid fertilizers. This practice assures more uniform micronutrient application to soils. Data related to critical deficiency concentrations are available, but data related to critical toxic concentrations in soils and plants are not as readily available. Further, experimental data related to interactions of micronutrients with other essential nutrients are limited. More research is required to cover these areas for plants grown in different agroecological regions and cropping systems. Developing micronutrient-efficient crop genotypes for absorption and utilization is another research area that needs special consideration. Genetic control is generally complex (polygenic) for the macronutrients N, P, and K, and relatively simple (monogenic in many cases) for micronutrients depending on species.

Silicon is important for disease and insect control in crops such as flooded rice and sugarcane so that strong culms prevail to resist lodging. In addition, Si reduces Fe and Mn toxicity and improves P metabolism of many crop plants. Calcium silicate is a major source of Si to supplement soils deficient in this element. Aluminum is a toxic element to many crop plants and its concentration in acidic soils is pH dependent. When pH of acidic soils is below 5, Al toxicity can generally be expected. Toxicity of Al is associated with reductions in plant growth, including roots, and nutritional imbalances are common. Liming is an effective and practical solution to improve crop yields for crops grown on acidic soils. However, use of Al-tolerant crop species or cultivars may be a complementary solution for twenty-first century crop production on acidic soils.

Appendix

Plant Species

Alfalfa	*Medicago sativa* L.
Barley	*Hordeum vulgare* L.
Bean (adzuki)	*Vigna anguaris* L.
Bean (common, dry, pinto)	*Phaseolus vulgaris* L.
Bean (faba)	*Vicia faba* L.
Bean (mung)	*Vigna radiata* (L.) R. Wilczek
Bean (urd/black gram)	*Vigna mungo* (L.) Hepper
Bentgrass (creeping)	*Agrostis palustris* Huds.
Bluegrass	*Poa annua* L.
Cassava	*Manihot esculenta* Crantz
Chickpea	*Cicer arietinum* L.
Clover (subterranean)	*Trifolium subterraneum* L.
Clover (arrowleaf)	*Trifolium vesiculosium* L.
Cotton	*Gossypium hirsutum* L.
Cowpea	*Vigna unguiculata* (L.) Walp.
Cucumber	*Cucumis sativus* L.
Lentil	*Lens culinaris* L.
Lovegrass (boer) (weeping)	*Eragrostris curvula* Nees
Lupin (white)	*Lupinus albus* L.
Maize	*Zea mays* L.
Millet (pearl)	*Pennisetum glaucum* (L.) R. Br.
Mustard (white)	*Sinapsis alba* L.
Oat	*Avena sativa* L.
Oat (animated)	*Avena sterilis* L.
Pea	*Pisum sativum* L.
Pea (pigeon)	*Cajanus cajan* (L.) Huth.
Peanut	*Arachis hypogoaea* L.
Pine	*Pinus spp.*
Potato	*Solanum tuberosum* L.
Rape (summer)	*Brassica napus* L.

Rape (winter)	*Brassica napus* L.
Rice	*Oryza sativa* L.
Rye	*Secale cereale* L.
Ryegrass (perennial)	*Lolium perenne* L.
Safflower	*Carthamus tinctorius* L.
Sorghum	*Sorghum bicolor* (L.) Moench
Soybean	*Glycine max* (L.) Merr.
Sugarbeet	*Beta vulgaris* L. subsp. *vulgaris*
Sugarcane	*Saccharum officinarum* L.
Sunflower	*Helianthus annuus* L.
Taro	*Colocasia esculenta L* .
Tomato	*Lycopersicon esculentum* L.
Vetch (common)	*Vicia sativa* L.
Wheat	*Triticum aestivum* L.
Yam	*Dioscorea alata* L.

References

Chapter 1

Asana, R. D. 1968. Growth habits of crops of non-irrigated areas: Important characteristics of plant types. *Indian Farming* 18:25-27.

Austin, R. B., J. Bingham, R. D. Blackwell, L. T. Evans, M. A. Ford, C. L. Morgan, and M. Tylor. 1980. Genetic improvement in winter wheat yields since 1900 and associated physiological changes. *J. Agric. Sci.* 94:675-689.

Batch, J. J. 1981. Recent development in growth regulators for cereal crops, pp. 371-378. In: F. C. Peacock (ed.) *Outlook on agriculture*, Vol. 10. Berkshire, UK: Imperial Chemical Industries.

Blum, A., C. Y. Sullivan, and H. T. Nguyen. 1997. The effect of plant size on wheat response to agents of drought stress: II. Water deficit, heat and ABA. *Aust. J. Plant Physiol.* 24:43-48.

Braun, H. J., S. Rajaram, and M. V. Ginkel. 1997. CIMMYT's approach to breeding for wide adaptation, pp. 197-205. In: P. M. A. Tigerstedt (ed.) *Adaptation in plant breeding*. Dordrecht, the Netherlands: Kluwer Academic Publishers.

Bridge, R. R., and W. R. Meredith. 1983. Comparative performance of obsolete and current cotton cultivars. *Crop Sci.* 23:949-952.

Bridge, R. R., W. R. Meredith, and J. F. Chism. 1971. Comparative performance of obsolete varieties and current varieties of upland cotton. *Crop Sci.* 11:29-32.

Brown, J. H., G. Paliyath, and J. E. Thompson. 1991. physiological mechanisms of plant senescence, pp. 227-275. In: F. C. Steward (ed.) *Plant physiology, a treatise*, Vol. 10, *Growth and development*. New York: Academic Press.

Byerlee, D., and B. C. Curtis. 1988. Wheat: A crop transformed. *Span* 30:110-113.

Calderini, D. F., and G. A. Slafer. 1998. Changes in yield and yield stability in wheat during the 20th century. *Field Crops Res.* 57:335-347.

Campbell, G. S., and J. M. Norman. 1990. The description and measurement of plant canopy structure, pp. 1-20. In: G. Russell, B. Marshall, and P. G. Jarvis (eds.) *Plant canopies: Their growth, form and function*. Cambridge, UK: Cambridge University Press.

Chang, T. T. 1976. Rice, pp. 98-104. In: N. W. Simmonds (ed.) *Evolution of crop plants*. London: Longman.

Colombo, B., J. R. Kiniry, and P. Debaeke. 2000. Effect of soil phosphorus on leaf development and senescence dynamics of field grown maize. *Agron. J.* 92:428-435.

Counce, P. A., B. R. Wells, and K. A. Gravois. 1992. Yield and harvest index responses to preflood nitrogen fertilization at low rice plant populations. *J. Prod. Agric.* 5:492-497.

Craufurd, P. Q., T. R. Wheeler, R. H. Ellis, R. J. Summerfield, and J. H. Williams. 1999. Effect of temperature and water deficit on water-use efficiency, carbon isotope discrimination, and specific leaf area in peanut. *Crop Sci.* 39:136-142.

Cuevas-Perez, F. E., L. E. Berrio, D. I. Gonzalez, F. Correa-Victoria, and E. Tulande. 1995. Genetic improvement in yield of semidwarf rice cultivars in Colombia. *Crop Sci.* 35:725-729.

Denis, J. C., and M. W. Adams. 1978. A factor analysis of variables related to yield in dry beans: I. Morphological traits. *Crop Sci.* 18:74-78.

Dofing, S. M., and M. G. Karlsson. 1993. Growth and development of uniculm and conventional tillering barley lines. *Agron. J.* 85:58-61.

Donald, C. M. 1968. The breeding of crop ideotypes. *Euphytica* 17:385-403.

Duncan, W. G. 1975. Maize, pp. 23-50. In: L. T. Evans (ed.) *Crop physiology: Some case histories.* Cambridge, UK: Cambridge University Press.

Eberhart, S. A., and W. A. Russell. 1966. Stability parameters for comparing varieties. *Crop Sci.* 6:357-360.

Edmeades, G. O., and H. R. Lafitte. 1993. Defoliation and plant density effects on maize selected for reduced plant height. *Agron. J.* 85:850-857.

Evans, L. T. 1976. Physiological adaptation to performance as crop plants. *Phil. Trans. R. Soc. London.* 275:71-83.

Evans, L. T. 1980. The natural history of crop yield. *Am. Sci.* 68:388-397.

Evans, L. T., R. M. Visperas, and B. S. Vergara. 1984. Morphological and physiological changes among rice varieties used in the Philippines over the last seventy years. *Field Crops Res.* 8:105-124.

Fageria, N. K. 1998. *Annual report of the project "The study of liming and fertilization for rice and common bean in cerrado region."* National Rice and Bean Research Center of EMBRAPA, Goiania, Brazil.

Fageria, N. K., and V. C. Baligar. 1997. Upland rice genotypes evaluation for phosphorus use efficiency. *J. Plant Nutr.* 20:499-509.

Fageria, N. K., V. C. Baligar, and C. A. Jones. 1997. Growth and mineral nutrition of field crops. 2nd ed. New York: Marcel Dekker.

Fageria, N. K., A. B. Santos, and V. C. Baligar. 1997. Phosphorus soil test calibration for lowland rice on an Inceptisol. *Agron. J.* 89:737-742.

Feyerherm, A. M., K. E. Kemp, and G. M. Paulsen. 1988. Wheat yield analysis in relation to advancing technology in the Midwest United States. *Agron. J.* 80:998-1001.

Finlay, K.W., and G.W. Wilkinson. 1963. The analysis of adaptation in a plant breeding program. *Aust. J. Agric. Res.* 14:742-754.

Fischer, K. S., and A. F. E. Palmer. 1984. Tropical maize, pp. 231-248. In: P. R. Goldsworthy and N. M. Fisher (eds.) *The physiology of tropical field crops.* New York: John Wiley & Sons.

Gan, S., and R. M. Amasino. 1997. Making sense of senescence. *Plant Physiol.* 113:313-319.

Gebbing, T., and H. Schnyder. 1999. Pre-anthesis reserve utilization for protein and carbohydrate synthesis in grains of wheat. *Plant Physiol.* 121:871-878.

Goldsworthy, P. R., A. F. E. Palmer, and D. W. Sperling. 1974. Growth and yield of lowland tropical maize in Mexico. *J. Agric. Sci.* (Cambridge). 83:223-230.

Gouis, J. L., and P. Pluchard. 1997. Genetic variation for nitrogen use efficiency in winter wheat (*Triticum aestivum* L.), pp. 243-246. In: P. M. A. Tigerstedt (ed.) *Adaptation in plant breeding.* Dordrecht, the Netherlands: Kluwer Academic Publishers.

Handa, K. 1995. Differentiation and development of tiller buds, pp. 61-65. In: T. Matsuo, K. Kumazawa, R. Ishii, K. Ishihara, and H. Hirata (eds.) *Science of the rice plant: Physiology,* Vol. 2. Tokyo: Food and Agriculture Policy Research Center.

Heitholt, J. J., W. T. Pettigrew, and W. R. Meredith Jr. 1992. Light interception and lint yield of narrow-row cotton. *Crop Sci.* 32:728-733.

Hoshikawa, K. 1989. *The growing rice plant.* Tokyo: Nobunkyo.

International Rice Research Institute (IRRI). 1988. *Standard evaluation system for rice.* 3rd ed., Los Baños, Philippines: IRRI.

International Rice Research Institute (IRRI). 1989. *IRRI toward 2000 and beyond.* Los Baños, Philippines: IRRI.

International Rice Research Institute (IRRI). 1993. *IRRI program report for 1993.* Los Baños, Philippines: IRRI.

Jennings, P. R., W. R. Coffman, and H. E. Kauffman. 1979. *Rice improvement.* Los Baños, Philippines: IRRI.

Johnson, E. C., K. S. Fischer, G. O. Edmeades, and A. F. E. Palmer. 1986. Recurrent selection for reduced plant height in lowland tropical maize. *Crop Sci.* 26: 253-260.

Jordan, W. R. 1983. Whole plant response to water deficits: An overview, pp. 283-317. In: H. Tylor, W. R. Jordan, and T. R. Sinclair (eds.) *Limitations to efficient water use in crop production.* Madison, WI: American Society of Agronomy.

Kang, M. S. 1998. Using genotype-by-environment interaction for crop cultivar development. *Adv. Agronomy.* 35:199-240.

Khan, A., and L. Spilde. 1992. Agronomic and economic response of spring wheat cultivars to ethephon. *Agron. J.* 84:399-402.

Khush, G. S. 1995. Increased genetic potential of rice yield: Methods and perspectives, pp. 13-29. In: B. S. Pinheiro and E. P. Guimarais (eds.) *Rice in Latin America: Perspectives to increase production and yield potential.* Document No. 60. Goiânia, Brazil: EMBRAPA-CNPAF.

Kozlowski, T .T. 1973. Extent and significance of shedding plant parts, pp. 1-37. In: T. T. Kozlowski (ed.) *Shedding of plant parts.* New York: Academic Press.

Lauer, J. G., and S. R. Simmons. 1985. Photoassimilate partitioning of main shoot leaves in field grown spring barley. *Crop Sci.* 25:851-855.

Lauer, J. G., and S. R. Simmons. 1989. Canopy light and tiller mortality in spring barley. *Crop Sci.* 29:420-424.

Liang, G. S., J. R. Muo, and C. F. Ran. 1986. Rice physiology and ecology, pp. 159-220. In: Chinese Academy of Agric. Sciences (ed.) *Rice cultivation in China.* Beijing: Chine Agric. Press.

Lu, Z., and P. M. Neumann. 1999. Low cell-wall extensibility can limit maximum leaf growth rates in rice. *Crop Sci.* 39:126-130.

Lynch, J. 1995. Root architecture and plant productivity. *Plant Physiol.* 109:7-13.

Meredith, W. R., and R. Wells. 1989. Potential for increasing cotton yields through enhanced partitioning to reproductive structures. *Crop Sci.* 29:636-639.

Miller, B. C., J. E. Hill, and S. R. Roberts. 1991. Plant population effects on growth and yield in water-seeded rice. *Agron. J.* 83:291-297.

Mock, J. J., and R. B. Pearce. 1975. An ideotype of maize. *Euphytica* 24:613-623.

Morais, O. P. 1998. Annual report of the project, "Breeding upland rice cultivars." Goiania, Brazil: National Rice and Bean Research Center of EMBRAPA.

Neild, R. E., and M. W. Seeley. 1977. Growing degree days predictions for corn and sorghum development and some applications to crop production in Nebraska. Agric. Exp. Stn. Res. Bull. 280. Lincoln: Univ. of Nebraska.

Nooden, L. D. 1980. Senescence in the whole plant, pp. 220-258. In: K.V. Thimann (ed.) *Senescence in plants.* Boca Raton, FL: CRC Press.

Nooden, L. D. 1988. Whole plant senescence, pp. 281-327. In: L. D. Nooden and A. C. Leopold (eds.) *Senescence and aging in plants.* San Diego, CA: Academic Press.

Nooden, L. D., J. J. Guiamet, and I. John. 1997. Senescence mechanisms. *Physiol. Plant.* 101:746-753.

Okatan, Y., G. M. Kahanak, and L. D. Nooden. 1981. Characterization and kinetics of maturation and monocarpic senescence. *Physiol. Plant.* 52:330-338.

Ori, N., M. T. Juarez, D. Jackson, D. Yamaguchi, and G. M. Banowetz. 1999. Leaf senescence is delayed in tobacco plants expressing the maize homeobox gene *knotted1* under the control of a senescence-activated promoter. *Plant Cell* 11: 1073-1080.

Ortiz, R., and H. Langie. 1997. Path analysis and ideotypes for plantain breeding. *Agron. J.* 89:988-994.

Palta, J. A., T. Kobata, N. C. Turner, and I. R. Fillery. 1994. Remobilization of carbon and nitrogen in wheat as influenced by post-anthesis water deficits. *Crop Sci.* 43:118-124.

Pendleton, J. W., G. E. Smith, S. R. Winter, and T. J. Johnson. 1968. Field investigations of the relationships of leaf angle in corn to grain yield and apparent photosynthesis. *Agron. J.* 6:422-424.

Perkins, J. M., and J. L. Jinks. 1968. Environmental and genotype-environmental components of variability: III. Multiple lines and crosses. *Heredity* 23:339-356.

Rao, M. S. S., B. G. Mullinix, M. Rangapa, E. Cebert, A. S. Bhagsari, V. T. Sapra, J. M. Joshi, and R. B. Dadson. 2002. Genotype × environment interactions and yield stability of food-grade soybean genotypes. *Agron. J.* 94:72-80.

Rasmusson, D. C. 1987. An evaluation of ideotype breeding. *Crop Sci.* 27:1140-1146.

Redfearn, D. D., K. J. Moore, K. P. Vogel, S. S. Waller, and R. B. Mitchell. 1997. Canopy architecture and morphology of switch grass populations differing in forage yield. *Agron. J.* 89:262-269.

Rees, D., K. Sayre, E. Acevedo, T. N. Sanchez, Z. Lu, E. Zeiger, and L. Limon. 1993. *Canopy temperatures of wheat: Relationship with yield and potential as a technique for early generation selection.* Wheat Special Report No 10. Mexico City: CIMMYT.

Reta-Sánchez, D. G., and J. L. Fowler. 2002. Canopy light environment and yield of narrow-row cotton as affected by canopy architecture. *Agron. J.* 94:1317-1323.

Rhodes, I. 1975. The relationship between productivity and some components of canopy structure in ryegrass: IV. Canopy characters and their relationship with sward yield in some intra population selection. *J. Agric. Sci.* (Cambridge). 84:345-351.

Saeed, M., and C. A. Francis. 1984. Association of weather variables with genotype × environment interactions in grain sorghum. *Crop Sci.* 24:13-20.

Schmidt, J. W. 1984. Genetic contributions to yield gains in wheat, pp. 89-101. In: W. R. Fehr (ed.) *Genetic contributions to yield gains of five major crop plants.* Spec. Publ. 7. Madison, WI: Crop Sci. Soc. Am.

Siegenthaler, V. L., J. E. Stepanich, and L. W. Briggle. 1986. Distribution of the varieties and classes of wheat in the United States, 1984. *USDA Sta. Bull.* 739. Washington, DC: U.S. Gov. Print. Office.

Simmons, S. R., and J. G. Lauer. 1986. Shoot photoassimilate partitioning patterns during the tillering phase in spring barley, pp. 519-526. In: J. Cronshow, W. J. Lucas, and R. T. Giaquinta (eds.) *Phloem transport.* New York: Alan R. Liss.

Simmons, S. R., D. C. Rasmusson, and J. V. Wiersma. 1982. Tillering in barley: Genotype, row spacing and seedling rate effects. *Crop Sci.* 22:801-805.

Sinha, S. K., and M. S. Swaminathan. 1984. New parameters and selection criteria in plant breeding, pp. 1-31. In: P. B. Vose and S. G. Blixt (eds.) *Crop breeding: A contemporary basis.* Oxford, UK: Pergamon Press.

Slafer, G. A., D. F. Calderini, and D. J. Miralles. 1996. Generation of yield components and compensation in wheat: Opportunities for further increasing yield potential, pp. 101-133. In: M. Reynolds (ed.) *Increasing yield potential in wheat: Breaking the barriers.* CIMMYT Int. Symp. CIANO, Cd. Obregon, Mexico. Mexico City: CIMMYT.

Slafer, G. A., E. H. Satorre, and F. H. Andrade. 1994. Increase in grain yield in bread wheat from breeding and associated physiological changes, pp. 1-68. In: G. A Slafer (ed.) *Genetic improvement in field crops.* New York: Marcel Dekker.

Spiertz, J. H., and J. Ellen. 1972. The effect of light intensity on some morphological and physiological aspects of the crop perennial ryegrass (*Lolium perenne* L.) and its effect on seed production. *Neth. J. Agric. Sci.* 20:232-246.

Stoskopf, N. C. 1981. *Understanding crop production*. Reston, VA: Reston Publishing Co.

Thomas, H., and J. L. Stoddart. 1980. Leaf senescence. *Annu. Rev. Plant Physiol.* 31:83-111.

Tollenaar, M., and A. Aguilera. 1992. Radiation use efficiency of an old and a new maize hybrid. *Agron. J.* 84:536-541.

Tomlinson, P. B. 1987. Architecture of tropical plants. *Annu. Rev. Ecol.* 18:1-21.

Vandenberg, A., and T. Nleya. 1999. Breeding to improve plant type, pp. 167-183. In: S. Singh (ed.) *Common bean improvement in the twenty-first century*. Dordrecht, the Netherlands: Kluwer Academic Publishers.

Welles, J. M., and J. M. Norman. 1991. Instrument for indirect measurement of canopy architecture. *Agron. J.* 83:818-825.

Wells, R., and W. R. Meredith. 1984. Comparative growth of obsolete and modern cotton cultivars: III. Relationship of yield to observed growth characteristics. *Crop Sci.* 24:868-872.

Williams, J. F., S. R. Roberts, J. E. Hill, S. C. Scardaci, and G. Tibbitts. 1990. Managing water for weed control in rice. *California Agric.* 44(5):7-10.

Willmot, D. B., G. E. Pepper, and D. Nafziger. 1989. Random stand deficiency and replanting delay effects on soybean yield, yield components, canopy, and morphological responses. *Agron. J.* 81:425-430.

Wolfe, D. W., D. W. Henderson, T. C. Hsiao, and A. Alvino. 1988a. Interactive water and nitrogen effects on senescence of maize: I. Leaf area duration, nitrogen distribution, and yield. *Agron. J.* 80:859-864.

Wolfe, D. W., D. W. Henderson, T. C. Hsiao, and A. Alvino. 1988b. Interactive water and nitrogen effects on senescence of maize: II. Photosynthetic decline and longevity of individual leaves. *Agron. J.* 80:865-870.

Wu, G., L. T. Wilson, and A. M. McClung. 1998. Contribution of rice tillers to dry matter accumulation and yield. *Agron. J.* 90:317-323.

Yang, J., J. Zg, L. Liu, Z. Wang, and Q. Zhu. 2002. Carbon remobilization and grain filling in Japonica/Indica hybrid rice subjected to postanthesis water deficits. *Agron. J.* 94:102-109.

Yang, J., Z. Zhang, Z. Huang, Q. Zhu, and L. Wang. 2000. Remobilization of carbon reserves is improved by controlled soil drying during grain filling of wheat. *Crop Sci.* 40:1645-1655.

Yang, J., J. Zhang, Z. Wang, Q. Zhu, and L. Liu. 2001. Water deficit induced senescence and its relationship to the remobilization of pre-stored carbon in wheat during grain filling. *Agron. J.* 93:196-206.

Yoshida, S. 1972. Physiological aspects of grain yield. *Annu. Rev. Plant Physiol.* 23:437-464.

Yoshida, S. 1981. *Fundamentals of rice crop science*. Los Baños, Philippines: International Rice Research Institute.

Zelitch, I. 1982. The close relationship between net photosynthesis and crop yield. *BioScience* 32:796-802.

Zhang, J., X. Sui, B. Li, B. Su, J. Li, and D. Zhou. 1998. An improved water use efficiency for winter wheat grown under reduced irrigation. *Field Crops Res.* 59:91-98.

Chapter 2

Adepetu, J. A., and L. K. Akapa. 1977. Root growth and nutrient uptake characteristics of some cowpea varieties. *Agron. J.* 69:940-943.

Alva, A. K., F. P. C. Blamey, D. G. Edwards, and C. J. Asher. 1986. An evaluation of aluminum indices to predict aluminum toxicity to plants grown in nutrient solutions. *Commun. Soil Sci. Plant Anal.* 17:1271-1280.

Anderson, E. L. 1987. Corn root growth and distribution as influenced by tillage and nitrogen fertilization. *Agron. J.* 79:544-549.

Armstrong, J. 1979. Aeration in higher plants, pp. 226-332. In: H. W. Woolhouse (ed.) *Advances in botanical research*, Vol. 7. London: Academic Press.

Aung, L. H. 1974. Root-shoot relationships, pp. 29-61. In: E. W. Carson (ed.) *The plant root and its environment.* Charlottesville: University Press of Virginia.

Baligar, V. C. 1986. Interrelationships between growth and nutrient uptake in alfalfa and corn. *J. Plant Nutr.* 9:1391-1404.

Baligar, V. C., N. K. Fageria, and M. Elrashidi. 1998. Toxicity and nutrient constraints on root growth. *HortSci.* 33:960-965.

Baligar, V. C., R. J. Wright, K. D. Ritchey, J. L. Ahlrichs, and B. K. Woolum. 1991. Soil and solution properties effects on root growth of aluminum tolerant and intolerant wheat cultivars, pp. 245-252. In: R. J. Wright, V. C. Baligar, and R. P. Murrmann (eds.) *Plant-soil interactions at low pH.* Dordrecht, the Netherlands: Kluwer.

Barber, S. A. 1995. *Soil nutrient bioavailability: A mechanistic approach.* 2nd ed. New York: John Wiley & Sons.

Barley, K. P. 1970. The configuration of the root system in relation to nutrient uptake. *Adv. Agron.* 22:159-198.

BassiriRad, H., J. W. Radin, and N. Matsuda. 1991. Temperature-dependent water and ion transport properties of barley and sorghum roots: I. Relationship to leaf growth. *Plant Physiol.* 97:426-432.

Bauhus, J., and C. Messier. 1999. Evaluation of fine root length and diameter measurements obtaining using RHIZO image analysis. *Agron. J.* 91:142-147.

Belford, R. K., and F. K. G. Henderson. 1985. Measurement of growth of wheat roots using a TV camera system in the field, pp. 99-105. In: W. Day and R. K. Atrin (eds.) *Wheat growth and modeling.* New York: Plenum Publ.

Beyrouty, C. A., B. R. Wells, R. J. Norman, J. N. Marvel, and J. A. Pillow. 1987. Characterization of rice roots using a minirhizotron technique, pp. 99-108. In: H. M. Taylor (ed.) *Minirhizotron observation tubes: Methods and applications for measuring rhizosphere dynamics.* Spec. Publ. 50. Madison, WI: Am. Soc. Agron.

Beyrouty, C. A., B. R. Wells, R. J. Norman, J. N. Marvel, and J. A. Pillow. 1988. Root growth dynamics of a rice cultivar grown at two locations. *Agron. J.* 80:1001-1004.

Bland, W. L. 1993. Cotton and soybean root system growth in three soil temperature regimes. *Agron. J.* 85:906-917.

Bland, W. L., and W. A. Dugas. 1988. Root length density from minirhizotron observations. *Agron. J.* 82:1024-1026.

Bohm, W. 1979. *Methods of studying root systems.* New York: Springer-Verlag.

Boote, K. J., J. R. Stansell, A. M. Schubert, and J. F. Stone. 1982. Irrigation, water use, and water relationships, pp. 164-205. In: H. E. Pattee and C. T. Yong (eds.) *Peanut science and technology.* Yoakum, TX: Am. Peanut Res. Educ. Soc.

Borg, H., and D. W. Grimes. 1986. Depth development of roots with time: An empirical description. *Trans. Am. Soc. Agric. Eng.* 29:194-197.

Bosemark, N. O. 1954. The influence of nitrogen on root development. *Physiol. Plant.* 7:497-502.

Bragg, P. L., G. Govi, and R. Q. Cannell. 1983. A comparison of methods, including angled and vertical minirhizotrons for studying root growth and distribution in a spring oat crop. *Plant Soil* 73:435-440.

Brar, G. S., J. F. Gomez, B. L. McMichael, A. G. Matches, and H. M. Taylor. 1990. Root development of 12 forage legumes as affected by temperature. *Agron. J.* 82:1024-1026.

Bruce, R. C., L. A. Warrell, D. G. Edwards, and L. C. Bell. 1988. Effects of aluminum and calcium in the soil solution of acid soils on root elongation of *Glycine max* cv. Forrest. *Aust. J. Agric. Res.* 39:319-338.

Burke, J. J., J. R. Mahan, and J. L. Hontfield. 1988. Crop specific thermal kinetic windows in relation to wheat and cotton biomass production. *Agron. J.* 80:553-556.

Cakmak, I., C. Hengeler, and H. Marschner. 1994. Changes in phloem export of sucrose in leaves in response to phosphorus, potassium and magnesium deficiency in bean plants. *J. Exp. Bot.* 45:1251-1257.

Castillo, S. R., R. H. Dowdy, J. M. Bradford, and W. E. Larson. 1982. Effects of applied mechanical stress on plant growth and nutrient uptake. *Agron. J.* 74:526-530.

Chairidchai, P., and G. S. P. Ritchie. 1993. The effect of citrate and pH on zinc uptake by wheat. *Agron. J.* 85:322-328.

Champigny, M. L., and A. Talouizte. 1981. Photosynthate distribution and metabolic fate in relation to nitrogen metabolism in wheat seedlings, pp. 645-652. In: G. Akoyunoglou (ed.) *Photosynthesis: IV. Regulation of carbon metabolism.* Philadelphia: Balban Inf. Science Services.

Chanway, C. P., and L. M. Nelson. 1991. Tissue culture bioassay for plant growth promoting rhizobacteria. *Soil Biol. Biochem.* 24:331-333.

Clark, R. B., and S. K. Zeto. 2000. Mineral acquisition by arbuscular mycorrhizal plants. *J. Plant Nutr.* 23:867-902.

Clemens, J., and P. G. Jones. 1978. Modification of drought resistance by water stress conditioning in Acacia and Eucalyptus. *J. Exp. Bot.* 29:895-904.

Comfort, S. D., G. L. Malzer, and R. H. Busch. 1988. Nitrogen fertilization of spring wheat genotypes: Influence on root growth and soil water depletion. *Agron. J.* 80:114-120.

Costa, C., L. M. Dwyer, X. Zhou, P. Dutilleul, C. Hamel, L. M. Reid, and D. L. Smith. 2002. Root morphology of contrasting maize genotypes. *Agron. J.* 94:96-101.

Cutforth, H. W., C. F. Shaykewich, and C. M. Cho. 1986. Effect of soil water and temperature on corn root growth during emergence. *Can. J. Soil Sci.* 66:51-58.

Davidson, R. L. 1969. Effects of soil nutrients and moisture on root/shoot ratios in *Lolium perenne* L. and *Tifolium repens* L. *Ann. Bot.* (London) 33:571-577.

Devries, J. D., J. M. Bennett, S. L. Albrecht, and K. J. Boote. 1989. Water relations, nitrogenase activity and root development of three grain legumes in response to soil water deficits. *Field Crop Res.* 21:215-226.

Doss, B. D., W. T. Dumas, and Z. F. Lund. 1979. Depth of lime incorporation for correction of subsoil acidity. *Agron. J.* 71:541-544.

Drew, M. C., and J. M. Lynch. 1980. Soil anaerobiosis microorganisms and root function. *Annu. Rev. Phytopathol.* 8:37-66.

Durieux, R. P., E. J. Kamprath, W. A. Jackson, and R. H. Moll. 1994. Root distribution of corn: The effect of nitrogen fertilization. *Agron. J.* 86:958-962.

Eghball, B., and J. W. Maranville. 1993. Root development and nitrogen influx of corn genotypes grown under combined drought and nitrogen stresses. *Agron. J.* 85:147-152.

Eghball, B., J. R. Settimi, J. W. Maranville, and A. M. Parkhurst. 1993. Fractal analysis for morphological description of corn roots under nitrogen stress. *Agron. J.* 85:287-289.

Entz, M. H., K. G. Gross, and D. B. Fowler. 1992. Root growth and soil water extraction by winter and spring wheat. *Can. J. Plant Sci.* 72:1109-1120.

Evans, L. T., and I. F. Wardlaw. 1976. Aspects of the comparative physiology of grain yield in cereals. *Adv. Agron.* 28:301-359.

Fageria, N. K. 1992. *Maximizing crop yields.* New York: Marcel Dekker.

Fageria, N. K. 2000. Adequate and toxic levels of boron for rice, common bean, corn, soybean and wheat production in cerrado soil. *Brazilian J. Agric. Eng. Environ.* 4:57-62.

Fageria, N. K. 2001. Effect of liming on upland rice, common bean, corn, and soybean production in cerrado soil. *Pesq. Agropec. Bras.* 36:1419-1424.

Fageria, N. K., and V. C. Baligar. 1993. Screening crop genotypes for mineral stresses, pp. 142-159. In: J. W. Maranville, V. C. Baligar, R. R. Duncan, and J. M. Yohe (eds.) *Proceedings of the Workshop on Adaptation of Plants to Soil Stresses.* University of Nebraska, Lincoln. Lincoln, NE: INTSORMIL/USAID.

Fageria, N. K., V. C. Baligar, and R. B. Clark. 2002. Micronutrients in crop production. *Adv. Agron.* 77:185-268.

Fageria, N. K., V. C. Baligar, and D. G. Edwards. 1990. Soil-plant nutrient relationships at low pH stress, pp. 475-507. In: V. C. Baligar and R. R. Duncan (eds.) *Crops as enhancers of nutrient use.* San Diego, CA: Academic Press.

Fageria, N. K., V. C. Baligar, and R. J. Wright. 1997. Soil environment and root growth dynamics of field crops. *Recent Res. Devel. Agron.* 1:15-58.

Fageria, N. K., and C. M. R. Souza. 1991. Upland rice, common bean, and cowpea response to magnesium application on an Oxisols. *Commun. Soil Sci. Plant Anal.* 22:1805-1816.

Fageria, N. K., R. J. Wright, V. C. Baligar, and J. R. P. Carvalho. 1991. Response of upland rice and common bean to liming on an Oxisol, pp. 519-525. In: R. J. Wright, V. C. Baligar, and R. P. Murrmann (eds.) *Plant-soil interactions at low pH*. Dordrecht, the Netherlands: Kluwer Acad. Publ.

Fageria, N. K., and F. J. P. Zimmermann. 1996. Influence of pH on growth and nutrient uptake by crop species in an Oxisols. IV Int. Symposium on Plant-Soil Interactions at Low pH, March 17-24, Belo Horizonte, Brazil.

Fitter, A. H. 1982. Morphometric analysis of root systems: Application of the technique and influence of soil fertility on root system development in two herbaceous species. *Plant Cell Environ.* 5:313-322.

Foy, C. D. 1992. Soil chemical factors limiting plant root growth. *Adv. Soil Sci.* 19:97-149.

Fredeen, A. L., I. M. Rao, and N. Terry. 1989. Influence of phosphorus nutrition on growth and carbon partitioning in *Glycine max*. *Plant Physiol.* 89:225-230.

Garrity, D. P., C. Y. Sullivan, and D. G. Watts. 1983. Moisture deficits and grain sorghum performance: Drought stress conditions. *Agron. J.* 75:997-1004.

Gonzalez-Erico, E., E. J. Kamprath, G. C. Nederman, and W. V. Soares. 1979. Effect of depth of lime incorporation on the growth of corn on an Oxisol of central Brazil. *Soil Sci. Soc. Am. J.* 43:1155-1158.

Granato, T. C., and C. D. Raper. 1989. Proliferation of maize roots in response to localized supply of nitrate. *J. Exp. Bot.* 40:263-275.

Hackett, C., and D. A. Rose. 1972. A model of the extension and branching of a seminal root of barley and its use in studying relations between root dimensions: I. The model. *Aust. J. Biol. Sci.* 25:669-679.

Hallmark, W. B., and S. A. Barber. 1984. Root growth and morphology, nutrient uptake, and nutrient status of early growth of soybeans as affected by soil P and K. *Agron. J.* 76:209-212.

Hatlitligil, M. B., R. A. Olson, and W. A Compton. 1984. Yield, water use, and nutrient uptake of corn hybrids under varied irrigation and nitrogen regimes. *Fert. Res.* 5:321-333.

Hernandez-Armenta, R., W. C. Wien, and A R. J. Eaglesham. 1989. Carbohydrate partitioning and nodule function in common bean after heat stress. *Crop Sci.* 29:1292-1297.

Horiguchi, T. 1995. Rhizosphere and root oxidation activity, pp. 221-248. In: T. Matsuo, K. Kumazawa, R. Ishii, K. Ishihara, and H. Hirata (eds.) *Science of the rice plant*, Vol. 2. Tokyo: Food Agric. Policy Res. Center.

Hurd, E. A. 1974. Phenotype and drought tolerance in wheat. *Agric. Meteorol.* 14:39-55.

Jackson, M. B. 1982. Ethylene as a growth promoting hormone under flooded conditions, pp. 291-301. In: P. F. Waring (ed.) *Plant growth substances.* London: Academic Press.

Jackson, W. A., W. L. Pan, R. H. Moll, and E. J. Kamprath. 1986. Uptake, translocation, and reduction of nitrate, pp. 95-98. In: C. A. Neyra (ed.) *Biochemical basis of plant breeding, Vol. 2, Nitrogen metabolism.* Boca Raton, FL: CRC Press.

Jaffar, M. N., L. R. Stone, and D. E. Goodwin. 1993. Rooting depth and dry matter development of sunflower. *Agron. J.* 85:281-286.

Jama, A. O., and M. J. Ottman. 1993. Timing of the first irrigation in corn and water stress conditions. *Agron. J.* 85:1159-1164.

Jordan, W. R., and F. R. Miller. 1980. Genetic variability in sorghum root: systems: Implications for drought tolerance, pp. 383-399. In: N. C. Turner and P. J. Kramer (eds.) *Adaptation of plants to water and high temperature stress.* New York: John Wiley & Sons.

Kamprath, E. J., and C. D. Foy. 1985. Lime-fertilizer-plant interactions in acid soils, pp. 91-151. In: O. P. Engelstad (ed.) *Fertilizer technology and use,* 3rd Ed. Madison, WI: Soil Sci. Soc. Am.

Kaspar, T. C., H. M. Taylor, and R. M. Shibles. 1984. Taproot elongation rates of soybean cultivars in the glasshouse and their relation to field rooting depth. *Crop Sci.* 24:916-920.

Ketring, D. L. 1984. Root diversity among peanut genotypes. *Crop Sci.* 24:229-232.

Ketring, D. L., and J. L. Reid. 1993. Growth and peanut roots under field conditions. *Agron. J.* 85:80-85.

Kitchen, N. R., D. D. Buchholz, and C. J. Nelson. 1990. Potassium fertilizer and potato leafhopper effects on alfalfa growth. *Agron. J.* 82:1069-1074.

Kleeper, B., R. K. Belford, and R. W. Rickman. 1984. Root and shoot development in winter wheat. *Agron. J.* 76:117-122.

Kleeper, B., and T. C. Kaspar. 1994. Rhizotrons: Their development and use in agricultural research. *Agron. J.* 88:745-753.

Kludze, H. K., and R. D. DeLaune. 1995. Straw application effects on methane and oxygen exchange and growth in rice. *Soil Sci. Soc. Am. J.* 59:824-830.

Kucey, R. M. N., H. H. Janzen, and M. E. Leggett. 1989. Microbial mediated increases in plant availability P. *Adv. Agron.* 42:159-228.

Laan, P., M. J. Benevoets, S. Lythe, W. Armstrong, and. P. M. Bloom. 1989. Root morphology and aerenchyma formation as indicators of the flood-tolerance of rumex species. *J. Ecology* 77:693-703.

Lynch, J. 1995. Root architecture and plant productivity. *Plant Physiol.* 109:7-13.

Mackay, A. D., and S. A. Barber. 1985. Soil moisture effects on root growth and phosphorus uptake by corn. *Agron. J.* 77:519-523.

Mackay, A. D., and S. A Barber. 1986. Effect of nitrogen on root growth of two corn genotypes in the field. *Agron. J.* 78:659-703.

Marschner, H. 1991. Root-induced changes in the availability of micronutrients in the rhizosphere, pp. 503-528. In: Y. Waisel, A. Eshel, and U. Kafkafi (eds.) *Plant roots: The hidden half.* New York: Marcel Dekker.

Mason, W. K., H. R. Rowse, A. T. P. Bennie, T. C. Kaspar, and H. M. Taylor. 1982. Response of soybeans to two row spacing and two water levels: II. Water use, root growth and plant water status. *Field Crops Res.* 5:15-29.

McMichael, B. L., and H. M. Taylor. 1987. Applications and limitations of rhizotrons and minirhizotrons, pp. 1-13. In: H. M. Taylor (ed.) *Minirhizotron observation tube: Methods and applications for measuring rhizosphere dynamics.* Am. Soc. Agron. Spec. Publ. 50. Madison, WI: Am. Soc. Agron, Crop Sci. Soc. Am., and Soil Sci. Soc. Am.

McWilliam, J. R. 1986. The national and international importance of drought and salinity effects of agricultural production. *Aust. J. Plant Physiol.* 13:1-13.

Meisner, C. A., and K. J. Karnok. 1992. Peanut rot response to drought stress. *Agron. J.* 84:159-165.

Melhuish, F. M., and A. R. G. Lang. 1968. Quantitative studies of rots in soil: I. Length and diameters of cotton roots in a clay-loam soil by analysis of surface ground blocks of resin-impregnated soil. *Soil Sci.* 106:16-22.

Mengel, D. B., and S. A. Barber. 1974. Development and distribution of the corn root system under field conditions. *Agron. J.* 68:341-344.

Mengel, K. 1985. Dynamics and availability of major nutrients in soils. *Adv. Soil Sci.* 2:66-131.

Mengel, K., and E. A. Kirkby. 1978. *Principles of plant nutrition.* Bern, Switzerland: Int. Potash Inst.

Merrill, S. D. 1992. Pressurized-well minirhizotron for field observation of root growth. *Agron. J.* 84:755-758.

Meyer, W. S., and H. D. Baris. 1985. Nondestructive measurement of wheat roots in large undisturbed and replaced clay soil cores. *Plant Soil.* 85:237-247.

Miltner, E. D., K. J. Karnok, and R. S. Hussey. 1991. Root response of tolerant and intolerant soybean cyst nematodes. *Agron. J.* 83:571-576.

Miyasaka, S. C., and D. L. Grunes. 1990. Root temperature and calcium level effects on winter wheat forage: I. Shoot and root growth. *Agron. J.* 82:236-242.

Mullins, G. L., D. W. Reeves, C. H. Burmester, and H. H. Bryant. 1994. In-row subsoiling and potassium placement on root growth and potassium content of cotton. *Agron. J.* 86:136-139.

Murphy, J. A., M. G. Hendricks, P. E. Rieke, A. J. M. Smucker, and B. E. Branham. 1994. Turfgrass root systems evaluated using the minirhizotron and video recording methods. *Agron. J.* 86:247-250.

Narasimham, R. L., I. V. S. Rao, and M. S. Rao. 1977. Effect of moisture stress on response of groundnut to phosphate fertilization. *Ind. J. Agric. Sci.* 11:573-576.

Newman, E. I. 1966. A method of estimating the total length of root in a sample. *J. Appl. Ecol.* 3:139-145.

Onstad, D. W., C. A. Shoemaker, and B. C. Hansen. 1984. Management of potato leafhopper on alfalfa with the aid of systems analysis. *Environ. Entomol.* 13: 1046-1058.

O'Toole, J. C., and S. K. De Datta. 1983. Genotypic variation in epicuticular wax of rice. *Crop Sci.* 23:392-394.

Oussible, M., R. R. Allmaras, R. D. Wych, and R. K. Crookston. 1993. Subsurface compaction effects on tillering and nitrogen accumulation in wheat. *Agron. J.* 85:619-625.

Perrenoud, S. 1977. Potassium and plant health, pp. 1-118. In: Int. Potash Inst. (ed.) *Research topics, no. 3.* Bern, Switzerland: Int. Potash Institute.

Power, J. F., D. L. Grunes, G. A. Reichman, and W. O. Willis. 1970. Effect of soil temperature on rate of barley development and nutrition. *Agron. J.* 62:567-571.

Radcliffe, D. E., R. S. Hussey, and R. W. McClendon. 1990. Cyst nematode vs. tolerant and intolerant soybean cultivars. *Agron. J.* 82:855-860.

Reichman, G. A., and T. P. Trooien. 1993. Corn yield and salinity response to irrigation on slowly permeable subsoil. *Soil Sci. Soc. Am. J.* 57:1549-1554.

Robertson, W. K., L. C. Hammond, J. T. Johnson, and G. M. Prine. 1979. Root distribution of corn, soybeans, peanuts, sorghum and tobacco in fine sands. *Proc. Soil Crop Sci. Soc. Fl.* 38:54-59.

Roth, G. W., D. D. Calvin, and S. M. Lweloff. 1995. Tillage, nitrogen timing, and planting date effects on western corn rootworm injury to corn. *Agron. J.* 87:189-193.

Rovira, A. D., R. C. Foster, and J. K. Martin. 1983. Origin, nature and nomenclature of the organic materials in the rhizosphere, pp. 1-4. In: J. L. Harley and R. S. Russell (eds.) *Soil root interface.* London: Academic Press.

Rufty, T. W., Jr., D. W. Israel, R. J. Volk, J. Qui, and T. Sa. 1993. Phosphate regulation of nitrate assimilation in soybean. *J. Exp. Bot.* 44:879-891.

Saini, G. R., and T. L. Chow. 1982. Effect of compact subsoil and water stress on shoot and root activity of corn and alfalfa in a growth chamber. *Plant Soil* 66:291-298.

Schenk, M. K., and S. A. Barber. 1980. Potassium and phosphorus uptake by corn genotype grown in the field as influenced by root characteristics. *Plant Soil* 54:65-75.

Schomberg, H. H., and R. W. Weaver. 1992. Nodulation, nitrogen fixation, and early growth of arrowleaf clover in response to root temperature and starter nitrogen. *Agron. J.* 84:1046-1050.

Shalhevet, J., M. G. Huck, and B. P. Schroeder. 1995. Root and shoot growth responses to salinity in maize and soybean. *Agron. J.* 87:512-516.

Sharp, R. E., and W. J. Davies. 1985. Root growth and water uptake by maize plants in drying soil. *J. Exp. Bot.* 36:1441-1456.

Sharratt, B. S. 1991. Shoot growth, root length density, and water use of barley grown at different soil temperatures. *Agron. J.* 83:237-239.

Sharratt, B. S., and V. L. Cochran. 1993. Skip-row and equidistant-row barley with nitrogen placement: Yield, nitrogen uptake and root density. *Agron. J.* 85:246-250.

Silberbush, M., and S. A. Barber. 1984. Phosphorus and potassium uptake of field grown soybeans predicted by a simulation model. *Soil Sci. Soc. Am. J.* 48:592-596.

Slaton, N. A., and C. A. Beyrouty. 1992. Rice shoot response to root confinement within a membrane. *Agron. J.* 84:50-53.

Slaton, N. A., C. A. Beyrouty, B. R. Wells, R. J. Norman, and E. E. Gbur. 1990. Root growth and distribution of two short-season rice genotypes. *Plant Soil* 126:269-278.

Sponchiado, B. N., J. W. White, J. A. Castillo, and P. G. Jones. 1989. Root growth of four common bean cultivars in relation to drought tolerance in environments with contrasting soil types. *Exp. Agric.* 25:249-257.

Steponkus, P. J., J. M. Cutler, and J. C. O'Toole. 1980. Adaptation to water stress in rice, pp. 401-418. In: N. C. Turner and P. J. Kramer (eds.) *Adaptation of plants to water and high temperature stress.* New York: John Wiley & Sons.

Stone, L. F., and A. L. Pereira. 1994a. Rice-common bean rotation under sprinkler irrigation: Effects of row spacing, fertilization and cultivar on growth, root development and water consumption of common bean. *Pesq. Agropec. Bras., Brasilia* 29:939-954.

Stone, L. F., and A. L. Pereira. 1994b. Rice-common bean rotation under sprinkler irrigation: Effects of row spacing, fertilization and cultivar on growth, root development and water consumption of rice. *Pesq. Agropec. Bras., Brasilia* 29:1577-1592.

Suresh, R., C. D. Foy, and J. R. Weidner. 1989. Effects of water deficit and manganese toxicity on two cultivars of soybeans: Possible applications in remote sensing. *J. Plant Nutr.* 12:995-1003.

Talouizte, A. M., M. L. Champigny, E. Bismuth, and A. Moyse. 1984. Root carbohydrate metabolism associated with nitrate assimilation in wheat previously deprived of nitrogen. *Physiol. Vege.* 22:19-27.

Tatsumi, J., A. Yamauchi, and Y. Kono. 1989. Fractal analysis of plant root system. *Ann. Bot.* 64:499-503.

Taylor, H. M., M. G. Huck, and B. Klepper. 1972. Root development in relation to soil physical conditions, pp. 57-77. In: D. Hillel (ed.) *Optimizing the soil physical environment toward greater crop yields.* New York: Academic Press.

Tennant, D. 1975. A test of a modified line intersect method of estimating root length. *J. Ecol.* 63:955-1001.

Teo, Y. H., C. A. Beyrouty, and E. E. Gbur. 1995. Evaluation of a model to predict nutrient uptake by field grown rice. *Agron. J.* 87:7-12.

Teyker, R. H., and D. C. Hobbs. 1992. Growth and root morphology of corn as influenced by nitrogen. *Agron. J.* 84:694-700.

Thangraj, M., J. C. O'Toole, and S. K. De Datta. 1990. Root response to water stress in rainfed lowland rice. *Exp. Agric.* 28:287-296.

Thomas, A. L., and T. C. Kaspar. 1997. Maize nodal root response to time of soil ridging. *Agron. J.* 89:195-200.

Thompson, T. E., and G. W. Fick. 1981. Growth response of alfalfa to duration of soil flooding and temperature. *Agron. J.* 73:329-332.

Troughton, A. 1957. *The underground organs of herbage grass.* Bull. 44. Hurley, Berkshire, UK: Commonwealth Bur. Pasture Field Crops.

Troughton, A. 1962. *The roots of temperate cereals (wheat, barley, oats and rye).* Mimeographed Publication No. 2. Hurley, Berkshire, UK: Commonwealth Bur. Pasture Field Crops.

Tu, J. C., and C. S. Tan. 1991. Effect of soil compaction on growth, yield and root rots of white beans in clay loam and sandy loam soil. *Soil Biol. Biochem.* 23:233-238.

Tupper, G. R. 1992. Technologies to solve K deficiency: Deep placement, pp. 73-76. In: 1992 Proc. Memphis, TN: Beltwide Cotton Pred. Res. Conf. Natl. Cotton Council of Am.

University of Arkansas Cooperative Extension Service Rice Committee. 1990. *Rice production handbook.* Fayetteville: Univ. of Arkansas Coop. Ext. Serve. Misc., Pub. MP 192.

University of Kentucky. 1978. *Liming acid soils.* Leaflet AGR-19. Lexington: Author.

Upchurch, D. R. 1985. Relationship between observations in minirhizotrons and true root length density. PhD Diss., Texas Tech. Univ., Lubbock (Diss. Abstr. 85:28594).

Upchurch, D. R., and J. T. Ritchie. 1983. Root observations using a video recording system in mini-rhizotrons. *Agron. J.* 75:1009-1015.

U.S. Dept. of Interior, Bureau of Reclamation. 1978. *Drainage manual.* Washington, DC: U.S. Govt. Printing Office.

Wardlaw, I. A. 1990. The control of carbon partitioning in plants. *New Phytol.* 116:341-381.

Welbank, P. J., M. J. Gibb, P. J. Taylor, and E. D. Williams. 1974. Root growth of cereal crops, pp. 26-66. In: *Rothamsted Exper. Stn. Report for 1973, Part 2.* Hertshire, UK: Harpenden.

Welbank, P. J., and E. D. Williams. 1968. Root growth of a barley crop estimated by sampling with portable powered soil coring equipment. *J. Appl. Ecol.* 5:447.

Wiesler, F., and W. J. Horst. 1993. Differences among maize cultivars in the utilization of soil nitrate and the related losses of nitrate through leaching. *Plant Soil* 151:193-203.

Wraith, J. M., and R. J. Hanks. 1992. Soil thermal regime influence on water use and yield under variable irrigation. *Agron. J.* 84:529-536.

Wright, R. J. 1989. Soil aluminum toxicity and plant growth. *Commun. Soil Sci. Plant Anal.* 20:1479-1497.

Yamauchi, A., Y. Kono, and J. Tatsumi. 1987a. Comparison of root system structures of 13 species of cereals. *Japan. J. Crop Sci.* 56:618-631.

Yamauchi, A., Y. Kono, and J. Tatsumi. 1987b. Qualitative analysis of root system structure of upland rice and maize. *Japan J. Crop Sci.* 56:608-617.

Yibrin, H., J. W. Johnson, and D. J. Eckert. 1993. No-till corn production as affected by mulch, potassium placement, and soil exchangeable potassium. *Agron. J.* 85:639-644.

Yoshida, S. 1981. *Fundamentals of rice crop science.* Los Baños, Philippines: IRRI.

Yu, L. X., J. D. Ray, J. C. O'Toole, and H. T. Nguyen. 1995. Use of wax-petrolatum layers for screening rice root penetration. *Crop Sci.* 35:684-687.

Zhao, Z. R., G. R. Li, and G. Q. Hung. 1991. Promotive effect of potassium on adventitious root formation in some plants. *Plant Sci.* 79:47-50.

Chapter 3

Abdul-Faith, H. A., and F. A. Bazzaz. 1980. The biology of *Ambrosia trifida* L.: IV. Demography of plants and leaves. *New Phytol.* 84:107-111.

Afuakwa, J. J., R. K. Crookston, and R. J. Jones. 1984. Effects of temperature and sucrose availability on kernel black layer development in maize. *Crop Sci.* 24:285-288.

Akhter, M., and C. H. Sneller. 1996. Yield and yield components of early maturing soybean genotypes in the mid-south. *Crop Sci.* 36:877-882.

Akita, S. 1995. Dry matter production of rice population, pp. 648-690. In: T. Matsuo, K. Kumazawa, R. Ishii, K. Ishihara, and H. Hirata (eds.) *Science of the Rice Plant: Physiology,* Vol. 2. Tokyo: Food Agric. Policy Res. Center.

Austin, R. B., J. Bingham, R. D. Blackwell, L. T. Evans, M. A. Ford, C. L. Morgan, and M. Taylor. 1980. Genetic improvement in winter wheat yields since 1900 and associated physiological changes. *J. Agric. Sci.* 94:675-689.

Austin, R. B., J. A. Edrich, M. A. Ford, and R. D. Blackwell. 1977. The fate of dry matter, carbohydrates, and ^{14}C lost from leaves and stems during grain filling. *Ann. Bot.* 41:1309-1321.

Bange, M. P., G. L. Hammer, and K. G. Rickert. 1998. Temperature and sowing date after the linear increase of sunflower harvest index. *Agron. J.* 90:324-328.

Bazzaz, F. A., and J. L. Harper. 1977. Demographic analysis of the growth of *Linum usitatissimum. New Phytol.* 78:193-208.

Bhatt, G. M. 1977. Response to two-way selection for harvest index in two wheat (*Triticum aestivum* L.) crosses. *Aust. J. Agric. Res.* 28:29-36.

Bingham, J. 1969. The physiological determinants of grain yield in cereals. *Agric. Prog.* 44:30-42.

Board, J. E., M. Kamal, and B. G. Harville. 1992. Temporal importance of greater light interception to increased yield in narrow-row soybean. *Agron. J.* 84:575-579.

Board, J. E., M. S. Kang, and B. G. Harville. 1999. Path analyses of the yield formation process for late-planted soybean. *Agron. J.* 91:128-135.

Board, J. E., and Q. Tan. 1995. Assimilatory capacity effects on soybean yield components and on pod number. *Crop Sci.* 35:846-851.

Board, J. E., A. T. Wier, and D. J. Boethel. 1994. Soybean yield reductions caused by defoliation during mid to late seed filling. *Agron. J.* 86:1074-1079.

Board, J. E., A. T. Wier, and D. J. Boethel. 1997. Critical light interception during seed filling for insecticide application and optimum soybean grain yield. *Agron. J.* 89:369-374.

Board, J. E., W. Zhang, and B. G. Harville. 1996. Yield ranking for soybean cultivars grown in narrow and wide rows with late planting dates. *Agron. J.* 88:240-245.

Bos, I., and L. D. Sparnaaij. 1993. Component analysis of complex characters in plant breeding: II. The pursuit of heterosis. *Euphytica* 70:237-245.

Brown, R. H. 1984. Growth of the green plants, pp. 153-174. In: M. B. Tesar (ed.) *Physiological Basis of Crop Growth and Development*. Madison, WI: Am. Soc. Agron.

Bruckner, P. L., and R. C. Frohberg. 1987. Rate and duration of grain fill in spring wheat. *Crop Sci.* 27:451-455.

Cirilo, A. G., and F. H. Andrade. 1996. Sowing date and kernel weight in maize. *Crop Sci.* 36:325-331.

Colson, J., A. Bouniols, and J. Jones. 1995. Soybean reproductive development: Adapting a model for European cultivars. *Agron. J.* 87:1129-1139.

Corke, H., and L. W. Kannenberg. 1989. Selection for vegetative phase and actual filling period duration in short season maize. *Crop Sci.* 29:607-612.

Cregan, P. B., and R. W. Yaklich. 1986. Dry matter and nitrogen accumulation and partitioning in selected soybean genotypes of different derivation. *Theor. Appl. Gene.* 72:782-786.

Crosbie, T. M. 1982. Changes in physiological traits associated with long-term breeding effects to improve grain yield of maize, pp. 206-223. In: H. Loden and D. Wilkinson (eds.) *Proc. 37th Ann. Corn Sorghum Res. Conf.*, Chicago, IL. Washington, DC: Am. Seed Trade Assn.

Cross, H. Z. 1975. Diallel analysis of duration and rate of grain filling of seven inbred lines of corn. *Crop Sci.* 15:532-535.

Darroch, B. A., and R. J. Baker. 1990. Grainfilling in three spring wheat genotypes: Statistical analysis. *Crop Sci.* 30:525-529.

Daynard, T. B., and L. W. Kannenberg, 1976. Relationships between length of the actual and effective grain filling periods and grain yield of corn. *Can. J. Plant Sci.* 56:237-242.

Daynard, T. B., J. W. Tanner, and W. G. Duncan. 1971. Duration of the grain filling period and its relation to grain yield in corn. *Crop Sci.* 11:45-48.

Dofing, S. M. 1995. Phenological development-yield relationships in spring barley in a subarctic environment. *Can. J. Plant Sci.* 75:93-97.

Dofing, S. M. 1997. Ontogenetic evaluation of grain yield and time to maturity in barley. *Agron. J.* 89:685-690.

Dofing, S. M., and C. W. Knight. 1994. Yield component compensation in uniculm barley lines. *Agron. J.* 86:273-276.

Donald, C. M., and J. Hamblin. 1976. The biological yield and harvest index of cereals as agronomic and plant breeding criteria. *Adv. Agron.* 28:361-405.

Duncan, W. G. 1986. Planting patterns and soybean yields. *Crop Sci.* 26:584-588.

Egli, D. B. 1981. Species differences in seed growth characteristics. *Field Crops Res.* 4:1-12.

Egli, D. B. 1988. Plant density and soybean yield. *Crop Sci.* 28:977-981.

Egli, D. B. 1994. Seed growth and development, pp. 127-148. In: K. J. Boote (ed.) *Physiology and Determination of Crop Yield.* Madison, WI: Am. Soc. Agron., Crop Sci. Soc. Am., and Soil Sci. Soc. Am.

Egli, D. B., and I. F. Wardlaw. 1980. Temperature response of seed growth characteristics of soybeans. *Agron. J.* 72:560-564.

Evans, L. T. 1980. The natural history of crop yield. *Am. Sci.* 68:388-397.

Evans, L. T., and I. F. Wardlaw. 1976. Aspects of the comparative physiology of the grain yield in cereals. *Adv. Agron.* 28:301-359.

Fageria, N. K. 1992. *Maximizing Crop Yields.* New York: Marcel Dekker.

Fageria, N. K. 1998. Final report of the project, the study of liming and fertilization for rice and common bean in cerrado region. Goiania, Brazil: Nat. Rice Bean Res. Center of EMBRAPA.

Fageria, N. K., V. C. Baligar, and C. A. Jones. 1997. *Growth and Mineral Nutrition of Field Crops,* Second Edition. New York: Marcel Dekker.

Fageria, N. K., A. B. Santos, and V. C. Baligar. 1997. Phosphorus soil test calibration for lowland rice on an Inceptisols. *Agron. J.* 89:737-742.

Feil, B. 1992. Breeding progress in small grain cereals: A comparison of old and modern cultivars. *Plant Breed.* 108:1-11.

Fischer, R. A., and Z. Kertesz. 1976. Harvest index in spaced populations and grain weight in microplots as indicators of yielding ability in spring wheat. *Crop Sci.* 16:673-677.

Frederick, J. R. 1997. Winter wheat leaf photosynthesis, stomatal conductance, and leaf nitrogen concentration during reproductive development. *Crop Sci.* 37: 1819-1826.

Gbikpi, P. J., and R. K. Crookston. 1981. Effect of flowering date on accumulation of dry matter and protein in soybean seeds. *Crop Sci.* 21:652-655.

Gebeyehou, G., D. R. Knott, and R. J. Baker. 1982. Relationships among durations of vegetative and grain filling phases, yield components and grain yield in durum wheat cultivars. *Crop Sci.* 22:287-290.

Gent, M. P. N., and R. K. Kiyomoto. 1989. Assimilation and distribution of photosynthate in winter wheat cultivars differing in harvest index. *Crop Sci.* 29:120-125.

Goyne, P. J., H. Meinke, S. P. Milroy, G. L. Hammer, and J. M. Hare. 1996. Development and use of a barley crop simulation model to evaluate production management strategies in northeastern Australia. *Aust. J. Agric. Res.* 47:997-1015.

Grabau, L. J., D. A. Van Sanford, and Q. W. Meng. 1990. Reproductive characters of winter wheat cultivars subjected to postanthesis shading. *Crop Sci.* 30:771-774.

Guindo, D., B. R. Wells, and R. J. Norman. 1994. Cultivar and nitrogen rate influence on nitrogen uptake and partioning in rice. *Soil Sci. Soc. Am. J.* 58:840-845.

Guldan, S. J., and W. A. Brun. 1985. Relationship of cotyledon cell number and seed respiration to soybean seed growth. *Crop Sci.* 25:815-819.

Hardwick, R. C. 1976. Components of yield static contrivance or dynamic concept. *J. Sci. Food Agric.* 27:397-398.

Hartung, R. C., C. G. Poneleit, and P. L. Cornelius. 1989. Direct and correlated responses to selection for rate and duration of grain fill in maize. *Crop Sci.* 39:740-745.

Hay, R. K. M. 1995. Harvest index: A review of its use in plant breeding and crop physiology. *Ann. Appl. Biol.* 126:197-216.

Hellewell, K. B., D. D. Stuthman, A. H. Markhart, and J. E. Erwin. 1996. Day and night temperature effects during grain-filling in oat. *Crop Sci.* 36:624-628.

Huhn, M. 1990. Comments on the calculation of mean harvest indices. *J. Agron. Crop Sci.* 165:86-93.

Hunt, R. 1978. *Plant Growth Analysis.* London: Edward Arnold.

Johnson, R. R., D. E. Green, and C. W. Jordan. 1982. What is the best soybean row width: A U.S. perspective. *Crops Soils* 34:10-13.

Jolliffe, P. A., G. W. Eaton, and J. L. Doust. 1982. Sequential analysis of plant growth. *New Phytol.* 92:287-296.

Jones, R. J., S. Ouattar, and R. K. Crookston. 1984 Thermal environment during endosperm cell division and grain filling in maize: Effects on kernel growth and development in vitro. *Crop Sci.* 24:133-137.

Jones, R. J., B. M. N. Schreiber, and J. A. Roessler. 1996. Kernel size capacity in maize: Genotypic and maternal regulation. *Crop Sci.* 36:301-306.

Kato, T. 1993. Variation in grain-filling process among grain positions within a panicle of rice. *SABRAO J.* 25:1-10

Kato, T. 1997. Selection responses for the characters related to yield sink capacity of rice. *Crop Sci.* 37:1472-1475.

Kelly, J. D., J. M. Kolkman, and K. Schneider. 1998. Breeding for yield in dry bean (*Phaseolus vulgaris* L.). *Euphytica* 102:343-356.

Kelly, J. D., K. A. Schneider, and J. M. Kolkman. 1999. Breeding to improve yield, pp. 185-222. In: S. P. Singh (ed.) *Common Bean Improvement in the Twenty-First Century.* Dordrecht, the Netherlands: Kluwer Acad. Publ.

Knott, D. R., and G. Gebeyehou. 1987. Relationship between the lengths of the vegetative and grain filling periods and agronomic characters in three durum wheat crosses. *Crop Sci.* 27:857-860.

Kulshrestha, V. P., and H. K. Jain. 1982. Eighty years of wheat breeding in India: Past selection pressures and future prospects. *Z. Pflanzenz.* 89:19-30.

Lambers, H. 1987. Does variation in photosynthetic rate explain variation in growth rate and yield. *Netherlands J. Agric. Sci.* 35:505-519.

Landes, A., and J. R. Porter. 1989. Comparison of scales used for categorizing the development of wheat, barley, rye and oats. *Ann. Appl. Biol.* 115:343-360.

Landsberg, J. J. 1977. Effects of weather on plant development, pp. 289-307. In: J. J. Landsberg and C. V. Cuttings (eds.) *Environmental Effects on Crop Physiology.* London: Academic Press.

Matsushima, S. 1976. *High-Yielding Rice Cultivation: A Method for Maximizing Rice Yield Through Ideal Plants.* Tokyo: University of Tokyo Press.

McCauley, G. N. 1990. Sprinkler vs. flood irrigation in traditional rice production regions of southeast Texas. *Agron. J.* 82:677-683.

McEwan, J. M., and R. J. Cross. 1979. Evolutionary changes in New Zealand wheat cultivars, pp. 198-203. In: S. Ramanujam (ed.) *Proc. 5th Inter. Wheat Genetics Symp., 23-28 Feb. 1978.* New Delhi, India: Indian Soc. Genetics Plant Breeding.

Meckel, L., D. B. Egli, R. E. Phillips, D. Radcliff, and J. E. Leggett. 1984. Effect of moisture stress on seed growth in soybean. *Agron. J.* 76:647-650.

Mou, B., and W. E. Kronstad. 1994. Duration and rate of grain filling in selected winter wheat populations: I. Inheritance. *Crop Sci.* 34:833-837.

Muchow, R. C. 1990a. Effect of high temperature on grain growth in field grown maize. *Field Crop Res.* 23:145-158.

Muchow, R. C. 1990b. Effect of high temperature on the rate and duration of grain growth in field grown *Sorghum bicolor* (L.) Moench. *Field Crops Res.* 18:31-43.

Murayama, N. 1995. Development and senescence of an individual plant, pp. 119-178. In: T. Matsuo, K. Kumazawa, R. Ishii, K. Ishihara, and H. Hirata (eds.) *Science of Rice Plant: Physiology,* Vol. 2. Tokyo: Food Agric. Policy Res. Center.

Nakamura, T., M. Osaki, T. Shinano, and T. Tadano. 1997. Different mechanisms of carbon-nitrogen interaction in cereal and legume crops, pp. 913-914. In: T. Ando, K. Fujita, T. Mae, H. Tsumoto, S. Mori, and J. Sekiya (eds.) *Plant nutrition for sustainable food production and environment.* Dordrecht, the Netherlands: Kluwer Academic Publishers.

Nass, H. G. 1980. Harvest index as a selection criterion for grain yield in two spring wheat crosses grown at two population densities. *Can. J. Plant Sci.* 60:1141-1146.

Osaki, M., Y. Fujisaki, K. Morikawa, M. Matsumoto, T. Shinano, and T. Tadano. 1993. Productivity of high yielding crops. IV. Parameters determining differences of productivity among field crops. *Soil Sci. Plant Nutr.* 39:605-615.

Osaki, M., K. Morikawa, M. Matsumoto, T. Shinano, M. Iyoda, and T. Tadano. 1993. Productivity of high yielding crops. III. Accumulation of ribulose-1, 5-bisphosphate carboxylase/oxygenase and chlorophyll in relation to productivity of high yielding crops. *Soil Sci. Plant Nutr.* 39:399-408.

Osaki, M., T. Shinano, M. Matsumoto, J. Ushiki, M. M. Shinano, M. Urayama, and T. Tadano. 1995. Productivity of high yielding crops: V. Root growth and specific absorption rate of nitrogen. *Soil Sci. Plant Nutr.* 41:635-647.

Osaki, M., T. Shinano, and T. Tadano. 1992. Carbon-nitrogen interaction in field crops production. *Soil Sci. Plant Nutr.* 38:553-564.

Peltonen-Sainio, P. 1991. Productive oat ideotype for northern growing conditions. *Euphytica* 54:27-32.

Peng, S., R. C. Laza, R. M. Visperas, A. L. Sanico, K. G. Cassman, and G. S. Khush. 2000. Grain yield of rice cultivars and lines developed in the Philippines since 1966. *Crop Sci.* 40:307-314.

Perry, M. W., and M. F. D'Antuono. 1989. Yield improvement and associated characteristics of some Australian spring wheat cultivars introduced between 1860 and 1982. *Aust. J. Agric. Res.* 40:457-472.

Peterson, C. M., C. H. Mosjidis, R. R. Dute, and M. E. Westgate. 1992. A flower and pod staging system for soybean. *Ann. Bot.* 69:59-67.

Piepho, H. P. 1995. A simple procedure for yield component analysis. *Euphytica* 84:43-48.

Pigeaire, A., C. Duthion, and O. Turc. 1986. Characterization of the final stage in seed abortion in indeterminate soybean, white lupin, and pea. *Agronomie* 6:371-378.

Radford, P. J. 1967. Growth analysis formulae: Their use and abuse. *Crop Sci.* 7:171-175.

Reddy, V. H., and T. B. Daynard. 1983. Endosperm characteristics associated with rate of grain filling and kernel size in corn. *Maydica* 28:339-355.

Riggs, T. J., P. R. Hanson, N. D. Start, D. M. Miles, C. L. Morgan, and M. A. Ford. 1981. Comparison of spring barley varieties grown in England and Wales between 1880 and 1980. *J. Agric. Sci.* 97:599-610.

Rosielle, A. A., and K. J. Frey. 1975. Estimates of selection parameters associated with harvest index in oat lines derived from a bulk population. *Euphytica* 24:121-131.

Russell, W. A. 1985. Evaluations for plant, ear, and grain traits of maize cultivars representing seven years of breeding. *Maydica* 30:85-96.

Sadras, V. O., M. P. Bange, and S. P. Milroy. 1997. Reproductive allocation of cotton in response to plant and environmental factors. *Ann. Bot.* 80:75-81.

Sayre, K. D., R. P. Singh, J. Huerta-Espino, and S. Rajaram. 1998. Genetic progress in reducing losses to leaf rust in CIMMYT-derived Mexican spring wheat cultivars. *Crop Sci.* 38:654-659.

Sedgley, R. H. 1991. An appraisal of the Donald ideotype after 21 years. *Field Crops Res.* 26:93-112.

Sharma, R. C. 1992. Duration of the vegetative and reproductive period in relation to yield performance of spring wheat. *European J. Agron.* 1:133-137.

Sharma, R. C., and E. L. Smith. 1986. Selection for high and low harvest index in three winter wheat populations. *Crop Sci.* 26:1147-1150.

Shinano, T., M. Osaki, K. Komatsu, and T. Tadano. 1991. Comparison of production efficiency of the harvesting organs among field crops: I. Growth efficiency of the harvesting organs. *Soil Sci. Plant Nutr.* 39:269-280.

Shinano, T., M. Osaki, and T. Tadano. 1993. Comparison and reconstruction of photosynthesized ^{14}C compounds incorporated into shoot between rice and soybean. *Soil Sci. Plant Nutr.* 37:409-417.

Shinano, T., M. Osaki, and T. Tadano. 1994. ^{14}C-Allocation of 14C-compounds introduced to a leaf to carbon and nitrogen components in rice and soybean during ripening. *Soil Sci. Plant Nutr.* 40:199-209.

Shinano, T., M. Osaki, and T. Tadano. 1995. Comparison of growth efficiency between rice and soybean at the vegetative growth stage. *Soil Sci. Plant Nutr.* 41:471-480.

Simmons, S. R. 1997. Growth, development, and physiology, pp. 77-113. In G. E. Heyne (ed.) *Wheat and Wheat Improvement,* Second Edition. Agron. Monograph 13. Madison, WI: Am. Soc. Agron., Crop Sci. Soc. Am., and Soil Sci. Soc. Am.

Sinclair, T. R. 1998. Historical changes in harvest index and crop nitrogen accumulation. *Crop Sci.* 38:638-643.

Singh, I. D., and N. C. Stoskopf. 1971. Harvest index in cereals. *Agron. J.* 63:224-226.

Sinha, S. K., and M. R. Swaminathan. 1984. New parameters and selection criteria in plant breeding, pp. 1-66. In: P. B. Vose and S. G. Blixt (eds.) *Crop Breeding: A Contemporary Basis.* New York: Pergamon Press.

Snyder, F. W., and G. E. Carlson. 1984. Selecting for partitioning of photosynthetic products in crops. *Adv. Agron.* 37:47-72.

Song, X. F., W. Agata, and Y. Kawamitsu. 1990. Studies on dry matter and grain production of F_1 hybrid rice in China: I. Characteristic of dry matter production. *Japan. J. Crop Sci.* 59:19-28.

Speath, S. C., and T. R. Sinclair. 1985. Linear increase in soybean harvest index during seed filling. *Agron. J.* 77:207-211.

Spiertz, J. H. J., B. A. Tem, and L. J. P. Kupers. 1971. Relation between green area duration and grain yield in some varieties of wheat. *Neth. J. Agric. Sci.* 19:211-222.

Spiertz, J. H. J., and J. Vos. 1985. Grain growth of wheat and its limitation by carbohydrate and nitrogen supply, pp. 129-141. In: W. Day and R. K. Atkin (eds.) *Wheat Growth and Modeling.* New York: Plenum Press.

Stansel, J. W. 1975. *The Rice Plant—Its Development and Yield in Six Decades of Rice Research in Texas.* Research Monograph 4. College Station: Texas Agric. Exp. Stan.

Subedi, K. D., C. B. Budhathoki, and M. Subedi. 1997. Variation in sterility among wheat genotypes in response to boron deficiency in Nepal. *Euphytica* 95:21-26.

Summerfield, R. J., R. H. Ellis, P. Q. Craufurd, Q. Aiming, E. H. Roberts, and T. R. Wheeler. 1997. Environmental and genetic regulation of flowering of tropical annual crops. *Euphytica* 96:83-91.

Syme, J. R. 1972. Single-plant characters as a measure of field plot performance of wheat cultivars. *Aust. J. Agric. Res.* 23:753-760.

Tanaka, A, and M. Osaki. 1983. Growth and behavior of photosynthesized ^{14}C in various crops in relation to productivity. *Soil Sci. Plant Nutr.* 29:147-158.

Tollenaar, M. 1989. Genetic improvement in grain yield of commercial maize hybrids grown in Ontario from 1959 to 1988. *Crop Sci.* 29:1365-1371.

Tollenaar, M., and T. B. Daynard. 1978. Effects of defoliation on kernel development in maize. *Can. J. Plant Sci.* 58:207-212.

Van Duivenbooden, N., C. T. De Wit, and H. Van Keulen. 1996. Nitrogen, phosphorus, and potassium relations in five major cereals reviewed in respect to fertilizer recommendations using simulation modeling. *Fert. Res.* 44:37-49.

Waddington, S. R., J. K. Ransom, M. Osmanzai, and D. A. Saunders. 1986. Improvement in the yield potential of bread wheat adapted to northwest Mexico. *Crop Sci.* 26:698-703.

Wallace, D. H., J. P. Baudoin, J. S. Beaver, D. P. Coyne, D. E. Halseth, P. N. Massaya, H. M. Munger, J. R. Mayers, M. Silbernagel, K. S. Yourstone, and R. W. Zobel. 1993. Improving efficiency of breeding for higher crop yield. *Theor. Appl. Genet.* 86:27-40.

Wallace, D. H., and W. Yan. 1998. *Plant Breeding and Whole System Crop Physiology, Improving Adaptation, Maturity and Yield.* Wallingford, UK: C. A. B. Int.

Wareing, P. F., and I. D. J. Phillips. 1981. *Growth and Differentiation in Plants,* Third Edition. New York: Pergamon Press.

Watson, D. J. 1958. The dependence of net assimilation rate on leaf area index. *Ann. Bot.* 22:37-54.

Wilhelm, W. W., and G. S. McMaster. 1995. Importance of the phyllochron in studying development and growth in grasses. *Crop Sci.* 35:1-3.

Wych, R. D., and D. C. Rasmusson. 1983. Genetic improvement in malting barley cultivars since 1920. *Crop Sci.* 23:1037-1040.

Wych, R. D., and D. D. Stuthman. 1983. Genetic improvement in Minnesota adapted oat cultivars since 1923. *Crop Sci.* 23:879-881.

Yamaguchi, J. 1978. Respiration and the growth efficiency in relation to crop productivity. *J. Fac. Agric. Hokkaido Univ. Japan* 59:59-129.

Yamamoto, Y., T. Yoshida, T. Enomoto, and G. Yoshikawa. 1991. Characteristics for the efficiency of spikelet production and the ripening in high yielding japonica-indica hybrid and semidwarf indica rice varieties. *Japan. J. Crop Sci.* 60:365-372.

Yamauchi, M. 1994. Physiological bases of higher yield potential in F_1 hybrids, pp. 71-80. In: S. S. Virmani (ed.) *Hybrid Rice Technology: New Developments and Future Prospects.* Los Baños, Philippines: Inter. Rice Res. Inst.

Yin, X., M. J. Kropff, and J. Goudriaan. 1997. Changes in temperature sensitivity of development from sowing to flowering in rice. *Crop Sci.* 37:1787-1794.

Yoshida, S. 1972. Physiological aspects of grain yield. *Annu. Rev. Plant Physiol.* 23:437-464.

Chapter 4

Al-Khatib, K., and G. M. Paulsen. 1984. Mode of high temperature injury to wheat during grain development. *Physiol. Plant.* 61:363-368.

Al-Khatib, K., and G. M. Paulsen. 1999. High-temperature effects on photosynthetic processes in temperate and tropical cereals. *Crop Sci.* 39:119-125.

Andrade, F. H., S. A. Uhart, and A. Cirilo. 1993. Temperature affects radiation use efficiency in maize. *Field Crops Res.* 32:17-25.

Bange, M. P., G. L. Hammer, and K. G. Rickert. 1997. Effect of radiation environment on radiation use efficiency and growth of sunflower. *Crop Sci.* 37:1208-1214.

Berry, J., and O. Bjorkman. 1980. Photosynthetic response and adaptation to temperature in higher plants. *Annu. Rev. Plant Physiol.* 31:491-543.

Black, C. C. 1971. Ecological implications of dividing plants into groups with distinct photosynthetic production capacities. *Adv. Ecol. Res.* 7:87-114.

Blacklow, W. M., and L. D. Incoll. 1981. Nitrogen stress of winter wheat changed the determinants of yield and the distribution of nitrogen and total dry matter during grain filling. *Aust. J. Plant Physiol.* 8:191-200.

Board, J. E., and B. G. Harville. 1993. Soybean yield component responses to a light interception gradient during the reproductive period. *Crop Sci.* 33:772-777.

Board, J. E., and B. G. Harville. 1994. A criterion for acceptance of narrow-row culture in soybean. *Agron. J.* 86:1103-1106.

Board, J. E., B. J. Harville, and A M. Saxton. 1990. Branch dry weight in relation to yield increases in narrow-row soybean. *Agron. J.* 82:540-544.

Board, J. E., M. Kamal, and B. G. Harville. 1992. Temporal importance of greater light interception to increased yield in narrow-row soybean. *Agron. J.* 84:575-579.

Board, J. E., A. T. Wier, and D. J. Boethel. 1997. Critical light interception during seed filling for insecticide application and optimum soybean grain yield. *Agron. J.* 89:369-374.

Bollero, G. A., D. G. Bullock, and S. E. Hollinger. 1996. Soil temperature and planting date effects on effects on corn yield, leaf area, and plant development. *Agron. J.* 88:385-390.

Bondada, B. R., D. M. Oosterhuis, R. J. Norman, and W. H. Baker. 1996. Canopy photosynthesis, growth, yield, and boll ^{15}N accumulation under nitrogen stress in cotton. *Crop Sci.* 36:127-133.

Botella, M. A., A. C. Cerda, and S. H. Lips. 1993. Dry matter production, yield, and allocation of carbon-14 assimilates by wheat as affected by nitrogen source and salinity. *Agron. J.* 85:1044-1049.

Caballero, R., A. Rebole, C. Barro, C. Alzueta, and L. T. Ortiz. 1998. Aboveground carbohydrate and nitrogen partitioning in common vetch during seed filling. *Agron. J.* 90:97-102.

Camp, P. J., S. C. Huber, J. J. Burke, and D. E. Moreland. 1982. Biochemical changes that occur during senescence of wheat leaves. *Plant Physiol.* 42:369-374.

Conocono, E. A., J. A. Egdane, and T. I. Setter. 1998. Estimation of canopy photosynthesis in rice by means of daily increases in leaf carbohydrate concentration. *Crop Sci.* 38:987-995.

Cooper, C. S., and P. W. MacDonald. 1970. Energetic of early seedlings growth in corn. *Crop Sci.* 10:136-139.

Cox, W. J. 1996. Whole plant physiological and yield responses of maize to plant density. *Agron. J.* 88:489-496.

Dale, R. F., D. T. Coelho, and K. P. Gallo. 1980. Prediction of daily green leaf area index for corn. *Agron. J.* 72:999-1005.

Daughtry, C. S. T., K. P. Gallo, and M. E. Bauer. 1983. Spectral estimates of solar radiation intercepted by corn canopies. *Agron. J.* 72:527-531.

De Datta, S. K. 1981. *Principles and Practices of Rice Production.* New York: John Wiley & Sons.

De Wit, C. T. 1967. Photosynthesis: Its relation to evaporation, pp. 315-320. In: A. San Pietro, F. A. Greer, T. J. Army (eds.) *Harvesting the Sun.* New York: Academic Press.

Downton, W. J. S. 1971. Adaptive and evolutionary aspects of C_4 photosynthesis, pp. 3-17. In: M. D. Hatch, C. B. Osmond, and R. O. Slayter (eds.) *Photosynthesis and Photorespiration.* New York: Wiley Interscience.

Downton, W. J. S. 1975. The occurrence of C_4 photosynthesis among plants. *Photosynthetica* 9:96-109.

Duncan, W. G. 1986. Planting patterns and soybean yield. *Crop Sci.* 26:584-588.

Dwyer, L., A. M. Anderson, D. W. Stewart, B. L. Ma, and M. Tollenaar. 1995. Changes in maize hybrid photosynthetic response to leaf nitrogen, from preanthesis to grain fill. *Agron. J.* 87:1221-1225.

Dwyer, L. M., M. Tollenaar, and D. W. Stewart. 1991. Changes in plant density dependence of leaf photosynthesis of maize (*Zea mays* L.) hybrid, 1959 to 1988. *Can. J. Plant Sci.* 71:1-11.

Eastin, J. A. 1969. Leaf position and leaf function in corn, pp. 81-89. In: J. I. Sutherland and R. J. Falasea (eds.) *Proc. 24th Ann. Corn and Sorghum Res. Conf.* Washington, DC: Am. Seed Trade Assoc.

Eastin, J. D., and C. Y. Sullivan. 1984. Environmental stress influence on plant persistence, physiology, and production, pp. 201- 236. M. B. Tesar (ed.) *Physiological Basis of Crop Growth and Development.* Madison, WI: Am. Soc. Am. and Crop Sci. Soc. Am.

Edwards, G. E. 1986. Carbon fixation and partitioning in the leaf, pp. 51-66. In: J. C. Shannon, D. P. Knievel, and C. D. Boyer (eds.) *Regulation of Carbon and Nitrogen Reduction and Utilization in Maize.* Rockville, MD: Am. Soc. Plant Physiol.

Egharevba, P. N. 1975. Planting pattern and light interception in maize, pp. 15-17. In: *Proc. Physiology Program Formulation Workshop,* Ibadan, Nigeria, April 1975. Ibadan, Nigeria: International Institute of Tropical Agriculture.

Egli, D. B. 1988. Plant density and soybean yield. *Crop Sci.* 28:977-981.

Egli, D. B. 1994. Mechanisms responsible for soybean yield response to equidistant planting patterns. *Agron. J.* 86:1046-1049.

Evans, J. R. 1983. Nitrogen and photosynthesis in the flag leaf of wheat. *Plant Physiol.* 72:297-302.

Evans, L. T., and I. F. Wardlaw. 1976. Aspects of the comparative physiology of the grain yield in cereals. *Adv. Agron.* 28:301-359.

Fageria, N. K. 1992. *Maximizing Crop Yields.* New York: Marcel Dekker.

Fageria, N. K., V. C. Baligar, and C. A. Jones. 1997. *Growth and Mineral Nutrition of Field Crops,* Second Edition. New York: Marcel Dekker.

Fehr, W. R., and C. E. Caviness. 1977. *Stages of Soybean Development.* Spec. Rep. 80. Ames: Iowa Agric. Exp. Stn.

Flenet, F., J. R. Kiniry, J. E. Board, M. E. Westgate, and D. C. Reicosky. 1996. Row spacing effects on light extinction coefficients of corn, sorghum, soybean, and sunflower. *Agron. J.* 88:185-190.

Fortin, M. C., F. J. Pierce, and M. Edwards. 1994. Corn leaf response to early-season soil temperature under crop residues. *Agron. J.* 86:355-359.

Frederick, J. R., and J. J. Camberato. 1994. Leaf net CO_2-exchange rate and associated leaf traits of winter wheat grown with various spring nitrogen fertilization rates. *Crop Sci.* 34:432-439.

Frederick, R., and J. J. Camberato. 1995. Water and nitrogen effects on winter wheat in the southeastern coastal plains: II. Physiological responses. *Agron. J.* 87:527-533.

Gallagher, J. N., and P. V. Biscoe. 1978. Radiation absorption, growth and yield of cereals. *J. Agric. Sci. Camb.* 91:47-60.

Gardner, F. P., R. B. Pearce, and R. L. Mitchell. 1985. *Physiology of Crop Plants.* Ames: Iowa Sate University.

Gifford, R. M., and L. T. Evans. 1981. Photosynthesis, carbon partitioning, and yield. *Annu. Rev. Plant Physiol.* 32:485-509.

Gwathmey, C. O., and D. D. Howard. 1998. Potassium effects on canopy light interception and earliness of no-tillage cotton. *Agron. J.* 90:144-149.

Hageman, R. H. 1986. Nitrate metabolism in roots and leaves, pp. 105-116. In: J. C. Shannon, D. P. Knievel, and C. D. Boyer (eds.) *Regulation of Carbon and Nitrogen Reduction and Utilization in Maize.* Rockville, MD: Am. Soc. Plant Physiol.

Hall, A. J., D. M. Whitfield, and D. J. Connor. 1990. Contribution of pre-anthesis assimilates to grain filling in irrigated and water-stressed sunflower crops: II. Estimates from a carbon budget. *Field Crops Res.* 20:95-112.

Hammer, G. I., and G. C. Wright. 1994. A theoretical analysis of nitrogen and radiation effects on radiation use efficiency in peanut. *Aust. J. Agric. Res.* 45:575-589.

Hatch, M. D., and C. R. Slack. 1970. Photosynthetic CO_2 fixation pathways. *Annu. Rev. Plant Physiol.* 21:141-162.

Heitholt, J. J. 1994. Canopy characteristics associated with deficient and excessive cotton plant population densities. *Crop Sci.* 34:1291-1297.

Heitholt, J. J., W. T. Pettigrew, and W. R. Meredith Jr. 1992. Light interception and lint yield of narrow-row cotton. *Crop Sci.* 32:728-733.

Hirose, T., and M. J. A. Werger. 1987. Nitrogen use efficiency in instantaneous and daily photosynthesis of leaves in the canopy of a *Solidago altissma* stand. *Physiol. Plant.* 70:215-222.

Hunt, I. A., and G. V. Poorten. 1985. Carbon dioxide exchange rates and leaf nitrogen contents during aging of the flag and penultimate leaves of five spring-wheat cultivars. *Can. J. Bot.* 63:1605-1609.

Imsande, J. 1989. Rapid dinitrogen fixation during soybean pod fill enhances net photosynthetic output and seed yield: A new perspective. *Agron. J.* 81:549-556.

Ishii, R. 1988. Varietal differences of photosynthesis and grain yield in rice. *Korean J. Crop Sci.* 33:315-321.

Johnson, R. R. 1987. Crop management, pp. 355-390. In: J. R. Wilcox (ed.) *Soybeans: Improvement, Production, and Uses.* Agron. Monogr. 16. Second Edition. Madison, WI: Am. Soc. Agron., Crop Sci. Soc. Am., and Soil Sci. Soc. Am.

Kelly, J. D., J. M. Kolkman, and K. Schneider. 1998. Breeding for yield in dry bean (*Phaseolus vulgaris* L.). *Euphytica* 102:343-356.

Kelly, M. O., and P. J. Davies. 1988. The control of whole plant senescence. *CRC Crit. Rev. Plant Sci.* 6:611-616.

Khanna, R., and S. K. Sinha. 1973. Change in the predominance from C_4 to C_3 pathway following anthesis in sorghum. *Biochem. Biophys. Res. Commun.* 52:121-124.

Kiniry, J. R., C. A Jones, J. C. O'Toole, R. Bouchet, M. Cabelguenne, and D. A. Spanel. 1989. Radiation-use efficiency in biomass accumulation prior to grain filling for five grain-crop species. *Field Crops Res.* 20:51-64.

Kisha, T. J., C. H. Sneller, and B. W. Diers. 1997. Relationship between genetic distance among parents and genetic variance in populations of soybean. *Crop Sci.* 37:1317-1325.

Kromer, S. 1995. Respiration during photosynthesis. *Annu. Rev Plant Physiol. Plant Mol. Biol.* 46:45-70.

Lambers, H. 1987. Does variation in photosynthetic rate explain variation in growth rate and yield. *Neth. J. Agric. Sci.* 35:505-519.

Loomis, R. S., and D. J. Connor. 1992. *Crop Ecology: Productivity and Management in Agricultural Systems.* New York: Cambridge University Press.

Maas, S. J., and J. R. Dunlap. 1989. Reflectance, transmittance, and absorbance of light by normal, etiolated, and albino corn leaves. *Agron. J.* 81:105-110.

Mae, T. 1997. Physiological nitrogen efficiency in rice: Nitrogen utilization, photosynthesis, and yield potential, pp. 51-60. In: T. Ando, K. Fujita, T. Mae, H. Tsumoto, S. Mori, and J. Sekiya (eds.) *Plant Nutrition for Sustainable Food Production and Environment.* Dordrecht, the Netherlands: Kluwer Acad. Publ.

Mahon, J. 1983. Limitations to the use of physiological variability in plant breeding. *Can. J. Plant Sci.* 63:11-21.

Major, D. J., B. W. Beasley, and R. I. Hamilton. 1991. Effect of maize maturity on radiation-use efficiency. *Agron. J.* 83:895-903.

Makino, A., and B. Osmond. 1991. Effects of nitrogen nutrition on nitrogen partitioning between chloroplasts and mitochondria in pea and wheat. *Plant Physiol.* 96:355-362.

McCree, K. J. 1974. Equations for the rate of dark respiration of white clover and grain sorghum as functions of dry weight, photosynthetic rate, and temperature. *Crop Sci.* 14:509-514.

McKee, G. W. 1964. A coefficient for computing leaf area in hybrid corn. *Agron. J.* 56:240-241.

Mendham, N. J., J. Russell, and N. K. Jarosz. 1990. Response to sowing time of three contrasting Australian cultivars of oilseed rape. *J. Agric. Sci. Camb.* 114:275-283.

Mendham, N. J., P. A. Shipway, and R. K. Scott. 1981. The effects of delayed sowing and weather on growth, development and yield of winter oil-seed rape. *J. Agric. Sci. Camb.* 96:389-416.

Milthorpe, F. L., and J. Moorby. 1988. *An Introduction to Crop Physiology*, Second Edition. Sydney: Cambridge University Press.

Mitchell, R. L. 1970. *Crop Growth and Culture*. Ames: Iowa State University Press.

Monteith, J. L. 1977. Climate and the efficiency of crop production in Britain. *Philos. Trans. R. Soc. London B* 281:277-294.

Morita, K. 1980. Release of nitrogen from chloroplasts during leaf senescence in rice. *Ann. Bot.* 46:297-302.

Morrison, M. J., P. B. E. McVetty, and R. Scarth. 1990. Effect of row spacing and seeding rates on summer rape in southern Manitoba. *Can. J. Plant Sci.* 70:127-137.

Morrison, M. J., and D. W. Stewart. 1995. Radiation-use efficiency in summer rape. *Agron. J.* 87:1139-1142.

Morrison, M. J., H. D. Voldeng, and E. R. Cober. 1999. Physiological changes from 58 years of genetic improvement of short-season soybean cultivars in Canada. *Agron. J.* 91:685-689.

Moss, D. N. 1984. Photosynthesis, respiration, and photorespiration in higher plants, pp. 131-152. In: M. B. Tesar (ed.) *Physiological Basis of Crop Growth and Development*. Madison, WI: Am. Soc. Agron. and Crop Sci. Soc. Am.

Muchow, R. C. 1989. Comparative productivity of maize, sorghum and pearl millet in a semi-arid tropical environment: I. Yield potential. *Field Crops Res.* 20:191-205.

Muchow, R. C., D. B. Coates, G. L. Wilson, and M. A. Foale. 1982. Growth and productivity of irrigated *Sorghum bicolor* (L.) Moench in Northern Australia: I. Plant density and arrangement effects on light interception and distribution and grain yield in the hybrid Texas 610SR in low and medium latitudes. *Aust. J. Agric. Res.* 33:773-784.

Muchow, R. C., and T. R. Sinclair. 1994. Nitrogen response of leaf photosynthesis and canopy radiation use efficiency in field-grown maize and sorghum. *Crop Sci.* 34:721-727.

Murata, Y. 1961. Studies on the photosynthesis of rice plant and culture significance. *Bull. Natl. Inst. Agric. Sci. (Tokyo) D* 9:1-169.

Nelson, C. J., and K. L. Larson. 1984. Seedling growth, pp. 93-129. In: M. B. Tesar (ed.) *Physiological Basis of Crop Growth and Development.* Madison, WI: Am. Soc. Agron. and Crop Sci. Soc. Am.

Nelson, T., and J. A. Langdale. 1992. Developmental genetics of C_4 photosynthesis. *Annu. Rev. Plant Physiol Plant Mol. Biol.* 43:25-47.

Novoa, R., and R. S. Loomis. 1981. Nitrogen and plant production. *Plant Soil* 58:177-204.

Orgaz, F., F. J. Villalobos, C. Gimenez, and E. Fereres. 1992. Radiation use efficiency of sunflower genotypes, pp. 268-273. In: *Proc. Int. Sunflower Conf.,* 13th, Pisa, Italy. Sept. Pisa: Int. Sunflower Assoc.

Osaki, M., and A. Tanaka. 1979. Current photosynthates and storage substances as the respiratory substances in the rice plant. *J. Sci. Soil Manure, Japan* 50:540-546.

Otegui, M. E., M. G. Nicolini, R. A. Ruiz, and P. A. Dodds. 1995. Sowing date effects on grain yield components for different maize genotypes. *Agron. J.* 87:29-33.

Parvez, A. O., F. P. Gardner, and K. J. Boote. 1989. Determinate and indeterminate type soybean cultivar responses to pattern, density and planting date. *Crop Sci.* 29:150-157.

Pearce, R. B., R. H. Brown, and R. E. Blaser. 1965. Relationship between leaf area index, light interception, and net photosynthesis in orchard grass. *Crop Sci.* 5:553-556.

Pepper, G. E. 1974. The effect of leaf orientation and plant density on the yield of maize (*Zea mays* L). Ph. D. Thesis (Diss. Abstr. 75-10,500) Ames: Iowa State Univ.

Rawson, H. M., R. M. Gifford, and P. M. Bremner. 1976. Carbon dioxide exchange in relation to sink demand in wheat. *Planta* 132:19-23.

Rochette, P., R. L. Desjardins, E. Pattey, and R. Lessard. 1995. Crop net carbon dioxide exchange rate and radiation use efficiency in soybean. *Agron. J.* 87:22-28.

Rochette, P., R. L. Desjardins, E. Pattey, and R. Lessard. 1996. Instantaneous measurement of radiation and water use efficiencies of a maize crop. *Agron. J.* 88:627-635.

Rosenthal, W. D., T. J. Gerik, and L. J. Wade. 1993. Radiation-use efficiency among grain sorghum cultivars and plant densities. *Agron. J.* 85:703-705.

Russell, G., P. G. Jarvis, and J. I. Monteith. 1989. Absorption of radiation by canopies and stand growth, pp. 21-40. In: G. Russell, B. Marshall, and P. G. Jarvis (eds.) *Plant Canopies: Their Growth, Form and Function.* Cambridge, England: Cambridge Univ. Press.

Shaver, D. L. 1983. Genetics and breeding of maize with extra leaves above the ear. *Proc. Annu. Corn Sorghum Res. Conf.* 38:161-180.

Sheehy, J. E., and L. C. Chapas. 1976. The measurement and distribution of irradiance in clear and overcast conditions in four temperate forage grass canopies. *J. Applied Ecol.* 13:831-840.

Shinano, T., M. Osaki, and T. Tadano. 1991. Comparison of reconstruction of photosynthesized ^{14}C compounds incorporated into shoot between rice and soybean. *Soil Sci. Plant Nutr.* 37:409-417.

Shiraiwa, T., and T. R. Sinclair. 1993. Distribution of nitrogen among leaves in soybean canopies. *Crop Sci.* 33:804-808.

Shopf, J. W. 1993. Microfossils of the early archean apec chert: New evidence of the antiquity of life. *Science* 260:640-646.

Sinclair, T. R., and T. Horie. 1989. Leaf nitrogen, photosynthesis, and crop radiation use efficiency review. *Crop Sci.* 29:90-98.

Sinclair, T. R., T. Shiraiwa, and G. L. Hammer. 1992. Variation in crop radiation use efficiency in response to increased proportion of diffuse radiation. *Crop Sci.* 32:1281-1284.

Snyder F. W., and G. E. Carlson. 1984. Selecting for partitioning of photosynthetic products in crops. *Adv. Agron.* 37:47-72.

Steer, B. T., S. P. Milroy, and R. M. Kamona. 1993. A model to stimulate the development, growth and yield of irrigated sunflower. *Field Crops Res.* 32:83-99.

Stoskopf, N. C. 1981. *Understanding Crop Production.* Reston, VA: Reston Publ. Co.

Swanson, S. P., and W. W. Wilhelm. 1996. Planting date and residue effects on growth, partitioning, and yield of corn. *Agron. J.* 88:205-210.

Tanaka, A., and M. Osaki. 1983. Growth and behavior of photosynthesized ^{14}C in various crops in relation to productivity. *Soil Sci. Plant Nutr.* 29:147-158.

Tetio-Kagho, F, and F. P. Gardner. 1988. Response of maize to plant population density: I. canopy development, light relationships, and vegetative growth. *Agron. J.* 80:930-935.

Tetio-Kagho, F., and F. P. Gardner. 1989a. Response of maize to plant population density: I. Canopy development, light relationships, and vegetative growth. *Agron. J.* 80:930-935.

Tetio-Kagho, F., and F. P. Gardner. 1989b. Reponse of maize to plant population density: II. Reproductive development, yield, and yield adjustment. *Agron. J.* 80:935-940.

Thornley, J. H. M. 1976. *Mathematical Models in Plant Physiology.* London: Academic Press.

Tolbert, N. E. 1997. The C_2 oxidative photosynthetic carbon cycle. *Annu. Rev. Plant Physiol. Plant Mol. Biol.* 48:1-25.

Tollenaar, M. 1991. Physiological basis of genetic improvement of maize hybrids in Ontario from 1959 to 1988. *Crop Sci.* 31:119-124.

Tollenaar, M., and A. Aguilera. 1992. Radiation use efficiency of an old and a new maize hybrid. *Agron. J.* 84:536-541.

Tollenaar, M., L. M. Dwyer, and D. W. Stewart. 1992. Ear and kernel formation in maize hybrids representing three decades of yield improvement in Ontario. *Crop. Sci.* 32:432-438.

Trapani, N., A. J. Hall, V. O. Sadras, and F. Vilella. 1992. Ontogenetic changes in radiation use efficiency of sunflower crops. *Field Crops Res.* 29:301-316.

Wardlaw, I. F., I. Sofield, and P. M. Cartwright. 1980. Factors limiting the rate of dry matter accumulation in the grain of wheat grown at high temperature. *Aust. J. Plant Physiol.* 7:387-400.

Wells, R. 1991. Soybean growth response to plant density: Relationships among canopy photosynthesis, leaf area, and light interception. *Crop Sci.* 31:755-761.

Wells, R., W. R. Meredith Jr., and J. R. Williford. 1986. Canopy photosynthesis and its relationship to plant productivity in nearisogenic cotton lines differing in leaf morphology. *Plant Physiol.* 82:635-640.

Wong, S. C., I. R. Cowan, and G. D. Farquhar. 1985. Lea conductance in relation to rate of CO_2 assimilation: I. Influence of nitrogen nutrition, phosphorus nutrition, photon flux density, and ambient partial pressure of CO_2 during ontogeny. *Plant Physiol.* 78:821-825.

Wullschleger, S. D., and D. M. Oosterhuis. 1992. Canopy leaf area development and age-class dynamics in cotton. *Crop Sci.* 32:451-456.

Yoshida, S. 1972. Physiological aspects of grain yield. *Annu. Rev Plant Physiol.* 23:437-464.

Yoshida, S. 1981. *Fundamentals of Rice Crop Science.* Los Baños, Philippines: Int. Rice Res. Inst.

Yoshida, S., D. A. Forno, J. H. Cook, and K. A. Gomez. 1976. *Laboratory Manual for Physiological Studies of Rice,* Third Edition. Los Baños, Philippines: Int. Rice Res. Inst.

Zaffaroni, E., and A. A. Schneiter. 1989. Water-use efficiency and light interception of semidwarf and standard-height sunflower hybrids grown in different row arrangements. *Agron. J.* 81:831-836.

Zelitch, I. 1975. Improving the efficiency of photosynthesis. *Science* 188:626-633.

Chapter 5

Allison, J. C. S., J. H. H. Wilson, and J. H. Williams. 1975. Effect of defoliation after flowering on changes in stem and grain mass of closely and widely spaced maize. *Rhod. J. Agric. Res.* 13:145-147.

Amano, T., Q. Zhu, Y. Wang, N. Inoue, and H. Tanaka. 1993. Case studies on high yields of paddy rice in Jiangsu Provience, China: I. Characteristics of grain production. *Japan. J. Crop Sci.* 62:267-274.

Anderson, L. S., and J. E. Dale. 1983. The sources of carbon for developing leaves of barley. *J. Exp. Bot.* 34:405-414.

Barnett, K. H., and R. B. Pearce. 1983. Source-sink ratio alteration and its effect on physiological parameters in maize. *Crop Sci.* 23:294-299.

Blum, A., H. Poiarkova, G. Golan, and J. Mayer. 1983. Chemical desiccation of wheat plants as a simulator of post-anthesis stress: I. Effects on translocation and kernel growth. *Field Crops Res.* 6:51-58.

Board, J. E., and B. G. Harville. 1993. Soybean yield component responses to a light interception gradient during the reproductive period. *Crop Sci.* 33:772-777.

Board, J. E., and B. G. Harville. 1998. Late-planted soybean yield response to reproductive source/sink stress. *Crop Sci.* 38:763-771.

Board, J. E., and Q. Tan. 1995. Assimilatory capacity effects on soybean yield components and pod number. *Crop Sci.* 35:846-851.

Boyer, C. D., R. R. Daniels, and J. C. Shannon. 1976. Abnormal starch granule formation in *Zea mays* L. endosperms possessing the amylose-extender mutant. *Crop Sci.* 16:298-301.

Bruckner, P. L., and R. C. Frohberg. 1991. Source-sink manipulation as a post-anthesis stress tolerance screening technique in wheat. *Crop Sci.* 31:326-328.

Bruns, H. A., and C. A. Abel. 2003. Nitrogen fertility effects on Bt δ-endotoxin and nitrogen concentrations of maize during early growth. *Agron. J.* 95:207-211.

Capitanio, R., E. Gentinetta, and M. Motto. 1983. Grain weight and its components in maize inbred lines. *Maydica* 28:365-379.

Christensen, L. E., F. E. Below, and R. H. Hageman. 1981. The effects of ear removal on senescence and metabolism of maize. *Plant Physiol.* 68:1180-1181.

Cook, M. G., and L. T. Evans. 1983. The roles of sink size and location in the partitioning of assimilates in wheat ears. *Aust. J. Plant Physiol.* 10:313-327.

Crop Science Society of America (CSSA). 1992. *Glossary of Crop Science Terms.* Madison, WI: CSSA.

Daie, J. 1985. Carbohydrate partitioning and metabolism in crops. *Hortic. Rev.* 7:69-108.

Evans, L. T., and R. L. Dunstone. 1970. Some physiological aspects of evolution in wheat. *Aust. J. Biol. Sci.* 23:725-741.

Fageria, N. K. 1992. *Maximizing Crop Yields.* New York: Marcel Dekker.

Farrar, J. F. 1985. Fluxes of carbon in roots of barley plants. *New Phytol.* 99:57-69.

Faville, M. J., W. B. Silvester, T. G. Allan Green, and W. A. Jermyn. 1999. Photosynthetic characteristics of three asparagus cultivars differing in yield. *Crop Sci.* 39:1070-1077.

Food and Agriculture Organizations of the United Nations (FAO). 1994. *Rice Production Year Book.* Rome, Italy: Author.

Gifford, R. M., and L. T. Evans. 1981. Photosynthesis, carbon partitioning, and yield. *Annu. Rev. Plant Physiol.* 32:485-509.

Gifford, R. M., J. H. Yhorne, W. D. Hitz, and R. T. Giaquinta. 1984. Crop productivity and photoassimilate partitioning. *Science* 225:801-808.

Goldsworthy, P. R., and M. Coleogrove. 1974. Growth and yield of highland maize in Mexico. *J. Agric. Sci.* 83:213-221.

Ho, L. C. 1988. Metabolism and compartmentation of imported sugars in sink organs in relation to sink strength. *Annu. Rev. Plant Physiol. Plant Mol. Biol.* 39:355-378.

Imaizumi, N., E. Kiyota, and K. Ishihara. 1988. Photosynthetic characteristics of the ear of rice plants. *Japan J. Crop Sci.* 57:175-176.

Inoue, H., and A. Tanaka. 1978. Comparison of source and sink potentials between wild cultivated potatoes. *J. Sci. Soil and Man. Japan.* 49:321-328.

International Rice Research Institute (IRRI). 1987. *Annual Report for 1986.* Los Baños, Philippines IRRI.

International Rice Research Institute (IRRI). 1988. *Annual Report for 1987.* Los Baños, Philippines IRRI.

International Rice Research Institute (IRRI). 1989. *Annual Reports for 1988.* Los Baños, Philippines IRRI.

Jones, R. J., J. A. Roessler, and S. Ouattar. 1985. Thermal environment during endosperm cell division in maize: Effects on number of endosperm cells and starch granules. *Crop Sci.* 25:830-834.

Jones, R. J., B. M. N. Schreiber, and J. A Roessler. 1996. Kernel sink capacity in maize: Genotypic and maternal regulation. *Crop Sci.* 36:301-306.

Jones, R. J., and S. R. Simmons. 1983. Effect of altered source-sink ratio on growth of maize kernels. *Crop Sci.* 23:129-134.

Khush, G. S. 1995. Increased genetic potential of rice yield: Methods and perspectives, pp. 13-29. In: B. S. Pinheiro and E. P. Guimarais (eds.) *Rice in Latin America: Perspectives to Increase Production and Yield Potential.* Document No. 60. Goiânia, Brazil: EMBRAPA-CNPAF.

Kokubun, M., and K. Watanable. 1983. Analysis of the yield determining process of field grown soybeans in relation to canopy structure: VII. Effects of source and sink manipulation during reproductive growth on yield and yield components. *Japan J. Crop Sci.* 52:215-219.

Kowles, R. W., and R. L. Phillips. 1988. Endosperm development in maize. *Int. Rev. Cytol.* 112:97-136.

Kropff, M. J., K. G. Cassman, S. Peng, R. B. Matthews, and T. L. Setter. 1994. Quantitative understanding of yield potential, pp. 21-38. In: K. G. Cassman (ed.) *Breaking the Yield Barrier.* Los Baños, Philippines: International Rice Research Institute.

Kumura, A. 1995. Physiology of high-yielding rice plants from the viewpoint of dry matter production and its partitioning, pp. 704-736. In: T. Matsuo, K. Kumazawa, R. Ishii, K. Ishihara, and H. Hirata (eds.) *Science of the Rice Plant: Physiology,* Vol. 2. Tokyo: Food Agric. Policy Res. Center.

Li, Z., R. M. P. Shannon, J. W. Stansel, and A. H. Paterson. 1998. Genetic dissection of the source-sink relationship affecting fecundity and yield in rice (*Oryza sativa* L.). *Molecular Breeding* 4:419-426.

Olson, R. A., and D. H. Sander. 1999. Corn production, pp. 639-686. In G. F. Sprague and J. W. Dudley (eds.) *Corn and Corn Improvement.* Agron. Monogr. 18,

Third Edition. Madison, WI: Am. Soc. Am., Crop Sci. Soc. Am., and Soil Sci. Soc. Am.

Osaki, M., M. Matsumoto, T. Shinano, and T. Tadano. 1996. A root-shoot interaction hypothesis for high productivity of root crops. *Soil Sci. Plant Nutr.* 42:289-301.

Osaki, M., T. Shinano, M. Matsumoto, J. Ushiki, M. M. Shinano, M. Urayama, and T. Tadano. 1995. Productivity of high-yielding crops: V. Root growth and root activity. *Soil Sci. Plant Nutr.* 41:635-647.

Osaki, M., T. Shinano, M. Matsumoto, T. Zheng, and T. Tadano. 1997. A root-shoot interaction hypothesis for high productivity of field crops, pp. 669-674. In: T. Ando, K. Fujita, T. Mae, H. Tsumoto, S. Mori, and J. Sekiya (eds.) *Plant Nutrition for Sustainable Food Production and Environment.* Tokyo, Japan: Kluwer Acad. Publ.

Pace, P. F., H. T. Cralle, J. T. Cothren, and S. A. Senseman. 1999. Photosynthate and dry matter partitioning in short and long season cotton cultivars. *Crop Sci.* 39:1065-1069.

Penning de Varies, F. W. T., D. M. Jansen, H. F. M. Berge, and A. Bakema. 1989. *Simulation of Ecophysiological Processes of Growth in Several Annual Crops.* Wageningen, the Netherlands: Pudoc.

Peterson, D. M. 1983. Effects of spikelet removal and post-heading thinning on distribution of dry matter and N in oats. *Field Crops Res.* 7:41-50.

Reed, A. J., G. W. Singletary, J. R. Schussler, D. R. Williamson, and A. I. Cristy. 1988. Shading effects on dry matter and nitrogen partitioning, kernel number, and yield of maize. *Crop Sci.* 28:819-825.

Shinano, T., M. Osaki, K. Komatsu, and T. Tadano. 1993. Comparison of production efficiency of the harvesting organs among field crops: I. Growth efficiency of the harvesting organs. *Soil Sci. Plant Nutr.* 39:269-280.

Simmons, S. R., R. K. Crookston, and J. E. Kurle. 1982. Growth of spring wheat kernels as influenced by reduced kernel number per spike and defoliation. *Crop Sci.* 22:983-988.

Sugiharto, B., K. Miyata, H. Nakamoto, H. Sasakawa, and T. Sugiyama. 1990. Regulation of expression of carbon-assimilation enzymes by nitrogen in maize leaf. *Plant Physiol.* 92:936-969.

Tanaka, A. 1972. *The Relative Importance of the Source and the Sink As the Yield Limiting Factors of Rice.* Tech. Bull. No. 6. Taipei City, Taiwan: Food Fertil. Tech. Center.

Tanaka, A. 1980. *Source and Sink Relationship in Crop Production.* Tech. Bull. No. 52. Taipei City, Taiwan: Food Fertil. Tech. Center.

Tanaka, A., and J. Yamaguchi. 1968. The growth efficiency in relation to the growth of the rice plant. *Soil Sci. Plant Nutr.* 14:110-116.

Tanaka, A., and J. Yamaguchi. 1972. Dry matter production, yield components and grain yield of the maize plant. *J. Fac. Agri. Hokkaido Univ., Japan* 57:71-132.

Tollenaar, M., and T. B. Daynard. 1978. Relationship between assimilate source and reproductive sink in maize grown in a short season environment. *Agron. J.* 70:219-223.

Turgeon, R. 1989. The sink-source transition in leaves. *Annu. Rev Plant Physiol. Plant Mol. Biol.* 40:119-138.

Uhart, S. A., and F. H. Andrade. 1991. Source-sink relationship in maize grown in a cool temperate area. *Agronomie* 11:863-875.

Uhart, S. A., and F. H. Andrade. 1995. Nitrogen and carbon accumulation and remobilization during grain filling in maize under different source/sink ratios. *Crop Sci.* 35:183-190.

Walker, A. J., and L. C. Ho. 1977. Carbon translocation in the tomato: Carbon import and fruit growth. *Ann. Bot.* 41:813-823.

Wells, R., and W. R. Meredith. 1984. Comparative growth of obsolete and modern cotton cultivars: I. Vegetative dry matter partitioning. *Crop Sci.* 24:858-862.

Yamaguchi, J. 1978. Respiration and the growth efficiency in relation to crop productivity. *Res. Bull. Fac. Agric. Hokkaido Univ., Japan* 59: 59-129.

Yamaguchi, J., K. Kawachi, and A. Tanaka. 1975. Studies on the growth efficiency of crop plants: V. Growth efficiency of the soybean at successive growth stages and grain productivity in comparison with rice and maize. *J. Sci. Soil Manure Japan* 46:120-125.

Yamaguchi, J., K. Watanabe, and A. Tanaka. 1975. Studies on the growth efficiency of crop plants: IV. Respiratory rate and the growth efficiency of various organs of rice and maize. *J. Sci. Soil Manure Japan* 41:73-77.

Yamazaki, M., A. Watanabe, and T. Sugiyama. 1986. Nitrogen-regulated accumulation of mRNA and protein for photosynthetic carbon-assimilating enzymes in maize. *Plant Cell Physiol.* 27:443-452.

Yoshida, S. 1981. *Fundamentals of Rice Crop Science.* Los Baños, Philippines: International Rice Research Institute.

Chapter 6

Allen, L. H., Jr., J. T. Baker, S. L. Albrecht, K. J. Boote, D. Pan, and J. V. C. Vu. 1995. Carbon dioxide and temperature effects on rice, pp. 258-277. In: S. Peng, K. T. Ingram, H. U. Neue, and L. H. Ziska (eds.) *Climate Change and Rice.* New York: Springer-Verlag, Int.; Los Baños, Philippines: Rice Res. Inst.

Allen, L. H., J. C. V. Vu, R. R. Valle, K. J. Boote, and P. Jones. 1988. Nonstructural carbohydrates and nitrogen of soybean grown under carbon dioxide enrichment. *Crop Sci.* 28:84-94.

Alvarez, R., M. E. Russo, P. Prystupa, J. D. Scheiner, and L. Blotta. 1998. Soil carbon pools under conventional and no-tillage systems in the Argentine rolling pampa. *Agron. J.* 90:138-143.

Amthor, J. S. 1998. Perspective on the relative insignificance of increasing atmospheric CO_2 concentration to crop yield. *Field Crops Res.* 58:109-127.

Baker, J. M., E. J. A. Spaans, and C. F. Reece. 1996. Conductimetric measurement of CO_2 concentration: Theoretical basis and its verification. *Agron. J.* 88:675-682.

Baker, J. T., L. H. Allan Jr., K. J. Boote, P. Jones, and J. W. Jones. 1990. Rice photosynthesis and evapotranspiration in subambient, ambient, and superambient carbon dioxide concentration. *Agron. J.* 82:834-840.

Batjes, N. H. 1996. Total carbon and nitrogen in the soils of the world. *European J. Soil Sci.* 47:151-163.

Bauer, A., and A. L. Black. 1994. Quantification of the effects of soil organic matter content on soil productivity. *Soil Sci. Soc. Am. J.* 58:185-193.

Bazzaz, F. A. 1990. The response of natural ecosystem to the rising global CO_2 levels. *Annu. Rev. Ecol. Syst.* 21:167-196.

Bohn, H. L. 1976. Estimate of organic carbon in world soils. Soil *Sci. Soc. Am. J.* 40:468-469.

Bohn, H. 1982. Estimate of organic carbon in world soils. *Soil Sci. Soc. Am. J.* 46:1118-1119.

Bowler, J. M., and M. C. Press. 1996. Effects of elevated CO_2, nitrogen form and concentration on growth and photosynthesis of a fast and slow growing grass. *New Phytol.* 132:391-401.

Brooks, A., and G. D. Farquhar. 1985. Effect of temperature on the CO_2/O_2 specificity of ribulose-1,5-bisphosphate carboxylase/oxygenase and the rate of respiration in the light. *Planta* 165:397-406.

Bunce, J. A. 1993. Effects of doubled atmospheric carbon dioxide concentration on the responses of assimilation and conductance to humidity. *Plant Cell Environ.* 16:189-197.

Bunce, J. A. 1998. The temperature dependence of the stimulation of photosynthesis by elevated carbon dioxide in wheat and barley. *J Exp. Bot.* 49:1555-1561.

Buringh, P. 1984. Organic carbon in soils of the world. *SCOPE* 23:91-109.

Burk, J. C., C. M. Yonker, W. J. Parton, C. V. Cole, K. Flach, and D. S. Schimel. 1989. Texture, climate, and cultivation effects on soil organic matter content in U.S. grassland soils. *Soil Sci. Soc. Am. J.* 53:800-805.

Campbell, C. A., and R. P. Zentner. 1993. Soil organic matter as influenced by crop rotations and fertilizations. *Soil Sci. Soc. Am. J.* 57:1034-1040.

Chaudhuri, U. N., M. B. Kirkham, and E. T. Kanemasu. 1990a. Carbon dioxide and water level effects on yield and water use of winter wheat. *Agron. J.* 82:637-641.

Chaudhuri, U. N., M. B. Kirkham, and E. T. Kanemasu. 1990b. Root growth of winter wheat under elevated carbon dioxide and drought. *Crop Sci.* 30:853-857.

Cure, J. D., and B. Acock. 1986. Crop response to carbon dioxide doubling: A literature survey. *Agric. For. Meteorol.* 38:127-145.

Curtin, D., F. Selles, H. Wang, C. A. Campbell, and V. O. Biederbeck. 1998. Carbon dioxide emissions and transformation of soil carbon and nitrogen during wheat straw decomposition. *Soil Sci. Soc. Am. J.* 62:1035-1041.

Curtis, P. S., L. M. Balduma, P. S. Drake, and D. F. Whigham. 1990. Elevated atmospheric CO_2 effects on below-ground processes in C_3 and C_4 estuarine marsh communities. *Ecology* 71:2001-2006.

Drake, B. G., M. A. Gonzalez-Meler, and S. P. Long. 1997. More efficient plants: A consequence of rising atmospheric CO_2. *Annu. Rev. Plant Physiol. Plant Mol. Biol.* 48:609-639.

Duchein, M., A. Bonicel, and T. Betsche. 1993. Photosynthetic net CO_2 uptake and leaf phosphate concentrations in CO_2 enriched clover at three levels of phosphate nutrition. *J. Exp. Bot.* 44:17-22.

Eswaran, H., E. V. D. Berg, and P. Reich. 1993. Organic carbon in soils of the word. *Soil Sci. Soc. Am. J.* 57:192-194.

Ferris, R., T. R. Wheeler, R. H. Ellis, and P. Hadley. 1999. Seed yield after environmental stress in soybean grown under elevated CO_2. *Crop Sci.* 39:710-718.

Frank, A. B., and A. Bauer. 1996. Temperature, nitrogen, and carbon dioxide effects on spring wheat development and spikelet numbers. *Crop Sci.* 36:659-665.

Gallo, K. P., C. S. T. Daughtry, and C. L. Wiegand. 1993. Errors in measuring absorbed radiation and computing crop radiation use efficiency. *Agron. J.* 85:1222-1228.

Ginkel, J. H. V., and A. Gorissen. 1998. In situ decomposition of grass roots as affected by elevated atmospheric carbon dioxide. *Soil Sci. Soc. Am. J.* 62:951-958.

Goudriaan, J. 1995. Global carbon cycle, pp. 207-217. In: S. Peng, K. T. Ingram, H. U. Neue, and L. H. Ziska (eds.) *Climate Change and Rice.* New York: Springer-Verlag; Los Baños, Philippines: Int. Rice Res. Inst.

Grayston, S. J., C. D. Campbell, J. L. Lutze, and R. M. Gifford. 1998. Impact of elevated CO_2 on the metabolic diversity of microbial communities in N-limited grass swards. *Plant Soil* 203:289-300.

Grotenhuis, T. P., and B. Bugbee. 1997. Super optimal CO_2 reduces seed yield but not vegetative growth in wheat. *Crop Sci.* 37:1215-1222.

Halvorson, A. D., C. A. Reule, and R. F. Follett. 1999. Nitrogen fertilization effects on soil carbon and nitrogen in a dryland cropping system. *Soil Sci. Soc. Am. J.* 63:912-917.

Heagle, A. S., J. E. Miller, and F. L. Booker. 1998. Influence of ozone stress on soybean response to carbon dioxide enrichment: I. Foliar properties. *Crop Sci.* 38:113-121.

Heagle, A. S., J. E. Miller, and W. A. Pursley. 1998. Influence of ozone stress on soybean response to carbon dioxide enrichment: I. Yield and seed quality. *Crop Sci.* 38:128-134.

Hogan, K. P., A. P. Smith, and L. H. Ziska. 1991. Potential effects of elevated CO_2 and changes in temperature on tropical plants. *Plant Cell Environ.* 14:763-778.

Hontoria, C., J. C. Rodriguez-Murillo, and A. Saa. 1999. Relationship between soil organic carbon and site characteristics in peninsular Spain. *Soil Sci. Soc. Am. J.* 63:614-621.

Houghton, J. T., G. J. Jenkins, and J. J. Ephraums. 1990. *Climate Change: The IPCC (Intergovernmental Panel on Climate Change) Scientific Assessment.* IPCC Rep. Working Group 1. Cambridge, England: Cambridge Univ. Press.

Houghton, J. T., L. G. Meira, B. A. Callander, N. Harris, A. Kattenberg, and K. Maskell. 1996. *Climate Change: The Science of Climate Change.* Cambridge, England: Cambridge University Press.

Idso, K. E., and S. B. Idso. 1994. Plant responses to atmospheric CO_2 enrichment in the face of environmental constraints: A review of the past 10 years research. *Agric. For. Meteorol.* 69:153-203.

Idso, S. B., B. A. Kimball, and J. R. Mauney. 1988. Effects of atmospheric CO_2 enrichment on root: shoot ratios of carrot, radish, cotton and soybean. *Agric. Ecosyst. Environ.* 21:293-299.

Ingvardsen, C., and B. Veierskov. 1994. Response of young barley plants to CO_2 enrichment. *J. Exp. Bot.* 45:1373-1378.

Intergovernmental Panel on Climate Change (IPCC). 1990. *Climate Change: The IPCC Assessment.* J. T. Houghton, L. G. Meira Filho, B. A. Callander, N. Harris, A. Kattenberg, and K. Maskell (eds.). Cambridge, England: Cambridge University Press.

Jackson, R. B., O. E. Sala, C. B. Field, and H. A. Mooney. 1994. CO_2 alters water use, carbon gain, and yield for the dominant species in a natural grassland. *Oecologia* 98:257-262.

Johnson, M. G., and J. S. Kerns. 1991. *Sequestering Carbon in Soils: A Workshop to Explore the Potential for Mitigating Global Climate Change.* Environ. Res. Lab. Rep. 600/3-91/031. Corvallis, OR: U.S. Environmental Protection Agency.

Jones, P., L. H. Allen Jr., J. W. Jones, K. J. Boote, and W. J. Campbell. 1984. Soybean canopy growth, photosynthesis, and transpiration responses to whole-season carbon dioxide enrichment. *Agron. J.* 76:633-637.

Jones, P., J. W. Jones, and L. H. Allen Jr. 1985. Carbon dioxide effects on photosynthesis and transpiration during vegetative growth in soybeans. *Proc. Soil Crop Sci. Soc. Fla.* 44:129-143.

Karl, T. R., N. Nicholls, and J. Gregory. 1997. The coming climate. *Sci. Am.* 276:78-83.

Keeling, C. D., T. P. Whorf, M. Wahlen, and J. V. Plicht. 1995. Inter annual extremes in the rate of rise atmospheric dioxide since 1980. *Nature* 375:666-670.

Kempe, S. 1979. Carbon in the rock cycle. *SCOPE* 13:343-378.

Kern, J. S., and M. G. Johnson. 1993. Conservation tillage impacts on national soil and atmospheric carbon levels. *Soil Sci. Soc. Am. J.* 57:200-210.

Kimball, B. A. 1983. Carbon dioxide and agricultural yield: An assessment and analysis of 430 prior observations. *Agron. J.* 75:779-788.

Kimball, B. A., K. Kobayashi, and M. Bindi. 2002. Responses of agricultural crops to free-air CO_2 enrichment. *Adv. Agron.* 77:293-368.

Klimov, V. V., S. I. Allakhverdiev, Y. M. Feyziev, and S. V. Baranov. 1995. Bicarbonate requirement for the donor side of photosystem II. *FEBS Lett.* 363:251-255.

Krenzer, E. G., Jr., and D. N. Moss. 1975. Carbon dioxide enrichment effects upon yield and yield components in wheat. *Crop Sci.* 15:71-74.

Lal, R., J. M. Kimble, R. F. Follett, and C. V. Cole. 1998. *The Potential of U. S. Cropland to Sequester Carbon and Mitigate the Greenhouse Effect.* Chelsea, MI: Ann Arbor Press.

Lee, J. J., and R. Dodson. 1996. Potential carbon sequestration by afforestation of pasture in the South-Central United States. *Agron. J.* 88:381-384.

Lutze, J. L., and R. M. Gifford. 1995. Carbon storage and productivity of a carbon dioxide enriched nitrogen limited grass sward after one years growth. *J. Biogeog.* 22:227-233.

Mahieu, N., D. S. Powlson, and E. W. Randall. 1999. Statistical analysis of published carbon-13 CPMAS NMR spectra of soil organic matter. *Soil Sci. Soc. Am. J.* 63:307-319.

Merckx, R., J. H. V. Ginkel, J. Sinnaeve, and A. Cremers. 1986. Plant induced changes in the rhizosphere of maize and wheat: I. Production and turnover of root-derived material in the rhizosphere of maize. *Plant Soil* 96:85-93.

Miller, J. E., A. S. Heagle, and W. A. Pursley. 1998. Influence of ozone stress on soybean response to carbon dioxide enrichment: I. Biomass and development. *Crop Sci.* 38:122-128.

Morison, J. I. L. 1985. Sensitivity of stomata and water use efficiency under high CO_2. *Plant Cell Environ.* 8:467-474.

Neales, T. F., and A. O. Nicholls. 1978. Growth response of young wheat plants to a range of ambient CO_2 levels. *Aust. J. Plant Physiol.* 5:45-49.

Newbery, R. M., and J. Wolfenden. 1996. Effects of elevated CO_2 and nutrient supply on the seasonal growth and morphology of *Agrostis capillaris*. *New Phytol.* 132:403-411.

Newton, P. C. D., H. Clark, C. C. Bell, E. M. Glasgow, and B. D. Campbell. 1994. Effects of elevated CO_2 and simulated changes in temperature on the species composition and growth rate of pasture turves. *Ann. Bot.* 73:53-59.

Newton, P. C. D., H. Clark, C. C. Bell, E. M. Glasgow, D. J. Ross, and G. W. Yeates. 1995. Plant growth and soil processes in temperate grassland communities at elevated CO_2. *J. Biogeogr.* 22:235-240.

Nie, D., H. He, G. Mo, M. B. Kirkham, and E. T. Kanemasu. 1992. Canopy photosynthesis and evapotranspiration of rangeland plants under doubled carbon dioxide in closed-top chambers. *Agric. For. Meteorol.* 61:205-217.

Norby, R. J., J. Pastor, and J. M. Melillo. 1986. Carbon-nitrogen interactions in CO_2-enriched white oak: Physiological and long-term perspectives. *Tree Physiol.* 2:233-241.

Parks, P. J. 1992. Opportunities to increase forest area and timber growth on marginal crop and pasture lands, pp. 97-121. In R. N. Sampson and D. Hair (eds.) *Forest and Global Change,* Vol. 1, *Opportunities for Increasing Forest Cover.* Washington, DC: Am. For. Assn.

Patterson, D. T. 1986. Responses of soybean and three C4 grass weeds to CO_2 enrichment during drought. *Weed Sci.* 34:203-210.

Pinter, P. J. J., B. A. Kimball, R. L. Garcia, G. W. Wall, D. J. Hunsake, and R. L. Lamorte. 1996. Free-air CO_2 enrichment responses of cotton and wheat crops, pp. 215-249. In: G. W. Koch and H. A Mooney (eds.) *Carbon Dioxide and Terrestrial Ecosystems*. San Diego: Academic Press.

Polley, H. W., H. B. Johnson, B. D. Marino, and H. S. Mayeux. 1993. Increase in C_3 plant water-use efficiency and biomass over glacial to present CO_2 concentrations. *Nature* 361:61-64.

Polley, H. W., H. B. Johnson, and H. S. Mayeux. 1994. Increasing CO_2: Comparative responses of the C_4 grass *Schizachyrium* and grassland invader *Prosopis*. *Ecology* 75:976-988.

Polley, H. W., H. B. Johnson, H. S. Mayeux, and S. R. Malone. 1993. Physiology and growth of wheat across a subambient carbon dioxide gradient. *Ann. Bot.* 71:347-356.

Poorter, H. 1993. Interspecific variation in the growth response of plants to an elevated ambient CO_2 concentration. *Vegetation* 104/105:77-97.

Portis, A. R. 1995. The regulation of Rubisco by Rubisco activase. *J. Exp. Bot.* 46:1285-1291.

Prinn, R. G. 1995. Global change: Problems and uncertainties, pp. 3-7, In: S. Peng, K. T. Ingram, H. U. Neue, and L. H. Ziska (eds.) *Climate Change and Rice*. New York: Springer-Verlag; Los Baños, Philippines: Int. Rice Res. Inst.

Qian, Y., and R. F. Follett. 2002. Assessing soil carbon sequestration in turfgrass systems using long-term soil testing data. *Agron. J.* 94:930-935.

Rawson, H. M. 1995. Yield responses of two wheat genotypes to CO_2 and temperature in field studies using temperature gradient tunnels. *Aust. J. Plant Physiol.* 22:23-32.

Reddy, K. R., G. H. Davidonis, A. S. Johnson, and B. T. Vinyard. 1999. Temperature regime and carbon dioxide enrichment alter cotton boll development and fiber properties. *Agron. J.* 91:851-858.

Rodhe, H. 1990. A comparison of the contribution of various gases to the greenhouse effects. *Science* 248:1217-1219.

Rogers, H. H., G. E. Bingham, J. D. Cure, J. M. Smith, and K. A. Surano. 1983. Responses of selected plant species to elevated CO_2 in the field. *J. Environ. Quality* 12:569-574.

Rogers, H. H., J. D. Cure, and J. M. Smith. 1986. Soybean growth and yield response to elevated carbon dioxide. *Agric. Ecosyst. Environ.* 16:113-128.

Rogers, H. H., J. D. Cure, J. R. Thomas, and J. M. Smith. 1984. Influence of elevated CO_2 on growth of soybean plants. *Crop Sci.* 24:361-366.

Rogers, H. H., G. B. Runion, and S. V. Krupa. 1994. Plant response to atmospheric CO_2 enrichment with emphasis on roots and the rhizosphere. *Environ. Pollut.* 83:155-189.

Rogers, H. H., N. Sionit, J. D. Cure, J. M. Smith, and G. E. Bingham. 1984. Influence of carbon dioxide on water relations of soybean. *Plant Physiol.* 74:233-238.

Rosenzweig, C., and D. Hillel. 1998. *Climate Change and the Global Harvest.* Oxford, England: Oxford University Press.

Schroeder, P. E., and L. Ladd. 1991. Slowing the increase of atmospheric carbon dioxide: A biological approach. *Climate Change* 19:283-290.

Seneweera, S., and J. P. Conroy. 1997. Growth, grain yield and quality of rice in response to elevated CO_2 and phosphorus nutrition, pp. 873-878. In: T. Ando, K. Fujita, T. Mae, H. Matsumoto, S. Morie, and J. Sekiya (eds.) *Plant Nutrition for Sustainable Food Production and Environment.* Dordrecht, the Netherlands: Kluwer Acad. Publ.

Siegenthaler, U. 1990. El Niño and atmospheric CO_2. *Nature* 345:295-296.

Sionit, N., D. A. Mortensen, B. R. Strain, and H. Hellmers. 1981. Growth response of wheat to CO_2 enrichment and different levels of mineral nutrition. *Agron. J.* 73:1023-1027.

Sionit, N., H. H. Rogers, G. E. Bingham, and B. R. Strain. 1984. Photosynthesis and stomatal conductance with CO_2 enrichment of container and field grown soybeans. *Agron. J.* 76:447-451.

Swinnen, J., J. A. V. Veen, and R. Merckx. 1995. Carbon fluxes in the rhizosphere of winter wheat and spring barley with conventional vs. integrated farming. *Soil Biol. Biochem.* 27:811-820.

Thomas, J. F., and C. N. Harvey. 1983. Leaf anatomy of four species grown under continuous CO_2 enrichment. *Bot. Gaz.* 144: 303-309.

Tubiello, F. N., C. Rosenzweig, B. A. Kimball, P. J. Pinter Jr., G. W. Wall, D. H. Hunsaker, R. L. La Morte, and R. L. Garcia. 1999. Testing CERES-wheat with free-air carbon dioxide enrichment (FACE) experiment data: CO_2 and water interactions. *Agron. J.* 91:247-255.

Turner, D. P., J. J. Lee, G. J. Koerper, and J. R. Barker. 1993. *The Forest Sector Carbon Budget of the United States: Carbon Pools and Flux Under Alternative Policy Options.* Environ. Protect. Agency 600/3-93/093. Washington, DC: U.S. Environ. Protect. Agency.

Warrick, R. A., and E. M. Barrow. 1990. Climate and sea level change: A perspective. *Outlook on Agriculture* 19:5-8.

Watson, R. T., H. Rohde, H. Oeschger, and U. Siegenthaler. 1990. Greenhouse gases and aerosols, pp. 7-40. In: J. T. Houghton, B. A. Callandar, and S. K. Varney (eds.) *Climate Change: The IPCC Scientific Assessment.* Cambridge, England: Cambridge University Press.

Wilks, D. S., and S. J. Riha. 1996. High frequency climatic variability and crop yields. *Climatic Change* 32:231-235.

Woodrow, I. E., and J. A Berry. 1988. Enzymatic regulation of photosynthetic CO_2 fixation in C_3 plants. *Annu. Rev. Plant Physiol. Plant Mol. Biol.* 39:533-594.

Wyse, R. 1980. Growth of sugarbeet seedlings in various atmospheres of oxygen and carbon dioxide. *Crop Sci.* 20:456-458.

Ziska, L. H., K. P. Hogan, A. P. Smith, and B. G. Drake. 1991. Growth and photosynthetic response of nine tropical species with long term exposure to elevated carbon dioxide. *Oecologia* 86:383-389.

Chapter 7

Aase, J. K., and G. M. Schaefer. 1996. Economics of tillage practices and spring wheat and barley crop sequence in the northern Great Plains. *J. Soil Water Conserv.* 51:167-170

Adams, S. S., and W. R. Stevenson. 1990. Water management, disease development and potato production. *Am. Potato J.* 67:3-11.

Alessi, J., J. F. Power, and D. C. Zimmerman. 1977. Sunflower yield and water use as influenced by planting date, population, and row spacing. *Agron. J.* 69:465-469.

Allard, R. W., P. Garcia, L. E. Saenz de Miera, and V. M. Perez. 1993. Evolution of multilocus structure in *Avena hirtula* and *Avena barbata. Genetics* 135:1125-1139.

Amir, J., and T. R. Sinclair. 1996. A straw mulch system to allow continuous wheat production in an arid climate. *Field Crops Res.* 27:365-376.

Araghi, S. G., and M. T. Assad. 1998. Evaluation of four screening techniques for drought resistance and their relationship to yield reduction ratio in wheat. *Euphytica* 103:293-299.

Ashley, D. A., and W. J. Ethridge. 1978. Irrigation effects on vegetative and reproductive development of three soybean cultivars. *Agron. J.* 70:467-471.

Aspinall, D. 1984. Water deficit and wheat, pp. 91-110. In: C. J. Pearson (ed.), *Control of Crop Productivity.* Sydney, Australia: Academic Press.

Asseng, S., N. C. Turner, T. Botwright, and A. G. Condon. 2003. Evaluating the impact of a trait for increased specific leaf area on wheat yields using a crop simulation model. *Agron. J.* 95:10-19.

Austin, R. B. 1989. Maximizing crop production in water-limited environments, pp. 13-25. In: F. W. G. Baker (ed.), *Drought Resistance in Cereals.* London: Common. Agric. Bur.

Austin, R. B., C. L. Morgan, M. A. Ford, and R. D. Blackwell. 1980. Contributions to the grain yield from pre-anthesis assimilation in tall dwarf barley phenotypes in two contrasting seasons. *Ann. Bot.* 45:309-319.

Badaruddin, M., and D. W. Meyer. 1989. Water use by legumes and its effect on soil water status. *Crop Sci.* 29:1212-1216.

Baligar, V. C., N. K. Fageria, and M. A. Elrashidi. 1998. Toxicity and nutrient constraints on root growth. *Hort. Sci.* 33:960-965.

Barker, R., D. Dawe, T. P. Tuong, S. I. Bhuiyan, and L. C. Guerra. 1999. The outlook for water resources in the year 2020: Challenges for research on water management in rice production, pp. 96-109. In: D. V. Tran (ed.), *Assessment and*

Orientation Towards the 21st Century. Proceedings of 19th Session of International Rice Commission, September 7-9, 1998, Cairo, Egypt. Rome: FAO.

Bartels, D., and D. Nelson. 1994. Approaches to improve stress tolerance using molecular genetics. *Plant Cell Environ.* 17:659-667.

Bauder, J. W., and M. J. Ennen. 1979. Crop water use—How does sunflower rate? *The Sunflower* 5:10-11.

Baver, L. D., W. H. Gardner, and W. R. Gardner. 1972. *Soil Physics,* Fourth Edition. New York: John Wiley.

Benjamin, J. G., L. K. Porter, H. R. Duke, and L. R. Ahuja. 1997. Corn growth and nitrogen uptake with furrow irrigation and fertilizer bands. *Agron. J.* 89:609-612.

Bennett, J. M., T. R. Sinclair, R. E. Muchow, and S. R. Costello. 1987. Dependence of stomatal conductance on leaf water potential, turgor potential, and relative water content in field-grown soybean and maize. *Crop Sci.* 27:984-990.

Bezdicek, D. F., and D. Granatstein. 1989. Crop rotation efficiencies and biological diversity in farming systems. *Am. J. Alt. Agric.* 4:111-118.

Black A. L., P. L. Brown, A. D. Halvorson, and F. H. Siddoway. 1981. Dryland cropping strategies for efficient water use to control saline seeps in the northern Great Plains, U. S. A. *Agric. Water Manage.* 4:295-311.

Blum, A. 1983. Genetic and physiological relationships in plant breeding for drought resistance. *Agric. Water Manage.* 7:195-205.

Blum, A. 1988. *Plant Breeding for Stress Environments.* Boca Raton, FL: CRC Press.

Blum, A. 1989. Osmotic adjustment and growth of barley genotypes under stress. *Crop Sci.* 29:230-233.

Blum, A., J. Mayer, and G. Golan. 1983. Chemical desiccation of wheat plants as a simulator of post-anthesis stress: II. Relations to drought stress. *Field Crops Res.* 6:149-155.

Blum, A., J. Mayer, and G. Golan. 1989. Agronomic and physiological assessments of genotypic variation for drought resistance in sorghum. *Aust. J. Agric. Res.* 40:49-61.

Blum, A., H. Poiarkova, G. Golan, and J. Mayer. 1983. Chemical desiccation of wheat plants as a simulator of post-anthesis stress: I. Effects on translocation and kernel growth. *Field Crops Res.* 6:51-58.

Bolton, F. 1981. Optimizing the use of water and nitrogen through soil and crop management. *Plant Soil.* 58:231-247.

Bonfil, D., I. Mufradi., S. Klitman, and S. Asido. 1999. Wheat grain yield and soil profile water distribution in a no-till arid environment. *Agron. J.* 91:368-373.

Boyer, J. S. 1982. Plant productivity and environment. *Science* 218:443-448.

Boyer, J. S. 1985. Water transport. *Annu. Rev Plant Physiol.* 36:473-516.

Boyer, J. S. 1996. Advances in drought tolerance in plants. *Adv. Agron.* 56:187-218.

Boyle, M. G., J. S. Boyer, and P. W. Morgan. 1991. Stem infusion of liquid culture medium prevents reproductive failure of maize at low water potential. *Crop Sci.* 31:1246-1252.

Bremner, P. M., G. K. Preston, and G. C. Fazekas. 1986. A field comparison of sunflower and sorghum in a long drying cycle. *Aust. J. Agric. Res.* 37:483-493.

Campbell, C. A., F. Selles, R. P. Zentner, and B. G. McConkey. 1993. Available water and nitrogen effects on yield components and grain nitrogen of zero-till spring wheat. *Agron. J.* 85:114-120.

Carter, T. E., and T. W. Rufty. 1993. Soybean plant introductions exhibiting drought and aluminum tolerance, pp. 335-346. In: C. G. Kuo (ed.), *Adaptation of Food Crops to Temperature and Water Stress.* Proc. Int. Symp., Paipei, Taiwan. August 13-18, 1992. Taipei, Taiwan: Asian Vegetables Res. and Dev. Center.

Ceccarelli, S., and S. Grando. 1996. Drought as a challenge for the plant breeder. *Plant Growth Regulation* 20:149-155.

Connor, D. J., and Hall, A. J. 1997. Sunflower physiology, pp. 113-182. In: A. A. Schneiter (ed.), *Sunflower Technology and Production.* Agron. Monogr. No. 35. Madison, WI: Am. Soc. Agron., Crop Sci. Soc. Am., and Soil Sci. Soc. Am.

Cooper, P. J. M., and P. J. Gregory. 1987. Soil water management in the rainfed farming systems of the Mediterranean region. *Soil Use Manag.* 3:57-62.

Cooper, P. J. M., J. D. H. Keating, and H. Hughes. 1983. Crop evapotranspiration: A technique for calculation of its components by field measurements. *Field Crop Res.* 7:299-312.

Cosgrove. D. J. 1993. Wall extensibility: Its nature, measurement and relationship to plant cell growth. *Transley Review* 46. *New Phytol.* 124:1-23.

Crabtree, R. J., A. A. Yassin, I. Kargougou, and R. W. McNew. 1985. Effects of alternate-furrow irrigation: Water conservation on the yields of two soybean cultivars. *Agric. Water Manage.* 10:253-264.

Crookston, R. K., J. E. Kurle, P. J. Copeland, J. H. Ford, and W. E. Lueschen. 1991. Rotational cropping sequence affects yield of corn and soybean. *Agron. J.* 83:108-113.

Cruz, R. T., W. R. Jordan, and M. C. Drew. 1992. Structural changes and associated reduction of hydraulic conductance in roots of *Sorghum bicolor* L. following exposure to water deficit. *Plant Physiol.* 99:203-212.

Cruz, R. T., and J. C. O'Toole. 1984. Dryland rice response to an irrigation gradient at flowering stage. *Agron. J.* 76:178-183.

Darusman, A., H. Khan, L. R. Stone, W. E. Spurgeon, and F. R. Lamm. 1997. Water flux below the root zone vs. irrigation in drip-irrigated corn. *Agron. J.* 89:375-379.

Davidson, J. L., and J. W. Birch. 1978. Response of a standard Australian and Mexican wheat to temperature and water stress. *Aust. J. Agric. Res.* 29:1091-1106.

Day, A. D., and S. Intalap. 1970. Some effects of soil moisture on the growth of wheat. *Agron. J.* 62:27-29.

Day, W., B. J. Legg, B. K. French, A E. Johnston, D. W. Lawlor, and W. C. Jeffers. 1978. A drought experiment using mobile shelters: The effect of drought on barley yield, water use and nutrient uptake. *J. Agric. Sci., Cambridge* 91:599-623.

Denmead, O. T., and B. D. Miller. 1976. Field studies of the conductance of wheat leaves and transpiration. *Agron. J.* 68:307-311.

Dent, F. J. 1980. Major production systems and soil-related constraints in Southeast Asia, pp. 79-106. In: Int. Rice Rrs. Inst. (IRRI) (ed.), *Properties for Alleviating Food Production in the Tropics*. Los Baños, Philippines: IRRI.

De Souza, J. G., and J. V. Da Silva. 1987. Partitioning of carbohydrates in annual and perennial cotton (*Gossypium hirsutum* L.). *J. Exp. Bot.* 38:1211-1218.

Dingkuhn, M., G. D. Farquhar, S. K. De Datta, and J. C. O'Toole. 1991. Discrimination of C-13 among upland rice having different water use efficiency. *Aust. J. Agric. Res.* 42:1123-1131.

Doorenbos, J., and W. O. Pruitt. 1992. *Crop Water Requirements*. Food Agric. Org. (FAO) Irrigation and Drainage, Paper 24. Rome: FAO.

Downton, W. J. S. 1983. Osmotic adjustment during water stress protects the photosynthetic apparatus against photoinhibition. *Plant Mol. Biol.* 40:503-537.

Duncan, R. R., and R. N. Carrow. 1999. Turfgrass molecular genetic improvement for abiotic/edaphic stress resistance. *Adv. Agron.* 67:233-306.

Dunphy, E. J. 1985. *Soybean on Farm Test Report*. Raleigh, NC: North Carolina Agric. Ext. Serv.

Durand, J. L., J. E. Sheehy, and F. R. Minchin. 1987. Nitrogenase activity, photosynthesis and water potential in soybean plants experiencing water deprivation. *J. Exp. Bot.* 38:311-321.

Eck, H. V. 1984. Irrigated corn yield response to nitrogen and water. *Agron. J.* 76:421-428.

Egli, D. B., and S. J. Crafts-Brandner. 1996. Soybean, pp. 595-623. In: E. Zamski and A. A. Schaffer (eds.), *Photoasimilate Distribution in Plants and Crops: Source-Sink Relationships*. New York: Marcel Dekker.

Ehdaie, B. 1995. Variation in water-use efficiency and its components in wheat: II. Pot and field experiments. *Crop Sci.* 35:1617-1626.

Ehdaie, B., and J. G. Waines. 1993. Variation in water use efficiency and its components in wheat: I. Well watered pot experiment. *Crop Sci.* 33:294-299.

Fageria, N. K. 1992. *Maximizing Crop Yields*. New York: Marcel Dekker.

Fageria, N. K., O. P. Morais, V. C. Baligar, and R. J. Wright. 1988. Response of rice cultivars to phosphorus supply on an Oxisol. *Fert. Res.* 16:195-206.

Farquhar, G. D., J. R. Ehleringer, and K. Hubic. 1989. Carbon isotope discrimination and photosynthesis. *Annu. Ver. Plant Physiol. Plant Mol. Biol.* 40:503-537.

Farquhar, G. D., and R. A. Richards. 1984. Isotopic composition of plant carbon correlates with water-use efficiency of wheat genotypes. *Aust. J. Plant Physiol.* 11:539-552.

Fernandez, G. C. J. 1993. Effective selection criteria for assessing plant stress tolerance, pp. 257-270. In: C. G. Kuo (ed.), *Adaptation of Food Crops to Temperature and Water Stress*. Shanhua, Taiwan: AVRDC.

Fischer, R. A. 1973. The effect of water stress at various stages of development on yield processes in wheat, pp. 233-241. In: R. O. Slatyer (ed.), *Plant Response to Climatic Factors*. Paris: U.N. E S C Org. (UNESCO).

Fischer, R. A. 1979. Growth and water limitation to dryland wheat yield in Australia: A physiological framework. *J. Aust. Inst. Agric. Sci.* 45:83-94.

Fischer, R. A., and R. Maurer. 1978. Drought resistance in spring wheat cultivars: I. Grain yield responses. *Aust. J. Agric. Res.* 29:897-912.

French, R. J., and J. E. Schultz. 1984. Water use efficiency of wheat in a Mediterranean type environment: I. The relation between yield, water use and climate. *Aust. J. Agric. Res.* 35:743-764.

Frensch, J., and T. C. Hsiao. 1994. Transient responses of cell turgor and growth of maize roots as affected by changes in water potential. *Plant Physiol.* 104:247-254.

Fukai, S., and M. Cooper. 1996. Stress physiology in relation to breeding for drought resistance: A case study of rice, pp.123-149. In: V. P. Singh, R. K. Singh, B. B. Singh, and R. S. Ziegler (eds.), *Physiology of Stress Tolerance in Rice*. Proc. Int. Conf. Stress Physiol. Rice. Los Baños, Philippines: Int. Rice Res. Inst.

Garrity, D. P., and J. C. O'Toole. 1995. Selection for reproductive stage drought avoidance in rice, using infrared thermometry. *Agron. J.* 87:773-779.

Gerik, T. J., K. L. Faver, P. M. Thaxton, and K. M. El-Zik. 1996. Late season water stress in cotton: Plant growth, water use, and yield. *Crop Sci.* 36:914-921.

Gifford, R. M. 1986. Partitioning of photoassimilate in the development of crop yield, pp. 535-549. In: J. Cronshaw, W. J. Lucas, and R. T. Giaquinta (eds.), *Phloem Transport*. New York: A. R. Liss.

Ginkel, M. V., D. S. Calhoun, G. Gebeyehu, A. Miranda, C. Tian-you, R. P. Lara, R. M. Trethowan, K. Sayre, J. Crossa, and S. Rajaram. 1998. Plant traits related to yield of wheat in early, late, or continuous drought conditions. *Euphytica* 100:109-121.

Giunta, F., R. Motzo, and M. Deidda. 1993. Effect of drought on yield and yield components of durum wheat and triticale in a Mediterranean environment. *Field Crops Res.* 33:399-409.

Gleick, P. H. 1993. *Water in Crisis: A Guide to the World's Fresh Water Resources*. New York: Oxford University Press.

Greenland, D. J. 1997. *The Sustainability of Rice Farming*. Los Baños, Philippines: Int. Rice Res. Inst.

Guerra, L. C., S. I. Bhuiyan, T. P. Tuong, and R. Barker. 1998. *Producing More Rice with Less Water from Irrigated Systems*. Discussion Paper Series No 29. Los Baños, Philippines: Int. Rice Res. Inst.

Hale, M. G., and D. M. Orcutt. 1987. *The Physiology of Plants Under Stress*. New York: John Wiley.

Halvorson, A. D., A. L. Black, J. M. Krupinsky, S. D. Merrill, and D. L. Tanaka. 1999. Sunflower response to tillage and nitrogen fertilization under intensive cropping in a wheat rotation. *Agron. J.* 91:637-642.

Halvorson, A. D., and C. A. Reule. 1994. Nitrogen fertilizer requirements in an annual dryland cropping systems. *Agron. J.* 86:315-318.

Halvorson, A. D., Wienhold, B. J., and Black, A. L. 2001. Tillage and nitrogen fertilization influence grain and soil nitrogen in an annual cropping system. *Agron. J.* 93:836-841.

Hamblin, A., and D. Tennant. 1987. Root length density and water uptake in cereals and grain legumes: How well are they correlated? *Aust. J. Agric. Res.* 38:513-527.

Hansen, A. D., and W. D. Hitz. 1982 Metabolic responses of mesophytes to plant water deficits. *Annu. Rev. Plant Physiol.* 33:163-203.

Hanson, B. R. 1987. A systems approach to drainage reduction. *Calif. Agric.* 41:19-24.

Harris, H. C., P. J. M. Cooper, and M. Pala. 1991. *Soil and Crop Management for Improved Water Use Efficiency in Rainfed Areas*. Proc. Int. Workshop, Ankra, Turkey. May 15-19, 1989. Aleppo, Syria: ICARDA.

Hattendorf, M. J., M. S. Redelfs, B. Amos, L. R. Stone, and R. E. Groin Jr. 1988. Comparative water use characteristics of six row crops. *Agron. J.* 80:80-85.

Hodgson, A. S., and K. Y. Chan. 1984. Deep moisture extraction and crack formation by wheat and safflower in a vertisol following irrigated cotton rotations. *Rev. Rural Sci.* 5:299-304.

Howell, T. A., J. A. Tolk, A. D. Schneider, and S. R. Evett. 1998. Evapotranspiration, yield, and water use efficiency of corn hybrids differing in maturity. *Agron. J.* 90:3-9.

Hsiao, T. C. 1973. Plant responses to water stress. *Annu. Rev. Plant Physiol.* 24:519-570.

Hsiao, T. C., and E. Acevedo. 1974. Plant responses to water deficits, water-use efficiency, and drought resistance. *Agric. Meteorol.* 14:59-84.

Hubick, K. T., and G. D. Farquhar. 1989. Carbon isotope discrimination and the ratio of carbon gained to water lost in barley cultivars. *Plant Cell Environ.* 12:795-804.

Hubick, K. T., G. D. Farquhar, and R. Shorter. 1986. Correlation between water-use efficiency and carbon isotope discrimination in diverse peanut germplasms. *Aust. J. Plant Physiol.* 13:803-816.

Hudak, C. M., and R. P. Patterson. 1996. Root distribution and soil moisture depletion pattern of a drought-resistant soybean plant introduction. *Agron. J.* 88:478-485.

Hull, H. M., L. N. Wright, and C. A. Bleckmann. 1978. Epicuticular wax ultrastructure among lines of *Eragrostis lehmanniana* Nees developed for seedling drought tolerance. *Crop Sci.* 18:699-704.

Hurd, E. A. 1974. Phenotype and drought tolerance in wheat. *Agr. Meterol.* 14:39-55.

Ishihara, K., and H. Saito. 1983. Relationships between leaf water potential and photosynthesis in rice plants. *JARQ* 17:81-86.

Ismail, M. A., and A. E. Hall. 1992. Correlation between water-use efficiency and carbon isotope discrimination in diverse cowpea genotypes and isogenic lines. *Crop Sci.* 32:7-12.

Jensen, S. D., and A. J. Cavalieri. 1983. Drought tolerance in U.S. maize, pp. 223-236. In: J. F. Stone and W. O. Willis (eds.), *Plant Production and Management Under Drought Conditions.* Devel. Agric. Managed-Forest Ecol. 12. New York: Elsevier Sci. Publ.

Johnson, R. C., and E. T. Kanemasu. 1982. The influence of water availability on winter wheat yields. *Can. J. Plant Sci.* 62:831-838.

Joly, R. J., and D. T. Hahn. 1989. An empirical model for leaf expansion in cacao in relation to plant water deficit. *Ann. Bot.* 64:1-8.

Jones, C. A. 1985. *C_4 Grasses and Cereals: Growth, Development, and Stress Response.* New York: John Wiley.

Jones, M. M., and H. M. Rawson. 1979. Influence of rate of development of leaf water deficits upon photosynthesis, leaf conductance, water use efficiency, and osmotic potential in sorghum. *Physiol. Plant.* 45:103-111.

Jones, O. R., and W. C. Johnson. 1983. Cropping practices: Southern Great Plains, pp. 365-385. In: H. E. Dregne and W. O. Wills (eds.), *Dryland Agriculture.* Agron. Monogr. 23. Madison, WI: Am. Soc. Agron., Crop Sci. Soc. Am., and Soil Sci. Soc. Am.

Jordan, W. R., and F. R. Miller. 1980. Genetic variability in sorghum root systems: Implications for drought tolerance, pp. 383-399. In: N. C. Turner and P. J. Kramer (eds.), *Adaptation of Plants to Water and High Temperature Stress.* New York: John Wiley.

Jordan, W. R., R. L. Monk, F. R. Miller, D. T. Rosenow, R. E. Clark, and P. J. Shouse. 1983. Environmental physiology of sorghum: I. Environmental and genetic control of epicuticular wax load. *Crop Sci.* 23:552-555.

Kane, M. V., and L. J. Grabau. 1992. Early planted, early maturing soybean cropping system: Growth, development, and yield. *Agron. J.* 84:769-773.

Karlen, D. L., G. E. Varvel, D. G. Bullock, and R. M. Cruse. 1994. Crop rotations for the 21st century. *Adv. Agron.* 53:1-45.

Kauss, H. 1977. Biochemistry of osmotic regulation. *Plant Biochem.,* II 13:119-140.

Keatinge, J. D. H., M. D. Dennett, and J. Rodgers. 1986. The influence of precipitation regime on the management of dry areas in northern Syria. *Field Crops Res.* 13:239-249.

Kirda, C., S. K. A. Danso, and F. Zapata. 1989. Temporal water stress effects on nodulation, nitrogen accumulation and growth of soybean. *Plant Soil* 120:49-55.

Klaji, M. C., and G. Vachaud. 1992. Seasonal water balance of a sandy soil in Niger cropped with pearl millet, based on profile moisture measurements. *Agric. Water Manage.* 21:313-330.

Kobata, T. 1995. Drought resistance, pp. 474-483. In: T. Matsuo, K. Kumazawa, R. Ishii, K. Ishihara, and H. Hirata (eds.), *Science of the Rice Plant: Physiology,* Vol. 2. Tokyo: Food Agric. Policy Res. Center.

Kramer, P. J., and J. S. Boyer. 1995. *Water Relations of Plants and Soil.* San Diego, CA: Academic Press.

Labanauskas, C. K., P. Shouse, and L. H. Stolzy. 1981. Effects of water stress at various growth stages on seed yield and nutrient concentration of field grown cowpeas. *Soil Sci.* 131:249-256.

Larcher, W. 1995. *Physiological Plant Ecology,* Third Edition. New York: Springer-Verlag.

Lecoeur, J., and T. R. Sinclair. 1996. Field pea transpiration and leaf growth in response to soil water deficits. *Crop Sci.* 36:331-335.

Levitt, J. 1980. *Responses of Plants to Environmental Stresses,* Vol. 2. New York: Academic Press.

Lockhart, J. A. 1965. An analysis of irreversible plant cell elongation. *J. Theor. Biol.* 8:264-275.

López-Castañeda, C. D., and R. A. Richards. 1994. Variation in temperate cereals in rainfed environments: III. Water use and water use efficiency. *Field Crops Res.* 39:85-98.

Lorens, G. F., J. M. Bennett, and L. B. Loggale. 1987. Differences in drought resistance between two corn hybrids: II. Component analysis and growth rates. *Agron. J.* 79:808-813.

Ludlow, M. M., and R. C. Muchow. 1990. A critical evaluation of traits for improving crop yields in water-limited environments. *Adv. Agron.* 43:107-153.

Lynch, D. R., and G. C. C. Tai. 1989. Yield and yield component response of eight potato genotypes to water stress. *Crop Sci.* 29:1207-1211.

Mackerron, D. K. L., and R. A. Jefferies. 1986. The influence of early soil moisture stress on tuber number in potato. *Potato Res.* 29:299-312.

Mambani, B., and R. Lal. 1983. Response of upland rice cultivars to drought stress: III. Screening rice varieties by means of variable moisture along a toposequence. *Plant Soil* 73:73-94.

Mason, W. K., W. S. Meyer, R. C. G. Smith, and H. D. Bars. 1983. Water balance of three irrigated crops on fine-textured soils of the Riverine plain. *Aust. J. Agric. Res.* 34:183-191.

Matin, M. A., J. H. Brown, and H. Ferguson. 1989. Leaf water potential relative water content, and diffusive resistance as screening techniques for drought resistance in barley. *Agron. J.* 81:100-105.

Matyssek, R., A. C. Tang, and J. S. Boyer. 1991. Plants can grow on internal water. *Plant Cell Environ.* 14:925-930.

Maurya, D. M., and J. C. O'Toole. 1986. Screening upland rice for drought tolerance, pp. 245-261. In: Int. Rice Res. Inst. (IRRI) (ed.), *Progress in Upland Rice Research.* Proc. 1985 Jakarta Conf. Los Baños, Philippines: IRRI.

Mayhew, W. L., and C. E. Caviness. 1994. Seed quality and yield of early-planted, short-season soybean genotypes. *Agron. J.* 89:459-464.

McCree, K. J., C. E. Kallsen, and S. G. Richardson. 1984. Carbon balance of sorghum plants during osmotic adjustment to water stress. *Plant Physiol.* 76:898-902.

McGee, E. A., G. A. Peterson, and D. G. Westfall. 1997. Water storage efficiency in no-till dryland cropping systems. *J. Soil Water Conserv.* 52:131-136.

McWilliam, J. R. 1986. The national and international importance of drought and salinity effects on agricultural production. *Aust. J. Plant Physiol.* 13:1-13.

Meckel, L., D. B. Egli, R. E. Phillips, D. Radcliffe, and J. E. Leggett. 1984. Effect of moisture stress on seed growth in soybean. *Agron. J.* 76:647-650.

Mian, M. A. R., D. A. Ashley, and H. R. Boerma. 1998. An additional QTL for water use efficiency in soybean. *Crop Sci.* 38:390-393.

Mian, M. A. R., M. A. Bailey, D. A. Ashley, R. Wells, T. E. Carter Jr., W. A. Parrott. and H. R. Boerma. 1996. Molecular markers associated with water use efficiency and leaf ash in soybean. *Crop Sci.* 36:1252-1257.

Miller, D. E., and D. W. Burke. 1983. Response of dry beans to daily deficit sprinkler irrigation. *Agron. J.* 75:775-778.

Miller, D. E., and M. W. Martin. 1985. Effect of water stress during tuber formation on subsequent growth and internal defects in Russet Burbank potatoes. *Am. Potto J.* 62:83-89.

Morgan, J. M. 1977. Differences in osmoregulation between wheat genotypes. *Nature* 270:235-236.

Morgan, J. M. 1980. Osmotic adjustment in the spikelets and leaves of wheat. *J. Exp. Bot.* 31:655-665.

Morgan, J. M. 1984. Osmoregulation and water stress in higher plants. *Annu. Rev Plant Physiol.* 35:299-319.

Morgan, J. M. 1991. A gene controlling differences in osmoregulation in wheat. *Aust. J. Plant Physiol.* 18:249-257.

Morgan, J. M., R. A. Hare, and R. J. Fletcher. 1986. Genetic variation in osmoregulation in bread and durum wheat and its relationship to grain yield in a range of field environments. *Aust. J. Agric. Res.* 37:449-457.

Morgan, J. M., B. Rodriguez Maribona, and E. J. Knights. 1991. Adaptation to water deficit in chickpea breeding lines by osmoregulation: Relationship to grain yield in the field. *Field Crops Res.* 27:61-70.

Moss, G. I., and L. A. Downey. 1971. Influence of drought stress on female gametophyte development in corn and subsequent grain yield. *Crop Sci.* 11:368-372.

Moustafa, M. A., L. Boersma, and W. E. Kronstad. 1996. Response of four spring wheat cultivars to drought stress. *Crop Sci.* 36:982-986.

Munns, R. 1988. Why measure osmotic adjustment? *Aust. J. Plant Physiol.* 15:717-726.

Musick, J. T., and D. A. Dusek. 1974. Alternate-furrow irrigating of fine textured soils. *Trans. Am. Soc. Agric. Eng.* 17:289-294.

Musick, J. T., O. R. Jones, B. A. Stewart, and D. A. Dusek. 1994. Water yield relationship for irrigated and dryland wheat in the U. S. southern plains. *Agron. J.* 86:980-986.

Musick, J. T., F. B. Pringle, W. L. Harman, and B. A. Stewart. 1990. Long-term irrigation trends: Texas High Plains. *Appl. Eng. Agric.* 6:717-724.

Musick, J. T., F. B. Pringle, and J. D. Walker. 1988. Sprinkler and furrow irrigation trends: Texas High Plains. *Appl. Eng. Agric.* 4:46-52.

Namuco, O. S., and J. C. O'Toole. 1986. Reproductive stage water stress and sterility: I. Effects of stress during meioses. *Crop Sci.* 26:317-321.

Nesmith, D. S., and J. T. Ritchie. 1992. Maize (*Zea mays* L.) response to a severe soil water-deficit during grain filling. *Field Crops Res.* 29:23-35.

Neumann, P. M. 1995. The role of cell wall adjustment in plant resistance to water deficits. *Crop Sci.* 35:1258-1266.

Neumann, P. M., H. Azaizeh, and D. Leon. 1994. Hardening of root cell walls: A growth inhibitory response to salinity stress. *Plant Cell Environ.* 17:303-309.

Nguyen, H. T., R. C. Babu, and A. Blum. 1997. Breeding for drought resistance in rice: Physiology and molecular genetics considerations. *Crop Sci.* 37:1426-1434.

Nielsen, D. C. 1998. Comparison of three alternative oilseed crops for the central Great Plains. *J. Prod. Agric.* 11:336-341.

Nonomura, A. M., and A. A. Benson. 1992. The path of carbon in photosynthesis: Improved crop yields with methanol. *Proc. Nat. Acad. Sci., USA* 89:9794-9798.

Norwood, C. A. 1995. Comparison of limited irrigated vs. dryland cropping systems in the U. S. Great Plains. *Agron. J.* 87:737-743.

Ober, E. S., and R. E. Sharp. 1994. Proline accumulation in maize primary roots at low water potential: I. Requirement for increased levels of abscisic acid. *Plant Physiol.* 105:981-987.

O'Toole, J. C. 1982. Adaptation of rice to drought-prone environments, pp. 195-213. In: IRRI (ed.), *Drought Resistance in Crops with Emphasis on Rice.* Los Baños, Philippines: Int. Rice Res. Inst.

O'Toole, J. C., and T. B. Moya. 1976. Genotypic variations in maintenance of leaf water potential rice. *Crop Sci.* 18:873-876.

Oweis, T., M. Pala, and J. Ryan. 1998. Stabilizing rainfed wheat yields with supplemental irrigation and nitrogen in a Mediterranean climate. *Agron. J.* 90:672-681.

Oweis, T., H. Zeidan, and A. Taimeh. 1992. Modeling approach for optimizing supplemental irrigation management. Proc. Int. Conf. Supplemental Irrigation and Drought Water Management, Bari, Italy. *1st Agron. Mediter. Type Environ. Exp. Agric.* 32:339-349.

Passioura, J. B. 1983. Roots and drought resistance, pp. 265-280. In: J. F. Stone and W. Wills (eds.), *Plant Production and Management under Drought Conditions.* New York: Elsevier Sci. Publ.

Passioura, J. B. 1994. The yield of crops in relation to drought, pp. 343- 359. In: K. J. Boote, J. M. Bennett, T. R. Sinclair, and G. M. Paulsen (eds.), *Physiology and Determination of Crop Yield.* Madison, WI: Am. Soc. Agron., Crop Sci. Soc. Am., and Soil Sci. Soc. Am.

Payne, W. A. 1997. Managing yield and water use of pearl millet in the Sahel. *Agron. J.* 89:481-490.

Peterson, G. A., A. J. Schlegel, D. L. Tanaka, and O. R. Jones. 1996. Precipitation use efficiency as affected by cropping and tillage systems. *J. Prod. Agric.* 9:180-186.

Porter, G. A., G. B. Opena, W. B. Bradbury, J. C. Mcburine, and J. A. Sisson. 1999. Soil management and supplemental irrigation effects on potato: I. Soil properties, tuber yield, and quality. *Agron. J.* 91:416-425.

Pritchard, J. 1994. The control of cell expansion in roots. Tansley Review No. 68. *New Phytol.* 127:3-26.

Purcell, L. C., M. Silva, C. A King, and W. H. Kim. 1997. Biomass accumulation and allocation in soybean associated with genotypic differences in tolerance of nitrogen fixation to water deficits. *Plant Soil* 196:101-113.

Ramirez-Vallejo, P., and J. D. Kelly. 1998. Traits related to drought resistance in common bean. *Euphytica* 99:127-136.

Rawson, H. M., and J. M. Clarke. 1988. Nocturnal transpiration in wheat. *Aust. J. Plant Physiol.* 15:397-406.

Richards, R. A., C. Lopez-Castaneda, H. Gomez-Macpherson, and A. G. Condon. 1993. Improving the efficiency of water use by plant breeding and molecular biology. *Irrig. Sci.* 14:93-104.

Ritchie, J. T. 1981. Water dynamics in the soil-plant atmosphere system. *Plant Soil* 58:81-96.

Robinson, S. 1987. The role of osmotic adjustment in maintaining chloroplast volume during stress. *Curr. Topics Plant Biochem. Physiol.* 6:74-87.

Rodriguez Maribona, B., J. L. Tenorio, J. R. Conde, and L. Ayerbe. 1992. Correlation between yield and osmotic adjustment of peas (*Pisum sativum* L.) under drought stress. *Field Crops Res.* 29:15-22.

Royo, C., and R. Blanco. 1998. Use of potassium iodide to mimic drought stress in triticale. *Field Crops Res.* 59:201-212.

Sadler, E. J., and N. C. Turner. 1994. Water relationships in a sustainable agricultural system, pp. 21-46. In: J. L. Hatfield and D. L. Karlen (eds.), *Sustainable Agricultural Systems*. Boca Raton, FL: CRC Press.

Salih, A. A., I. A. Ali, A. Lux, M. Luxova, Y. Cohen, Y. Sugimoto., and S. Inanaga. 1999. Rooting, water uptake, and xylem structure adaptation to drought of two sorghum cultivars. *Crop Sci.* 39:168-173.

Sall, K., and T. R. Sinclair. 1991. Soybean genotypic differences in sensitivity of symbiotic nitrogen fixation to soil dehydration. *Plant Soil* 133:31-37.

Sanchez, F. J., M. Manzanares, E. F. Andres, J. L. Tenorio, and L. Ayerbe. 1998. Turgor maintenance, osmotic adjustment and soluble sugar and proline accumulation in 49 pea cultivars in response to water stress. *Field Crops Res.* 59:225-235.

Santamaria, J. M., M. M. Ludlow, and S. Fukai. 1990. Contribution of osmotic adjustment to grain yield in *Sorghum bicolor* L. under water limited conditions: I. Water stress before anthesis. *Aust. J. Agric. Res.* 41:51-65.

Savoy, B. R., J. T. Cothren, and C. R. Shumway. 1992. Early-season production systems utilizing indeterminate soybean. *Agron. J.* 84:394-398.

Schnyder, H. 1993. The role of carbohydrate storage and redistribution in the source-sink relations of wheat and barley during grain filling: A review. *New Phytol.* 123:233-245.

Schulze, E. D. 1986. Carbon dioxide and water vapor exchange in response to drought in the atmosphere and in the soil. *Annu. Rev. Plant Physiol.* 37:247-274.

Schussler, J. R., and M. E. Westgate. 1994. Increasing assimilate reserves does not prevent kernel abortion at low water potential in maize. *Crop Sci.* 34:1569-1576.

Seemann, J. R., W. J. S. Downton, and J. A. Berry. 1986. Temperature and leaf osmotic potential as factors in the acclimation of photosynthesis to high temperature in desert plants. *Plant Physiol.* 80:926-930.

Serraj, R., S. Bona, L. C. Purcell, and T. R. Sinclair. 1997. Nitrogen fixation response to water deficit in field grown Jackson soybean. *Field Crops Res.* 52:109-116.

Serraj, R., and T. R. Sinclair. 1996. Processes contributing to N_2-fixation insensitivity to drought in the soybean cultivar Jackson. *Crop Sci.* 36:961-968.

Serraj, R., and T. R. Sinclair. 1997. Variation among soybean cultivars in dinitrogen fixation response to drought. *Agron. J.* 89:963-969.

Serraj, R., and T. R. Sinclair. 1998. Soybean cultivar variability for nodule formation and growth under drought. *Plant Soil* 202:159-166.

Sharply, A., J. J. Meisinger, A. Breeuwsma, J. T. Sims, T. C. Daniel, and J. S. Schepers. 1998. Impacts of animal manure management on ground and surface water quality, pp. 173-242. In: J. L. Hatfield and B. A. Stewart (eds.), *Animal Waste Utilization: Effective Use of Manure As a Soil Resource*. Chelsea, MI: Ann Arbor Press.

Shouse, P., S. Dasberg, W. A. Jury, and L. H. Stolzy. 1981. Water deficit effects on water potential, yield and water use of cowpeas. *Crop Sci.* 73:333-336.

Siddique, K. H. M., D. Tennant, M. W. Perry, and R. K. Belford. 1990. Water use and water use efficiency of old and modern wheat cultivars in a Mediterranean type environment. *Aust. J. Agric. Res.* 41:431-447.

Simane, B., J. M. Peacock, and P. C. Struik. 1993. Differences in developmental plasticity and growth rate among drought-resistant and susceptible cultivars of durum wheat (*Triticum aestivum* L.). *Plant Soil* 157:155-166.

Sims, A. L., J. S. Schepers, R. A. Olson, and J. F. Power. 1998. Irrigated corn yield and nitrogen accumulation response in a comparison of no-till and conventional till: Tillage and surface residue variable. *Agron. J.* 90:630-637.

Sinclair, T. R., and M. M. Ludlow. 1985. Who taught plants thermodynamics? The unfulfilled potential of plant water potential. *Aust. J. Plant Physiol.* 12:213-217.

Sinclair, T. R., and M. M. Ludlow. 1986. Influence of soil water supply on the plant water balance of four tropical grain legumes. *Aust. J. Plant Physiol.* 13:329-341.

Sinclair, T. R., R. C. Muchow, J. M. Bennett, and L. C. Hammond. 1987. Relative sensitivity of nitrogen and biomass accumulation to drought in field grown soybean. *Agron. J.* 79:986-991.

Sinclar, T. R., and R. Serraj. 1995. Dinitrogen fixation sensitivity to drought among grain legume species. *Nature* (London) 378:344.

Singh, B. P. 1998. Soil environment and root growth: Introduction to the colloquium. *Hort. Sci.* 33:946-947.

Singh, G. 1969. A review of soil-moisture relationships in potatoes. *Am. Potato J.* 46:398-403.

Singh, S. P. 1995. Selection for water stress tolerance in interracial populations of common bean. *Crop Sci.* 35:118-124.

Sloane, R. J., R. P. Patterson, and T. E. Carter. 1990. Field drought tolerance of a soybean plant introduction. *Crop Sci.* 30:118-123.

Smiciklas, K. D., R. E. Mullen, R. E. Carlson, and A. D. Knapp. 1989. Drought-induced stress effect on soybean seed calcium and quality. *Crop Sci.* 29:1519-1523.

Sobrado, M. A. 1986. Tissue water relations and leaf growth of tropical corn cultivars under water deficit. *Plant Cell Environ.* 9:451-457.

Sojka, R. E., L. H. Stolzy, and R. A. Fisher. 1981. Seasonal drought response of selected wheat cultivars. *Agron. J.* 73:838-845.

Souza, P. I., D. B. Egli, and W. P. Bruening. 1997. Water stress seed filling and leaf senescence in soybean. *Agron. J.* 89:807-812.

Spollen, W. G., and C. J. Nelson. 1994. Response of fructan to water deficit in growing leaves of tall fescue. *Plant Physiol.* 106:329-336.

Spomer, L. A. 1985. Techniques for measuring plant water. *Hort. Sci.* 20:1021-1028.

Sponchiado, B. N., J. W. White, J. A. Castillo, and P. G. Jones. 1989. Root growth of four common bean cultivars in relation to drought tolerance in environments with contrasting soil types. *Exp. Agr.* 25:249-257.

Stanca, A. M., V. Terzi, and L. Cattivelli. 1992. Biochemical and molecular studies of stress tolerance in barley, pp. 277-288. In: P. R. Shewry (ed.), *Barley: Genetics, Biochemistry, Molecular Biology and Biotechnology.* Wallingford, UK: Common. Agric. Bur.

Stapper, M., and H. C. Harris. 1989. Assessing the productivity of wheat genotypes in a Mediterranean climate, using a crop simulation model. *Field Crops Res.* 20:129-152.

Steponkus, P. L., K. W. Shahan, and J. M. Cutler. 1982. Osmotic adjustment in rice, pp. 181-194. In: Int. Rice Res Inst. (IRRI) (ed.), *Drought Resistance in Crops with Emphasis on Rice.* Los Baños, Philippines: IRRI.

Stone, L. R., D. E. Goodrum, A. J. Schlegel, M. N. Jaafar, and A. H. Khan. 2002. Water depletion depth of grain sorghum and sunflower in the central high plains. *Agron. J.* 94:936-943.

Sullivan, C. Y., and J. D. Eastin. 1974. Plant physiological responses to water stress. *Agric. Meteorol.* 14:113-127.

Taiz, L. 1984. Plant cell expansion: Regulation of cell wall mechanical properties. *Annu. Rev. Plant Physiol.* 35:585-657.

Takami, S., H. M. Rawson, and N. Turner. 1982. Leaf expansion of four sunflower cultivars in relation to water deficits: I. Diurnal patterns during stress and recovery. *Plant Cell Environ.* 5:279-286.

Tanaka, D. L., and R. L. Anderson. 1997. Soil water storage and precipitation storage efficiency of conservation tillage systems. *J. Soil Water Conserv.* 52:363-367.

Tanji, K. K., and B. R. Hanson. 1990. Drainage and return flows in relation to irrigation management, pp. 1057-1087. In: B. A. Stewart and D. R. Nielsen (eds.), *Irrigation of Agricultural Crops.* Agron. Monogr. 30. Madison, WI: Am. Soc. Agron., Crop Sci. Soc. Am., and Soil Sci. Soc. Am.

Taylor, H. M., and E. E. Terrell. 1982. Rooting pattern and plant productivity, pp. 185-200. In: M. Rechcigl (ed.), *Handbook of Agricultural Productivity.* Boca Raton, FL: CRC Press.

Thomas, G. A., G. Gibson, R. G. H. Nielsen, W. D. Martin, and B. J. Radford. 1995. Effects of tillage, stubble, gypsum, and nitrogen fertilizer on cereal cropping on a red-brown earth in south-west Queensland. *Aust. J. Exp. Agric.* 35:997-1008.

Thornthwaite, C. W. 1948. An approach towards a rational classification of climate. *Geogr Ver.* 38:55-94.

Troeh, F. R., and L. M. Thompson. 1993. *Soils and Soil Fertility,* Fifth Edition. Oxford, UK: Oxford University Press.

Tsuda, M. 1986. Effects of water stress on the panicle emergence in rice and sorghum plants. *Japan. J. Crop Sci.* 55:196-200.

Tumanov, I. I. 1927. Ungenugende wasserversorgung und das welken der pflanzen als mittel zur erhohung ihrer durreresistenz. *Planta* 3:391-480.

Tuong, T. P., and S. I. Bhuiyan. 1994. Innovations toward improving water use efficiency of rice. Paper Presented at World Water Resources Seminar, December 13-15, Lansdowne Conference Resort, VA.

Turk, K. J., and A. E. Hall. 1980. Drought adaptation of cowpea: II. Influence of drought on seed yield. *Agron. J.* 72:421-427.

Turner, N. C., and M. M. Jones. 1980. Turgor maintenance by osmotic adjustment: A review and evaluation, pp. 155-172. In: N. C. Turner and P. J. Kramer (eds.), *Adaptation of Plants to Water and High Temperature Stress.* New York: John Wiley Inter-Science.

Turner, N. C., J. C. O'Toole, R. T. Cruz, E. B. Yambao, S. Ahmad, O S. Namuco, and M. Dingkuhn. 1986. Response of seven diverse rice cultivars to water deficits: 2. Osmotic adjustment, leaf elasticity, leaf extension, leaf death, stomatal conductance and photosynthesis. *Field Crops Res.* 13:273-286.

UNESCO. 1979. Carte de la répartition mondiale des regions arides. In: *Notes Techniques.* MAB 7. Paris: UNESCO.

Unger, P. W. 1984. Tillage and residue effects on wheat, sorghum, and sunflower grown in rotation. *Soil Sci. Soc. Am. J.* 48:885-891.

Unger, P. A. 1994. Tillage effects on dryland wheat and sorghum production in the southern great plains. *Agron. J.* 86:310-314.

Vieira, R. D., D. M. Tekrony, and D. B. Egli. 1992. Effect of drought stress on soybean seed germination and vigor. *J. Seed Technol.* 15:12-21.

Viets, F. G., Jr. 1972. Water deficits and nutrient availability, pp. 217-239. In: T. T. Kozlowski (ed.), *Water Deficits and Plant Growth.* New York: Academic Press.

Virgona, J. M., K. T. Hubick, H. M. Rawson, G. D. Farquhar, and R. W. Downes. 1990. Genotypic variation in transpiration efficiency, carbon isotope discrimination, and carbon allocation during early growth in sunflower. *Aust. J. Plant Physiol.* 17:207-214.

Weinhold, B. J., T. P. Trooien, and G. Reichman. 1995. Yield and nitrogen use efficiency of irrigated corn in the northern Great Plains. *Agron. J.* 87:842-846.

Wells, R. 2002. Stem and root carbohydrate dynamics of two cotton cultivars bred fifty years apart. *Agron. J.* 94:876-882.

Wenkert, W., E. R. Lemon, and T. R. Sinclair. 1978. Water content-potential relationship in soybean: Changes in component potentials for mature and immature leaves under field conditions. *Ann. Bot.* 42:295-307.

Westgate, M. E. 1994. Seed formation in maize during drought, pp. 361-364. In: K. J. Boote, J. M. Bennett, T. R. Sinclair, and G. M. Paulsen (eds.), *Physiology and Determination of Crop Yield.* Madison, WI: Am. Soc. Agron., Crop Sci. Soc. Am., and Soil Sci. Soc. Am.

Westgate, M. E., and J. S. Boyer. 1985. Carbohydrate reserves and reproductive development at low leaf water potentials in maize. *Crop Sci.* 25:762-769.

Westgate, M. E., and J. S. Boyer. 1986. Reproduction at low silk and pollen water potentials in maize. *Crop Sci.* 26:951-956.

Wien, H. C., E. J. Littleton, and A. Ayanaba. 1979. Drought stress of cowpea and soybean under tropical conditions, pp. 283-301. In: H. Mussell and R. C. Staples (eds.), *Stress Physiology of Crop Plants.* New York: Wiley Interscience.

Wilhelm, W. W., J. D. Doran, and J. F. Power. 1986. Corn and soybean yield response to crop residue management under no-tillage production systems. *Agron. J.* 78:184-189.

Wilson, J. H. H., and C. S. Allison. 1978. Effects of water stress on the growth of maize (*Zea mays* L.). *Rhod. J. Agric. Res.* 16:175-192.

Wraith, J. M., and C. K. Wright. 1998. Soil water and root growth. *Hort. Sci.* 33:951-959.

Wright, G. C., R. C. N. Rao, and G. D. Farquhar. 1994. Water use efficiency and carbon isotope discrimination in peanut under water deficit conditions. *Crop Sci.* 34:92-97.

Wright, L. N., and G. L. Jordan. 1970. Artificial selection for seedling drought tolerance in boer lovegrass (*Eragrostis curvula* Nees). *Crop Sci.* 10:99-102.

Zablotowicz, R. M., D. D. Focht, and G. H. Cannell. 1981. Nodulation and N fixation and droughty conditions. *Agron. J.* 73:9-12.

Zeiher, C., D. B. Egli, J. E. Leggett, and D. A. Reicosky. 1982. Cultivar differences in nitrogen redistribution in soybeans. *Agron. J.* 74:375-379.

Zimmermann, U. 1978. Physics of turgor and osmoregulation. *Ann. Rev. Plant Physiol.* 29:121-148.

Chapter 8

Adatia, M. H., and R. T. Besford. 1986. The effect of silicon in cucumber plants grown in recirculating nutrient solution. *Ann. Bot.* 58:343-351; *Adv. Soil Sci.* 2:65-131.

Ahlrichs, L. L., M. C. Karr, V. C. Baligar, and R. J. Wright. 1990. Rapid bioassay of aluminum toxicity in soil. *Plant Soil* 122:279-285.

Alvarez, J. M., M. I. Rico, and A. Obrador. 1997. Leachibility and distribution of zinc applied to an acid soil as controlled-release zinc chelates. *Commun. Soil Sci. Plant Anal.* 28:1579-1590.

Anderson, E. L., E. J. Kamprath, and R. H. Moll. 1984. Nitrogen fertility effects on accumulation, remobilization and partitioning of N and dry matter in corn genotypes differing in prolificacy. *Agron. J.* 76:397-404.

Anderson, E. L., E. J. Kamprath, and R. H. Moll. 1985. Prolificacy and N fertilizer effects on yield and N utilization in maize. *Crop Sci.* 25:598-602.

Anderson, J. W. 1990. Sulfur metabolism in plants, pp. 327-381. In: B. J. Miflin and P. J. Lea (eds.) *The Biochemistry of Plants,* Vol. 16. San Diego, CA: Academic Press.

Arnesen, A. K. M., and B. R. Singh 1998. Plant uptake and DTPA-extractability of Cd, Cu, Ni and Zn in a Norwegian alum shale soil as affected by previous addition of dairy and pig manures and peat. *Can. J. Soil Sci.* 78:531-539.

Arnon, D. I., and P. R. Stout. 1939. Molybdenum as an essential element for higher plants. *Plant Physiol.* 14:599-602.

Asher, C. J., and P. G. Ozanne. 1967. Growth and potassium content of plants in solution cultures maintained at constant potassium concentrations. *Soil Sci.* 103: 155-161.

Austin, R. B., M. A. Ford, J. A. Edrich, and R. D. Blackwell. 1977. The nitrogen economy of winter wheat. *J. Agric. Sci.* 88:159-167.

Austin, R. B., C. L. Morgan, M. A. Ford, and R. D. Blackwell. 1980. Contributions to grain yield from pre-anthesis assimilation in tall and dwarf barley phenotypes in two constrasting seasons. *Ann. Bot.* (London). 45:309-319.

Baligar, V. C., R. R. Duncan, and N. K. Fageria. 1990. Soil-plant interaction on nutrient use efficiency in plants: An overview, pp. 351-373. In: V. C. Baligar and R. R. Duncan (eds.), *Crops As Enhancer of Nutrient Use.* San Diego, CA: Academic Press.

Baligar, V. C., and Fageria, N. K. 1997. Nutrient use efficiency in acid soils: Nutrient management and plant use efficiency, pp. 75-93. In: A. C. Moniz, A. M. C. Furlani, N. K. Fageria, C. A. Rosolem, and H. Cantarells (eds.), *Plant-Soil Interactions at Low pH: Sustainable Agriculture and Forestry Production.* Campinas, Brazil: Brazilian Soil Sci. Soc.

Baligar, V. C., N. K. Fageria, and M. A. Elrashidi. 1998. Toxicity and nutrient constraints on root growth. *Hort Sci.* 33:960-965.

Baligar, V. C., N. K. Fageria, and Z. L. He. 2001. Nutrient use efficiency in plants. *Commun. Soil Sci. Plant Anal.* 32:921-950.

Baligar, V. C., R. E. Schaffert, H. L. Dos Santos, G. V. E. Pitta, and A F. De C. Bahia Filho. 1993. Growth and nutrient uptake parameters in sorghum as influenced by aluminum. *Agron. J.* 85:1068-1074.

Barber, S. A. 1995. *Soil Nutrient Bioavailability: A Mechanistic Approach,* Second Edition. New York: John Wiley.

Barkla, B. J., and O. Pantoja. 1996. Physiology of ion transport across the tonoplast of higher plants. *Annu. Rev. Plant Physiol. Plant Mol. Biol.* 47:159-184.

Bennett, R. N., and R. M. Wallsgrove. 1994. Secondary metabolites in plant defense mechanisms. *New Phytol.* 127:617-633.

Bennett, W. F. 1993.Plant nutrient utilization and diagnostic plant symptoms, pp.1-7. In: W.F. Bennett (ed.), *Nutrient Deficiencies and Toxicities in Crop Plants.* St. Paul, MN: Am. Phytopath. Soc.

Bentrup, F. W. 1990. Potassium ion channels in the plasmalemma. *Physiol. Plant.* 79:705-711.

Ben Zioni, A., Y. Vaadia, and S. H. Lips. 1971. Nitrate uptake by roots as regulated by nitrate reduction products of the shoots. *Physiol. Plant.* 24:288-290.

Beevers, L., and R. H. Hagenman. 1983. Uptake and reduction of nitrate: Bacteria and higher plants, pp. 351-375. In: A. Lauchli and R. L. Bieleski (eds.), *Inorganic Plant Nutrition: Encyclopedia of Plant Physiology,* New Series, Vol. 15A. Berlin: Springer-Verlag.

Blair, G. J. 1987. Nitrogen-sulfur interactions in rice, pp. 195-203. In: Int. Rice Res. Inst. (IRRI) (ed.), *Efficiency of Nitrogen Fertilizers for Rice.* Los Baños, Philippines: IRRI.

Blevins, D. G. 1994. Uptake, translocation, and function of essential mineral elements in crop plants, pp. 259-275. In: G. A. Peterson (ed.), *Physiology and Determination of Crop Yield.* Madison, WI: Am. Soc. Agron, Crop Sci. Soc. Am., and Soil Sci. Soc. Am.

Blevins, D. G., and K. M. Lukaszewski. 1998. Boron in plant structure and function. *Annu. Rev. Plant Physiol. Plant Mol. Biol.* 49:481-500.

Bloom, A. J. and F. S. Chapin. 1981. Differences in steady-state net ammonium and nitrate influx by cold and warm adapted barley varieties. *Plant Physiol.* 68:1064-1067.

Bloom, P. R., and M. S. Erick. 1995. The quantification of aqueous aluminum, pp. 1-38. In: G. Sposito (ed.), *The Environmental Chemistry of Aluminum,* Second Edition. Boca Raton, FL: CRC Publ.

Bock, B. R., J. J. Camberato, F. E. Below, W. L. Pan, and R. T. Koenig. 1991. Wheat responses to enhanced ammonium nutrition, pp. 93-106. In: J. R. Huffman (ed.), *Effects of Enhanced Ammonium Diets on Growth and Yield of Wheat and Corn.* Atlanta, GA: Found. Agron. Res.

Brennan, R. F. 1994. The residual effectiveness of previously applied copper fertilizer for grain yield of wheat grown on soils of southwest Australia. *Fert. Res.* 39:11-18.

Brennan, R. F., and M. D. A. Bolland. 2003. Comparing copper requirements of faba bean, chickpea, and lentil with spring wheat. *J. Plant Nutr.* 26:883-899.

Brown, J. C. 1978. Mechanism of iron uptake by plants. *Plant Cell Environ.* 1:249-257.

Broyer, T. C., A. B. Carlton, A. B. Johnson, and P. R. Stout. 1954. Chlorine: A micronutrient element for higher plants. *Plant Physiol.* 29:526-532.

Bruckner, P. L., and D. D. Morey. 1988. Nitrogen effects on soft red winter wheat yield, agronomic characteristics, and quality. *Crop Sci.* 28:152-157.

Brye, K. R., J. M. Norman, E. V. Nordheim, S. T. Gower, and L. G. Dundy. 2002. Refinements to an in-situ core technique for measuring net nitrogen mineralization in moist, fertilized agriculture soil. *Agron. J.* 94:864-869.

Bufogle, A., P. K. Bollich, J. L. Kovar, C. W. Lindau, and R. R. Macchiavellid. 1998. Comparison of ammonium sulfate and urea as nitrogen sources in rice production. *J. Plant Nutr.* 21:1601-1614.

Burk, J. J., P. Holloway, and M. J. Dalling. 1986. The effect of sulfur deficiency on the organization and photosynthetic capability of wheat leaves. *J. Plant Physiol.* 125:371-375.

Cakmak, I., A. Yilmaz, M. Kalayci, H. Ekiz, B. Torun, B. Erenoglu, and H. J. Braun. 1996. Zinc deficiency as a critical problem in wheat production in central Antolia. *Plant Soil* 180:167-172.

Carver, B. F., and J. D. Ownby. 1995. Acid soil tolerance in wheat. *Adv. Agron.* 54:117-173.

Cataldo, D. A., L. E. Schrader, D. M. Peterson, and D. Smith. 1975. Factors affecting seed protein concentration in oats: I. Metabolism and distribution of n and carbohydrate in two cultivars that differ in groat protein concentration. *Crop Sci.* 15:19-23.

Chairidchai, P., and G. S. P. Ritchie. 1993. The effect of citrate and pH on zinc uptake by wheat. *Agron. J.* 85:322-328.

Cherif, M., A. Asselin, and R. R. Belanger. 1994. Defense responses induced by soluble silicon in cucumber roots infected by *Pythium spp. Phytopath.* 84:236-242.

Christensen, N. W., R. G. Taylor, T. L. Jackson, and B. L. Mitchell. 1981. Chloride effects on water potentials and yield of winter wheat infected with take-all root rot. *Agron. J.* 73:1053-1058.

Claasen, N., and S. A. Barber. 1976. Simulation model for nutrient uptake from soil by a growing plant root system. *Agron. J.* 68:961-964.

Clark, R. B. 1982. Plant response to mineral element toxicity and deficiency, pp. 71-73. In: M. N. Christiansen and C. F. Lewis (eds.), *Breeding Plants for Less Favorable Environment.* New York: John Wiley.

Clark, R. B. 1991. Physiology of cereals for mineral nutrient uptake, use, and efficiency, pp. 131-209. In: V. C. Baligar and R. R. Duncan (eds.), *Crops As Enhancers of Nutrient Use.* San Diego, CA: Academic Press.

Clark, R. B. 1993. Sorghum, pp. 21-26. In: W. F. Bennett (ed.), *Nutrient Deficiencies and Toxicities in Crop Plants.* St. Paul, MN: Am. Phytopath. Soc.

Clark, R. B. 2001. Role of silicon in plant nutrition, pp. 205-223. In: K. Singh, S. Mori, and R. M. Welch (eds.), *Perspectives on the Micronutrient Nutrition of Crops.* Jodhpur, India: Scientific Publ.

Clark, R. B., and V. C. Baligar. 2000. Acid and alkaline soil constraints on plant mineral nutrition, pp. 133-177. In: R. E. Wilkinson (ed.), *Plant-Environment Interactions.* New York: Marcel Dekker.

Clark, R. B., and S. K. Zeto. 2000. Mineral acquisition by arbuscular mycorrhizal plants. *J. Plant Nutr.* 23:867-902.

Clark, R. B., S. K. Zeto, and R. W. Zobel. 1999. Arbuscular mycorrhizal fungal isolate effectiveness on growth and root colonization of *Panicum virgatum* in acid soil. *Soil Biol. Biochem.* 31:1757-1763.

Clark, R. B., R. W. Zobel, and S. K. Zeto. 1999. Effects of mycorrhizal fungus isolates on mineral acquisition by *Panicum virgatum* in acid soil. *Mycorrhiza* 9:167-176.

Clarkson, D. T. 1974. *Ion Transport and Cell Structure in Plants.* London: McGraw-Hill.

Clarkson, D. T. 1984. Ionic relationships, pp. 319-353. In: M. B. Wilkins (ed.), *Advanced Plant Physiology.* London: Pitman Publ.

Clarkson, D. T., and J. B. Hanson. 1980. The mineral nutrition of higher plants. *Annu. Rev. Plant Physiol.* 31:239-298.

Clarkson, D. T., M. J. Hopper, and L. H. P. Jones. 1986. The effect of root temperature on the uptake of nitrogen and the relative size of the root system in *Lolium perenne:* I. Solutions containing both NH_4^+ and NO_3^-. *Plant Cell Environ.* 9:535-545.

Clarkson, D. T., L. H. P. Jones, and J. V. Purves. 1992. Absorption of nitrate and ammonium ions by *Lolium perenne* from flowing solution cultures at low root temperatures. *Plant Cell Environ.* 15:99-106.

Cocker, K. M., D. E. Evans, and M. J. Hodson. 1998. The amelioration of aluminum toxicity by silicon in higher plants: Solution chemistry or an in planta mechanism? *Physiol. Plantarum* 104:608-614.

Cooper, H. D., and D. T. Clarkson. 1989. Cycling of amino nitrogen and other nutrients between shoots and roots in cereals: A possible mechanism integrating shoot and root in the regulation of nutrient uptake. *J. Exp. Bot.* 40:753-762.

Cope, J. T. Jr. 1981. Effects of 50 years of fertilization with phosphorus and potassium on soil test levels and yields at location. *Soil Sci. Soc. Am. J.* 45: 342-347.

Craswell, E. T., and P. L.G. Vlek. 1979. Fate of fertilizer nitrogen applied to wetland rice, pp. 175-192. In: Int. Rice Res. Inst. (IRRI) (ed.), *Nitrogen and Rice.* Los Baños, Philippines: IRRI.

Dalal, R. C. 1977. Soil organic phosphorus. *Adv. Agron.* 29:83-117.

Dalling, M. J., G. Boland, and J. H. Wilson. 1976. Relation between acid proteinase activity and redistribution of nitrogen during grain development in wheat. *Aust. J. Plant Physiol.* 3:721-730.

Danielli, J. F., and H. A. Davson. 1935. A contribution to the theory of the permeability of thin films. *J. Cellular Comp. Physiol.* 5:495-508.

Dannehl, H., A. Herbik, and D. Godde. 1995. Stress-induced degradation of the photosynthetic apparatus is accompanied by changes in thylakoid protein turnover and phosphorylation. *Physiol. Plant.* 93:179-186.

Datnoff, L. E., C. W. Deren, and G. H. Snyder. 1997. Silicon fertilization for disease management of rice in Florida. *Crop Prot.* 16:525-531.

Dell, B. 1981. Male sterility and outer wall structure in copper deficient plants. *Ann. Bot.* 48:599-608.

Desai, R. M., and C. R. Bhatia. 1978. Nitrogen uptake and nitrogen harvest index in durum wheat cultivars varying in their grain protein concentration. *Euphytica* 27:561-566.

Dietz, K. J. 1989. Recovery of spinach leaves from sulfate and phosphate deficiency. *J. Plant Physiol.* 134:551-557.

Dingkuhn, M., H. F. Schnier, S. K. De Datta, K. Dorffling, and C. Javellana. 1991. Relationships between ripening-phase productivity and crop duration, canopy photosynthesis, and senescence in transplanted and direct-seeded lowland rice. *Field Crops Res.* 26:327-345.

Dudal, R. 1976. Inventory of the major soils of the world with special reference to mineral stress hazards, pp. 3-13. In: M. J. Wright (ed.), *Plant Adaption to Mineral Stress in Problem Soils.* Ithaca, NY: Cornell University Press.

Duncan, R. R. 1994. Genetic manipulation, pp. 1-38. In: R. E. Wilkinson (ed.), *Plant-Environment Interactions.* New York: Marcel Dekker.

Duncan, R. R., and R. N. Carrow. 1999. Turfgrass molecular genetic improvement for abiotic/edaphic stress resistance. *Adv. Agron.* 67:233-306.

Edwards, D. G., and C. J. Asher. 1974. The significance of solution flow rate in flowing culture experiments. *Plant Soil.* 41:161-175.

Elwad, S. H., and V. E. Green Jr. 1979. Silicon and the rice plant environment: A review of recent work. *Riso* 28:235-253.

Engel, R. E., P. L. Bruckner, and J. Eckhoff. 1998. Critical tissue concentration and chloride requirements for wheat. *Soil Sci. Soc. Am. J.* 62:401-405.

Engels, C., and H. Marschner. 1995. Plant uptake and utilization of nitrogen, pp. 41-81. In: P. E. Bacon (ed.), *Nitrogen Fertilization in the Environment.* New York: Marcel Dekker.

Epstein, E. 1966. Dual pattern of ion absorption by plant cells and by plants. *Nature* 212:1324-1327.

Epstein, E. 1972. *Mineral Nutrition of Plants: Principles and Perspectives.* New York: John Wiley.

Epstein, E. 1973. Mechanisms of ion transport through plant cell membranes. *Inter. Rev. Cytol.* 34:123-168.

Epstein, E., and C. E. Hagen. 1952. A kinetic study of the absorption of alkali cations by barley roots. *Plant Physiol.* 23:457-474.

Epstein, E., D. W. Rains, and O. E. Elzam. 1963. Resolution of dual mechanisms of potassium absorption by barley roots. *Proc. Nat. Acad. Sci.* 49:684-692.

Fageria, N. K. 1973. Uptake of Nutrient by the Rice Plant from Dilute Solutions. Doctoral thesis, Catholic University, Louvain, Belgium.

Fageria, N. K. 1976. Effect of P, Ca, and Mg concentrations in solution culture on growth and uptake of these ions by rice. *Agron. J.* 68:726-732.

Fageria, N. K. 1984. *Fertilization and Mineral Nutrition of Rice.* (EMBRAPA-CNPAF) (eds.), Editora Campus, Rio de Janeiro.

Fageria, N. K. 1989. *Tropical Soils and Physiological Aspects of Field Crops.* Brasilia, Brazil: EMBRAPA-CNPAF.

Fageria, N. K. 1992. *Maximizing Crop Yields.* New York: Marcel Dekker.

Fageria, N. K. 1998. Potassium use efficiency in upland genotypes, pp. 99-102. In: J. D. Costa and E. P. Guimarães (eds.), *Perspectives of Rice Crop in Upland and Lowland Ecosystems Annals,* Vol. 1. Goiânia, Brazil: EMBRAPA-CNPAF.

Fageria, N. K. 1999. Adequate base saturation and pH for upland rice, corn, soybean and common bean grown on an Oxisol in rotation. *Annual Report: Fertilizer and Lime Requirements of Cerrado Region.* Santo Antonio de Goias, Brazil: EMBRAPA-CNPAF.

Fageria, N. K. 2000. *Response of upland rice to nitrogen, phosphorus and potassium fertilization in soil of state of MatoGrosso, Brazil.* Paper XIII Soil Water Management Conservation Meeting, 6-11 August, 2000, Ilhéus, Bahia, Brazil.

Fageria, N. K. 2001a. Adequate and toxic levels of copper and manganese in upland rice, common bean, corn, soybean and wheat grown on an Oxisol. Commun. *Soil Sci. Plant Anal.* 32:1659-1676.

Fageria, N. K. 2001b. Response of upland rice, dry bean, corn and soybean to base saturation in cerrdao soil. *R. Bras. Eng. Agric. Ambiental.* 5:416-424.

Fageria, N. K., and V. C. Baligar. 1997. Response of common bean, upland rice, corn, wheat, and soybean to fertility of an Oxisol. *J. Plant Nutr.* 20:1279-1289.

Fageria, N. K., and V. C. Baligar. 1999a. Growth and nutrient uptake by common bean, lowland rice, corn, soybean, and wheat at different soil pH on an Inceptisol. *J. Plant Nutr.* 22:1495-1507.

Fageria, N. K., and V. C. Baligar. 1999b. Yield and yield components of lowland rice as influenced by timing of nitrogen fertilization. *J. Plant Nutr.* 22:23-32.

Fageria, N. K., and V. C. Baligar. 2001. Nitrogen management for lowland rice production on an Inceptisol. *Commun. Soil Sci. Plant Anal.* 32:1405-1429.

Fageria, N. K., and V. C. Baligar. 2005. Enhancing nitrogen use efficiency in crop plants. *Adv. Agron.* 80:97-185.

Fageria, N. K., V. C. Baligar, and R. P. Clark. 2002. Micronutrients in crop production. *Adv. Agron.* 77;185-268.

Fageria, N. K., V. C. Baligar, and D. G. Edwards. 1990. Soil-plant nutrient relationship at low ph stress, pp. 475-507. In. V. C. Baligar and R. R. Duncan (eds.), *Crops As Enhancers of Nutrient Use.* San Diego, CA: Academic Press.

Fageria, N. K., V. C. Baligar, and C. A. Jones. 1997. *Growth and Mineral Nutrition of Field Crops,* Second Edition. New York: Marcel Dekker.

Fageria, N. K., V. C. Baligar, and R. J. Wright. 1988. Aluminum toxicity in crop plants. *J. Plant Nutr.* 11:303-319.

Fageria, N. K., V. C. Baligar, and R. J. Wright. 1990. Iron nutrition of plants: An overview on the chemistry and physiology of its deficiency and toxicity. *Pesq. Agropec. Bras.* 25:553-570.

Fageria, N. K., and M. P. Barbosa Filho. 1994. *Nutritional deficiency in rice crop: Identification and correction,* Document No.42. Goiania, Brazil: EMBRAPA-CNPAF.

Fageria, N. K., I. P. Oliveira, and L. G. Dutra. 1996. *Nutritional deficiency in common bean and their correction,* Document no. 65. Goiania, Brazil: EMBRAPA-CNPAF.

Fageria, N. K., and A. B. Santos. 1998. Rice and common bean growth and nutrient concentration as influenced by aluminum on a lowland soil. *J. Plant Nutr.* 21:903-912.

Fageria, N. K., N. A. Slaton, and V. C. Baligar. 2003. Nutrient management for improved lowland rice productivity and sustainability. *Adv. Agron.* 80:63-152.

Fageria, N. K., F. J. P. Zimmermann, and V. C. Baligar. 1995. Lime and phosphorus interactions on growth and nutrient uptake by upland rice, common bean, wheat and corn in an Oxisol. *J. Plant Nutr.* 18:2519-2532.

Fawcett, J. A., and K. J. Frey. 1982. Nitrogen harvest index variation in *Avena sativa* and *Avena sterilis. Proc. Iowa Acad. Sci.* 89:155-159.

Fawcett, J. A., and K. J. Frey. 1983. Associations among nitrogen harvest index and other traits within two Avena species. *Proc. Iowa Acad. Sci.* 90:150-153.

Fixen, P. E. 1993. Crop responses to chloride. *Adv. Agron.* 50:107-150.

Flowers, T. J. 1988. Chloride as a nutrient and as an osmoticum. *Adv. Plant Nutr.* 3:55-78.

Foy, C. D. 1992. Soil chemical factors limiting plant root growth. *Adv. Soil Sci.* 19:97-149.

Friesen, D. K., I. M. Rao, R. J. Thomas, A. Oberson, and J. I. Sanz. 1997. Phosphorus acquisition and cycling in crop and pasture systems in low fertility tropical soils, pp. 493-498. In: T. Ando, K. Fujita, T. Mae, H. Tsumoto, S. Mori and J. Sekiya (eds.), *Plant Nutrition for Sustainable Food Production and Environment.* Dordrecht, The Netherlands: Kluwer Acad. Publ.

Galvez, L., R. B. Clark, L. M. Gourley, and J. W. Maranville. 1987. Silicon interactions with manganese and aluminum toxicity in sorghum. *J. Plant Nutr.* 10: 1139-1147.

Gardner, W. K., D. A. Barber, and D. G. Parbery. 1983. The acquisition of phosphorus by *Lupinus albus L.* III. The probable mechanism by which phosphorus movement in the soil/root interface is enhanced. *Plant Soil.* 70:107-124.

Gaudin, R., and J. Dupuy. 1999. Ammonical nutrition of transplanted rice fertilizer with large urea granules. *Agron. J.* 91:33-36.

Goos, R. J., J. A. Schimelfenig, B. R. Bock, and B. E. Johnson. 1999. Response of spring wheat to nitrogen fertilizers of different nitrification rates. *Agron. J.* 91:287-293.

Graham, R. D. 1984. Breeding for nutritional characteristics in cereals. *Adv. Plant Nutr.* 1:57-102.

Gupta, U. C. 1979. Boron nutrition of crop plants. *Adv. Agron.* 31:273-307.

Gupta, V. K. 1995. Zinc research and agricultural production, pp. 132-164. In: H. L. S. Tondon (ed.), *Micronutrit Research and Agricultural Production.* New Delhi, India: Fert. Dev. Consult. Org.

Hai, T. V., and H. Laudelout. 1966. Phosphate uptake by intact rice plants by continuous flow method at low phosphate concentration. *Soil Sci.* 101:408-417.

Hale, M. G., and D. M. Orcutt. 1987. *The Physiology of Plants Under Stress.* New York: John Wiley.

Halloran, G. M. 1981. Cultivar differences in nitrogen translocation in wheat. *Aust. J. Agric. Res.* 32:535-544.

Hallsworth, E. G., S. B. Wilsen, and E. A. Greenwood. 1960. Copper and cobalt in nitrogen fixation. *Nature* 187:79-80.

Haynes, R. J., and K. M. Goh. 1978. Ammonium and nitrate nutrition of plants. *Biol. Rev.* 53:465-510.

Heenan, D. P., and L. C. Campbell. 1981. Influence of potassium and manganese on growth and uptake of magnesium by soybeans (*Glycine max* L. Merr.). *Plant Soil* 61:447-456.

Heitholt, J. J., J. J. Sloan, C. T. Mackown, and R. I. Cabrera. 2003. Soybean growth on calcareous soil as affected by three iron sources. *J. Plant Nutr.* 26:935-948.

Hewitt, E. J. 1963. The essential nutrients: Requirements and interactions in plants, pp. 137-360. In: F. C. Steward (ed.), *Plant Physiology.* New York: Academic Press.

Hiatt, A. J., and J. E. Leggett. 1974. Ionic interactions and antagonisms in plants, pp. 101-134. In: E. W. Carson (ed.), *The Plant Root and its Environment.* Charlottesville, VA: Univ. Press Virginia.

Higinbotham, N., B. Etherton, and R. J. Foster. 1967. Mineral ion contents and cell transmembrane electropotentials of pea and oat seedling tissue. *Plant Physiol.* 42:37-46.

Hinsinger, P. 1998. How do plant roots acquire mineral nutrients? Chemical processes involved in the rhizosphere. *Adv. Agron.* 64:225-265.

Hocking, P. J., G. Keerthisinghe, F. W. Smith, and P. J. Randall. 1997. Comparison of the ability of different crop species to access poorly-available soil phosphorus, pp. 305-308. In: T. Ando, K. Fujita, T. Mae, H. Tsumoto, S. Mori, and J. Sekiya (eds.), *Plant Nutrition for Sustainable Food Production and Environment.* Dordrecth, The Netherlands: Kluwer Acad. Publ.

Hodges, T. K. 1973. Ion absorption by plant roots. *Adv. Agron.* 25:163-207.

Hodges, T. K., R. T. Leonard, C. E. Bracker, and T. W. Keenan. 1972. Purification of an ion stimulated adenosine triphosphatase from plant roots association with plasma membranes. *Proc. Nat. Acad. Sci.* 69:3307-3311.

Hodson, M. J., and D. E. Evans. 1995. Aluminum/silicon interactions in higher plants. *J. Exp. Bot.* 46:161-171.

Huber, D. M., and D. C. Arny. 1985. Interactions of potassium with plant diseases, pp. 467-488. In: R. D. Munson (ed.), *Potassium in Agriculture.* Madison, WI: Am. Soc. Agron, Crop Sci. Soc. Am., and Soil Sci. Soc. Am.

Huffman, J. R. 1989. Effects of enhanced ammonium nitrogen availability for corn. *J. Agron. Educ.* 18:325-339.

Ingestad, T. 1982. Relative addition rate and external concentration driving variables used in plant nutrition research. *Plant Cell Environ.* 5:443-353.

Ishizuka, Y. 1978. Nutrient deficiencies of crops. Taipei, Taiwan: ASPAC Food Fert. Tech. Center.

Israel, D. W., and T. W. Rufty. 1988. Influence of phosphorus nutrition on phosphorus and nitrogen utilization efficiencies and associated physiological responses in soybean. *Crop Sci.* 28:954-960.

Jarrell, W. M., and R. B. Beverly. 1981. The dilution effect in plant nutrition studies. *Adv. Agron.* 34:197-224.

Jones, M. B., W. E. Martin, and W. A. Williams. 1968. Behavior of sulfate sulfur and elemental sulfur in three California soils in Lysimeters. *Soil Sci. Soc. Am. Proc.* 32:535-540.

Kim, Y. S., and S. C. Park. 1973. effect of split potassium application on paddy rice grown on acid sulphate soils. *Potash Rev.* 7:1-9.

Korentajer, L., B. H. Byrnes, and D. T. Hellums. 1984. Leaching losses and plant recovery from various sulfur fertilizers. *Soil Sci. Soc. Am. J.* 48:671-676.

Korndörfer, G. K., L. E. Datnoff, and G. F. Corrêa. 1999. Influence of silicon on grain discoloration and upland rice grown on four savanna soils of Brazil. *J. Plant Nutr.* 22:93-102.

Kurvits, A., and E. A. Kirkby. 1980. The uptake of nutrients by sunflower plants *(Helianthus annuus)* growing in a continuous flowing culture system, supplied with nitrate or ammonium as nitrogen source. *Z. Pflanzenernaehr. Bodenkd.* 143:140-149.

Lang, A. 1983. Turger regulated translocation. *Plant Cell Environ.* 6:683-689.

Larcher, W. 1995. *Physiological Plant Ecology,* Third Edition. New York: Springer-Verlag.

Laties, G. G. 1969. Dual mechanisms of salt uptake in relation to compartmentation and long-distance transport. *Annu. Rev. Plant Physiol.* 20:89-116.

Lin, S.C., and L. P. Yuan. 1980. Hybrid rice breeding in China, pp. 35-37. In: Int. Rice Res. Inst. (IRRI) (ed.), *Innovative Approaches to Rice Breeding.* Manila, Philippines: IRRI.

Liu, J., R. J. Reid, and F. A. Smith. 1998. Mechanisms of cobalt uptake in plants: [60]Co uptake and distribution in chara. *Physiol. Plant.* 104:351-356.

Loffler, C. M., and R. H. Busch. 1982. Selection for grain protein, grain yield, and nitrogen partitioning efficiency in hard red spring wheat. *Crop Sci.* 22:591-595.

Loneragan, J. F. 1968. Nutrient requirements of plants. *Nature* 188:1307-1308.

Loneragan, J. F. 1997. Plant nutrition in the 20th and perspectives for the 21st century, pp. 3-14. In: T. Ando, K. Fujita, T. Mae, H. Tsumoto, S. Mori, and J. Sekiya (eds.), *Plant Nutrition for Sustainable Food Production and Environment.* Dordrecht, The Netherlands: Kluwer Acad. Publ.

Loneragan, J. F., and K. Snowball. 1969. Calcium requirements of plants. *Aust. J. Agric. Res.* 20:465-278.

Loomis, W. D., and R. W. Durst. 1992. Chemistry and biology of boron. *Bio. Factors* 3:229-239.

Ma, B. L., L. M. Dwyer, and D. L. Smith. 1994. Evaluation of peduncle perfusion for in vivo studies of carbon and nitrogen distribution in cereal crops. *Crop Sci.* 34:1584-1588.

Macduff, J. H., and M. J. Hopper. 1985. Effect of root temperature on uptake of nitrate and ammonium ions by barley grown in flowing solution culture. *Plant Soil* 91:303-306.

Macduff, J. H., M. J. Hopper, A. Wild, and F. E. Tim. 1987. Comparison of the effects of root temperature on nitrate and ammonium nutrition of oil seed rape in flowing solution culture. I. Growth and uptake of nitrogen. *J. Exp. Bot.* 38:1104-1120.

Mae, T. 1997. Physiological nitrogen efficiency in rice: Nitrogen utilization, photosynthesis, and yield potential, pp. 51-60. In: T. Ando, K. Fujita, T. Mae, H. Tsumoto, S. Mori, and J. Sekiya (eds.), *Plant Nutrition for Sustainable Food Production and Environment.* Dordrecht, The Netherlands: Kluwer Acad. Publ.

Marschner, H. 1986. *Mineral Nutrition of Higher Plants.* New York: Academic Press.

Marschner, H. 1995. *Mineral Nutrition of Higher Plants.* Second Edition, New York: Academic Press.

Marschner, H., Treeby, M., and V. Römheld. 1989. Role of root-induced changes in the rhizosphere for iron acquisition in higher plants. *Z. Pflanzenern. Bodenk.* 152:197-204.

Mateo, P. I. Bonilla, E. Fernandez-Valiente, and E. Sanchez-Maseo. 1986. Essentiality of boron for dinitrogen fixation in *Anabaena sp.* PCC 7119. *Plant Physiol.* 81:430-433.

McGrath S. P., and Zhao, F. J. 1995. A risk assessment of sulfur deficiency in cereals using soil and atmospheric deposition data. *Soil Use Manage.* 11:110-114.

Mengel, K. 1974a. Ion uptake and translocation, pp. 83-100. In: E. W. Carson (ed.), *The Plant Root and Its Environment.* Charlottesville, VA: University Press Virginia.

Mengel, K. 1974b. Plant ionic status, pp. 63-81. In: E. W. Carson (ed.), *The Plant Root and Its Environment.* Charlottesville, VA: University Press Virginia.

Mengel, K. 1985. Dynamics and availability of major nutrients in soils. *Adv. Soil Sci.* 2:65-131.

Mengel, K., and E. A Kirkby. 1978. *Principles of Plant Nutrition,* Fourth Edition. Basel, Switzerland: Int. Potash Inst.

Mengel, K., E. A. Kirkby, H. Kosegarten, and T. Appel. 2001. *Principles of Plant Nutrition,* Fifth Edition. Dordrecht, The Netherlands: Kluwer Acad. Publ.

Minotti, P. L., D. C. Williams, and W. A Jackson. 1968. Nitrate uptake and reduction as affected by calcium and potassium. *Soil Sci. Soc. Am. Proc.* 32:692-698.

Moll, R. H., W. A. Jackson, and R. L. Mikkelsen. 1994. Recurrent selection for maize grain yield: Dry matter and nitrogen accumulation and partitioning changes. *Crop Sci.* 34:874-881.

Moll, R. H., E. J. Kamprath, and W. A. Jackson. 1987. Development of nitrogen efficient prolific hybrids of maize. *Crop Sci.* 27:181-186.

Monge, E., E. Vale, and J. Abadia. 1987. Photosynthetic pigment composition of higher plants grown under iron stress. *Prog. Photosyn. Res.* 4:201-204.

Morison, J. I. L., and G. D. Batten. 1986. Regulation of mesophyll photosynthesis in intact wheat leaves by cytoplasmic phosphate concentrations. *Planta* 168:200-206.

Mortvedt, J. J. 1994. Needs for controlled availability micronutrient fertilizers. *Fertilizer Research* 38:213-221.

Murata, Y., and S. Matsushima. 1975. Rice, pp. 73-99. In: L. T. Evans (ed.), *Crop Physiology: Some Case Histories.* Cambridge, England: Cambridge University Press.

Nakamura, T., M. Osaki, T. Shinano, and T. Tadano. 1997. Different mechanisms of carbon-nitrogen interaction in cereal and legume crops, pp. 913-914. In: T. Ando, K. Fujita, T. Mae, H. Tsumoto, S. Mori, and J. Sekiya (eds.), *Plant Nutrition for Sustainable Food Production and Environment.* Dordrecht, The Netherlands: Kluwer Acad. Publ.

Nielsen, N. E., and S. A. Barber. 1978. Differences among genotypes of corn in the kinetics of phosphorus uptake. *Agron. J.* 70:695-698.

Nissen, P. 1971. Uptake of sulfate by roots and leaf slices of barley mediated by single, multiphasic mechanisms. *Physiol. Plant.* 24:315-324.

Nissen, P. 1973. Kinetics of ion uptake in higher plants. *Physiol. Plant.* 28:113-120.

Nissen, P. 1974. Uptake mechanisms: Inorganic and organic. *Annu. Rev. Plant Physiol.* 25:53-79.

Nissen, P., N. K. Fageria, A. J. Rayer, M. M. Hassan, and T. V. Hai. 1980. Multiphasic accumulation of nutrients by plants. *Physiol. Plant.* 49:222-240.

Nogurchi, Y., and T. Sugawara. 1966. *Potassium and Japonica Rice.* Bern, Switzerland: Int. Potash Inst.

Norman, R. J., B. R. Wells, and K. A. K. Moldenhauer. 1989. Effect of application method and dicyandiamide on urea-nitrogen-15 recovery in rice. *Soil Sci. Soc. Am J.* 53:1269-1274.

Obata, H. 1995. Physiological functions of micro essential elements, pp. 402-419. In: *Science of the Rice Plant: Physiology*, Vol. 2. K. Matsuo, K. Kumazawa, R. Ishii, K. Ishihara, and H. Hirata (eds.), Tokyo: Food and Agriculture Policy Research Center.

Oertli, J. J. 1979. Plant nutrients, pp. 382-385. In: R. W. Fairbridge and C. W. Finkl Jr. (eds.), *The Encyclopedia of Soil Science*, Part 1. Stroudsburg, PA: Dowden, Hutchinson and Ross.

Okajima, H., I. Uritani, and Huang, H. K. 1975. The significance of minor elements on plant physiology. Taipei, Taiwan: ASPAC Food Fert. Tech. Center.

Oosterhuis, D. M. 1995. Potassium nutrition of cotton in the USA with particular reference to foliar fertilization, pp. 133-146. In: C. A Constable and N. W. Forrester (eds.), *Proc. World Cotton Res.* Brisbane, Australia: CSIRO.

Oosterhuis, D. M., and C. W. Bednarz. 1997. Physiological changes during the development of potassium deficiency in cotton, pp. 347-351. In: T. Ando, K. Fujita, T. Mae, H. Tsumoto, S. Mori, and J. Sekiya (eds.), *Plant Nutrition for Sustainable Food Production and Environment.* Dordrecht, The Netherlands: Kluwer Acad. Publ.

Otani, T., N. Ae, and H. Tanaka. 1996. Phosphorus uptake mechanisms of crops grown in soils with low P status. II. Significance of organic acids in root exudates of pigeon pea. *Soil Sci. Plant Nutr.* 42:553-560.

Ou, S. H. 1985. *Rice Diseases*, Second Edition. Kew, Surrey, England: Commonwealth Mycol. Inst.

Pan, W. L., E. J. Kamprath, R. H. Moll, and W. A. Jackson. 1984. Prolificacy in corn: Its effects on nitrate and ammonium uptake and utilization. *Soil Sci Soc. Am. J.* 48:1101-1106.

Pearman, I., S. M. Thomas, and G. N. Thorne. 1978. Effect of nitrogen fertilizer on growth and yield of semi-dwarf and tall varieties of winter wheat. *J. Agric. Sci., Camb.* 91:31-45.

Peoples, T. R., and D. W. Koch. 1979. Role of potassium in carbon dioxide assimilation in *Medicago sativa* L. *Plant Physiol.* 63:878-881.

Peterson, G. A., and W. W. Frye. 1989. Fertilizer nitrogen management, pp. 183-219. In: R. F. Follet (ed.), *Nitrogen Management and Ground Water Protection.* Amsterdam: Elsevier.

Pitman, M. G. 1976. Ion uptake by plant roots, pp. 95-128. In: U. Lüttage and M. G. Pitman (eds.) *Transport in Plants II, Part B. Tissues and Organs. Encyclopedia of Plant Physiol.*, New Series, Vol. II. Berlin: Springer-Verlag.

Prasad, R., and S. K. De Datta. 1979. Increasing efficiency of fertilizer nitrogen in wetland rice, pp. 465-484. In: Int. Rice Res. Inst. (IRRI) (ed.), *Nitrogen and Rice.* Los Baños, Philippines: IRRI.

Qiao, K., and G. Ho. 1997. The effects of clay amendment and composting on metal speciation in digested sludge. *Wat. Res.* 31:951-964.

Randall, P. J. 1995. Genotypic differences in phosphate uptake, pp. 31-47. In: C. Johansen, K. K. Lee, K. K. Sharma, G. V. Subbarao, and E. A. Kueneman (eds.),

Genetic Manipulation of Crop Plants to Enhance Integrated Nutrient Management in Cropping Systems. Proc. FAO-ICRISAT (Food Aric. Org. – Int. Crops Res. Inst. Semi-Arid Tropics) Expert Consult. Workshop. Patancheru, Andhra Pradesh, India.

Rao, S., and V. Adinarayana. 1995. Molybdenum research and agricultural production, pp. 115-131. In: H. L. S. Tandon (ed.), *Micronutrient Research and Agricultural Production.* New Delhi, India: Fert. Dev. Consult. Org.

Rasmussen, P. E., R. W. Rickman, and B. L. Klepper. 1997. Residue and fertility effects on yield of no-till wheat. *Agron. J.* 89:563-567.

Rattunde, H. F., and K. J. Frey. 1986. Nitrogen harvest index in oats: Its repeatability and association with adaptation. *Crop Sci.* 26:606-610.

Raun, W. R., and G. V. Johnson. 1999. Improving nitrogen use efficiency for cereal production. *Agron. J.* 91:357-367.

Raun, W. R., B. S. Solie, G. V. Johnson, M. L. Stone, R. W. Mullen, K. W. Freeman, W. E. Thomason, and E. V. Lukina. 2002. Improving nitrogen use efficiency in cereal grain production with optical sensing and variable rate application. *Agron. J.* 94:815-820.

Richardson, M. D., and S. S. Croughan. 1989. Potassium influence on susceptibility of bermudagrass to *Helminthosporium cynodontis* toxin. *Crop Sci.* 29:1280-1282.

Römheld, V. 1987. Different strategies for iron acquisition in higher plants. *Physiol. Plant.* 70:231-234.

Römheld, V. 1991. The role of phytosiderophores in acquisition of iron and other micronutrients in graminaceous species: An ecological approach. *Plant Soil* 130:127-134.

Römheld, V., and H. Marschner. 1986. Mobilization of iron in the rhizosphere of different plant species. *Adv. Plant Nutr.* 2:155-204.

Sabrawat, A. K., and S. Chand. 1999. Effect of zinc on plant regeneration in indica rice. *Rice Biotechnol. Quarterly* 37:17.

Salsac, L., S. Chaillou, J. F. Morot-Gaudry, C. Lesaint, and E. Jolivoe. 1987. Nitrate and ammonium nutrition in plants. *Plant Physiol. Biochem.* 25:553-558.

Savant, N. K., G. H. Snyder, and L. E. Datnoff. 1997. Silicon management and sustainable rice production. *Adv. Agron.* 58:151-199.

Sawant, A. S., V. H. Patil, and N. K. Savant. 1994. Rice hull ash applied to seedbed reduces deadheart in transplanted rice. *Int. Rice. Res. Notes* 19(4):21-22.

Segel, I. H. 1968. *Biochemical Calculations.* New York: John Wiley.

Sexton, P. J., W. D. Batchelor, and R. Shibles. 1997. Sulfur availability, rubisco content, and photosynthetic rate of soybean. *Crop Sci.* 37:1801-1806.

Shinano, T., M. Osaki, and T. Tadano. 1994. ^{14}C-Allocation of ^{14}C-compounds introduced to a leaf to carbon and nitrogen components in rice and soybean during ripening. *Soil Sci. Plant Nutr.* 40:199-209.

Shinano, T., M. Osaki, and T. Tadano. 1995. Comparison of growth efficiency between rice and soybean at the vegetative growth stage. *Soil Sci. Plant Nutr.* 41:471-480.

Shuman, L. M. 1994. Mineral nutrition, pp. 149-182. In: R. E. Wilkinson (ed.) *Plant-Environment Interactions.* New York: Marcel Dekker.

Shuman, L. M. 1995. Effects of nitrilotriacetic acid on metal adsorption isotherms for two soils. *Soil Sci.* 160:92-100.

Siddiqi, M. Y., and A. D. M. Glass. 1981. Utilization index: A modified approach to the estimation and comparison of nutrient utilization efficiency in plants. *J. Plant Nutr.* 4:289-302.

Sillanpaa, M., and P. L.G. Vlek. 1985. Micronutrients and the agroecology of tropic and Mediterranean regions. *Fert. Res.* 7:151-167.

Singer, S. J. 1972. A fluid lipid globular protein mosaic model of membrane structure. *Ann. New York Acad. Sci.* 195:16-23.

Singh, M., and R. K. Singh. 1979. Split application of potassium in rice to maximize its utilization. *Indian J. Agron.* 24:193-198.

Singh, U., J. K. Ladha, E. G. Castillo, G. Punzalan, A. Tirol-Padre, and M. Duqueza. 1998. Genotypic variation in nitrogen use efficiency in medium and long duration rice. *Field Crops Res.* 58:35-53.

Smart, D. R., and A. J. Bloom. 1988. Kinetics of ammonium and nitrate uptake among wild and cultivated tomatoes. *Oecologia* 76:336-340.

Snowball, L., A. D. Robson, and J. F. Loneragan. 1980. The effect of copper on nitrogen fixation in subterranean clover *(Trifolium subterraneum). New Phytol.* 85:63-72.

Snyder, G. H., D. B. Jones, and G. J. Gascho. 1986. Silicon fertilization of rice on Everglades Histosols. *Soil Sci. Soc. Am. J.* 50:1259-1263.

Sommer, A L., and C. B. Lipman. 1926. Evidence on the indispensable nature of zinc and boron for higher green plants. *Plant Physiol.* 1:231-249.

Spiegler, K. S., and C. D. Coryell. 1953. Electromigration in cation exchange resins. III. *J. Physis. Chem.* 57:687-690.

Stevenson, F. J. 1986. *Cycles of Soil: Carbon, Nitrogen, Phosphorus, Sulfur, Micronutrients.* New York: John Wiley.

Stone, L. F., P. M. Silveira, J. A .A. Moreira, and L. P. Yokoyama. 1999. Rice nitrogen fertilization under supplemental sprinkler irrigation. *Pesq. Agropec. Brasileira, Brasilia* 34:927-932.

Su, N. R. 1976. Potassium fertilization of rice, pp. 117-148. In: ASPAC (ed.), *The Fertility of Paddy Soils and Fertilizer Applications for Rice.* Taipei, Taiwan: ASPAC Food Fert. Tech. Center.

Swank, J. C., F. E. Below, R. J. Lambert, and R. H. Hageman. 1982. Interaction of carbon and nitrogen metabolism in the productivity of maize. *Plant Physiol.* 70:1185-1190.

Takagi, S., K. Nomoto, and Takemoto. 1984. Physiological aspects of mugineiacid, a possible phytosiderophore of graminaceous plants. *J. Plant Nutr.* 7:469-477.

Takahashi, E. 1995. Uptake mode and physiological functions of silica. *Tokyo Sci. Rice Plant.* 2:99-122.

Tamhane, R. V., D. P. Motiramani, Y. P. Bali, and R. L. Donahue. 1966. *Soils: Their Chemistry and Fertility in Tropical Asia.* New Delhi, India: Prentice-Hall.

Tanner, W., and H. Beevers. 1990. Does transpiration have an essential function in long-distance ion transport in plants? *Plant Cell Environ.* 13:745-750.

Taylor, R. G., T. I. Jackson, R. I. Powelson, and N. W. Christensen. 1981. Chloride, nitrogen form, lime, and planting date effects on take-all root rot of winter wheat. *Plant Dis.* 67:1116-1120.

Terry, N. 1976. Effects of sulfur on the photosynthesis of intact leaves and isolated chloroplasts of sugar beets. *Plant Physiol.* 57:477-479.

Terry, N. 1980. Limiting factors in photosynthesis. I. Use of iron stress to control photochemical capacity in vivo. *Plant Physiol.* 65:114-120.

Terry, N., and J. Abadia. 1986. Function of iron in chloroplasts. *J. Plant Nutr.* 9:609-646.

Ting, I. P. 1982. Plant mineral nutrition and ion uptake, pp. 331-363. In: *Plant Physiology.* Reading, MA: Addison-Wesley Publ.

Torun, B., A. Yazici, I. Gultekin, and I. Cakmak. 2003. Influence of gyttja on shoot growth and shoot concentrations on zinc and boron of wheat cultivars grown on zinc deficient and boron toxic soil. *J. Plant Nutr.* 26:869-881.

Touraine, B., B. Muller, and C. Grignon. 1992. Effect of phloem translocated malate on NO_3^- uptake by roots of intact soybean plants. *Plant Physiol.* 99:1118-1123.

Uhart, S. A., and F. H. Andrade. 1995. Nitrogen deficiency in maize. I. Effects on crop growth, development, dry matter partitioning, and kernel set. *Crop Sci.* 35:1376-1383.

Ulrich, C. I., and A. J. Novacky. 1990. Extra and intracellular pH and membrane potential changes induced by K^+, Cl^-, $H_2PO_4^-$ and NO_3^- uptake and fusicoccin in root hairs of *Limnobium stoloniferum*. *Plant Physiol.* 94:1561-1567.

Voss, R. D. 1993. Corn, pp. 11-14. In: W. F. Bennett (ed.), *Nutrient Deficiencies and Toxicities in Crop Plants.* St. Paul, MN: Am. Phytopath. Soc.

Vreugdenhil, D. 1985. Source-sink gradient of potassium in the phloem. *Planta* 163:238-240.

Wada, G., S. Shoji, and T. Mae. 1986. Relation between nitrogen absorption and growth and yield of rice plants. *JARQ* 20:135-145.

Walker, N. A. 1976. The structure of biological membranes, pp. 3-11. In: U. Lüttage and M. G. Pitman (eds.), *Transport in Plants, Part A, Cells. Encyclopedia of Plant Physiology*, New Series, Vol. 2. New York: Springer-Verlag.

Walker, T. W., and J. K. Syers. 1976. The fate of phosphorus during pedogenesis. *Geoderma.* 15:1-19.

Wang, X., and F. E. Below. 1992. Root growth, nitrogen uptake, and tillering of wheat induced by mixed-nitrogen source. *Crop Sci.* 12:997-1002.

Warington, K. 1923. The effect of boric acid and borax on the broad bean and certain other plants. *Ann. Bot.* 37:629-672.

Wasaki, J., M. Ando, K. Ozawa, M. Omura, M. Osaki, H. Ito, H. Matsui, and T. Tadano. 1997. Properties of secretory acid phosphtase from lupin roots under phosphorus deficient conditions, pp. 295-300. In: T. Ando, K. Fujita, T. Mae, H. Tsumoto, S. Mori, and J. Sekiya (eds.), *Plant Nutrition for Sustainable Food Production and Environment.* Dordrecth, The Netherlands: Kluwer Acad. Publ.

Welch, R. W., and Y. Y. Yong. 1980. The effects of variety and nitrogen fertilizer on protein production in oats. *J. Sci. Food Agric.* 31:541-548.

Wells, B. R., B. A. Huey, R. J. Morman, and R. S. Helms. 1993. Rice, pp. 15-19. In: W. F. Bennett (ed.) *Nutrient Deficiencies and Toxicities in Crop Plants.* St. Paul, MN: Am. Phytopath. Soc.

Westcott, M. P., D. M. Brandon, C. W. Lindau, and W. H. Patrick, Jr. 1986. Effects of seeding method and time of fertilization on urea-nitrogen-15 recovery in rice. *Agron. J.* 78:474-478.

Williams, R. F. 1948. The effects of phosphorus supply on the rates of uptake of phosphorus and nitrogen upon certain aspects of phosphorus metabolism in gramineous plants. *Aust. J. Sci. Res. Ser.* B1:336-361.

Wiren, N. V., S. Gazzarrini, and W. B. Frommer. 1997. Regulation of mineral nitrogen uptake in plants, pp.41-49. In: T. Ando, K. Fujita, T. Mae, H. Tsumoto, S. Mori, and J. Sekiya (eds.), *Plant Nutrition for Sustainable Food Production and Environment.* Dordrecht, The Netherlands: Kluwer Acad. Publ.

Xu, H. L., J. Lopez, F. Rachidi, N. Tremblay, L. Gauthier, Y. Desjardins, and A. Gosselin. 1996. Effect of sulphate on photosynthesis in greenhouse-grown tomato plants. *Physiol. Plant.* 96:222-226.

Yang, X. 1987. *Physiological mechanisms of nitrogen efficiency in hybrid rice.* PhD Dissertation, Zhejiang Agric. Univ., Hangzhou, China.

Yang, X., and X. Sun. 1988. Physiological characteristics of F_1 hybrid rice roots, pp. 159-164. In: Int. Rice Res. Inst. (IRRI) (ed.), *Hybrid Rice.* Manila, Philippines: IRRI.

Yang, X., J. Zhang, and W. Ni. 1999. Characteristics of nitrogen nutrition in hybrid rice. *Int. Rice Res. Notes.* 245:8.

Yoshida, S. 1972. Physiological aspects of grain yield. *Annu. Rev. Plant Physiol.* 23:437-464.

Yoshida, S. 1981. *Fundamentals of Rice Crop Science.* Los Baños, Philippines: Int. Rice Res. Inst.

Zhao, F. J., P. J. A Withers, E. J. Evans, J. Monaghan, S. E. Salmon, P. R. Shewry, and S. P. McGrath. 1997. Sulphur nutrition: An important factor for the quality of wheat and rapeseed, pp. 917-922. In: T. Ando, K. Fujita, T. Mae, H. Tsumoto, S. Mori, and J. Sekiya (eds.), *Plant Nutrition for Sustainable Food Production and Environment.* Dordrecht, The Netherlands: Kluwer Acad. Publ.

Index

Page numbers followed by the letter "f" indicate figures and those followed by the letter "t" indicate tables.

CPSIA information can be obtained at www.ICGtesting.com
Printed in the USA
LVOW06s1215221215

467464LV00015B/145/P